DISCARDED

JUN 19 2025

Asheville-Buncombe
Technical Community College
Learning Resources Center
340 Victoria Road
Asheville, NC 28801

# Mechatronics and Machine Tools

# Mechatronics and Machine Tools

## HMT Limited

**McGraw-Hill**
New York  San Francisco  Washington, D.C.  Auckland  Bogotá
Caracas  Lisbon  London  Madrid  Mexico City  Milan
Montreal  New Delhi  San Juan  Singapore
Sydney  Tokyo  Toronto

# McGraw-Hill

*A Division of The McGraw-Hill Companies*

Copyright © 1999 by The McGraw-Hill Companies, Inc. All rights reserved. Printed in the United States of America. Except as permitted under the United States Copyright Act of 1976, no part of this publication may be reproduced or distributed in any form or by any means, or stored in a data base or retrieval system, without the prior written permission of the publisher.

Originally published with the title *Mechatronics* by Tata McGraw-Hill Publishing Company Limited, New Delhi, India. © 1998, Tata McGraw-Hill Publishing Company Limited.

1 2 3 4 5 6 7 8 9 0 DOC/DOC 9 0 4 3 2 1 0 9

ISBN 0-07-134634-1

*The sponsoring editor for this book was Linda Ludewig and the production supervisor was Tina Cameron.*

*Printed and bound by R. R. Donnelley & Sons Company.*

This book is printed on recycled, acid-free paper containing a minimum of 50% recycled, de-inked fiber.

McGraw-Hill books are available at special quantity discounts to use as premiums and sales promotions, or for use in corporate training programs. For more information, please write to the Director of Special Sales, McGraw-Hill, 11 West 19th Street, New York, NY 10011. Or contact your local bookstore.

---

Information contained in this work has been obtained by The McGraw-Hill Companies, Inc. ("McGraw-Hill") from sources believed to be reliable. However, neither McGraw-Hill nor its authors guarantee the accuracy or completeness of any information published herein, and neither McGraw-Hill nor its authors shall be responsible for any errors, omissions, or damages arising out of use of this information. This work is published with the understanding that McGraw-Hill and its authors are supplying information but are not attempting to render engineering or other professional services. If such services are required, the assistance of an appropriate professional should be sought.

# About HMT Limited

HMT Limited was established by the Government of India in 1953, in collaboration with Messrs Oerlikon of Switzerland to produce quality machine tools and catalyze industrial growth of India. Ever since, it has evolved as a multitechnology engineering complex with 16 units and 22 manufacturing divisions, enriched with state-of-the-art expertise from all over the world. Strongly backed by a highly motivated workforce of about 21,000 employees, the company produces a wide range of world class products—many of them import substitutes.

The company's product range is broadly divided into five major business groups:

- the *Machine Tool Business Group* covering general purpose machine tools, special purpose machine tools, CNC machine tools, horological machinery, advanced manufacturing systems, and controls;
- the *Industrial Machinery Business Group* comprising diecasting and plastic machinery, printing machinery, presses and dairy machinery;
- the *Tractors Business Group* consisting of tractors;
- the *Consumer Products Business Group* consisting of watches and lamps; and
- the *Engineering Components Business Group* consisting of bearings and ballscrews.

HMT has always focussed on consistent quality of products and performance, customer orientation and skill upgradation. The frontline technology, world class machinery and state-of-the-art equipment, updated manufacturing processes and professionally trained human resources are the major factors that have resulted in quality products meeting international standards in design, performance and safety. The sophisticated quality assurance facilities ensure products that are highly reliable and dependable. All manufacturing units at HMT are ISO 9001 accredited.

In addition to manufacturing and marketing activities, HMT also offers specialised project consultancy services in India and abroad. With a wide network of offices and representatives spread across the world, HMT (International), a wholly owned subsidiary of HMT, places HMT's products and services in the global arena. It has already executed several turnkey projects in Nigeria, Tanzania, Algeria, Malaysia, Singapore and Maldives.

Together with its national and international network of offices and service centres, HMT is a veritable giant among the public sector units.

During the course, HMT has also emerged as one of the leading disseminators of technology and technical information. *Production Technology* authored by HMT and published in 1981 by Tata McGraw-Hill Publishing Company Ltd is even today considered as one of the best reference books by the practising engineers and students. *Mechatronics* is designed to bridge the technical information gap further and cater to the long-felt need of engineers and students in the field of mechatronics.

# Foreword

In the increasingly competitive environment of the day, mechatronics has become the key to industrial prosperity. The rapid advancements in the fields of electronics engineering, information technology and systems engineering have been responsible for evolving new concepts aimed at developing highly sophisticated machine tools for enhanced productivity. In such an environment, a systematic programme to train its human resources, to prepare them to be competitive enough to face this challenge becomes a vital issue in any corporate sector. HMT, which has reposed a strong faith in its workforce, introduced the training program in mechatronics for its workforce as a first step to acquaint them with this science. Our customer engineers have also evinced keen interest in these training programmes and are constantly updating the technological capabilities of their human resources.

HMT has always believed in sharing its expertise and knowledge with others for industrial and technological prosperity of the nation. An earlier publication, *Production Technology* brought out with this intention has been very well received by the practising engineers and students. It is heartening that this belief is being reaffirmed with the publication of *Mechatronics*. The dedicated efforts of Mr J P Malik, Director, Machine Tool Business Group and his team of executives and engineers, particularly Mr P J Mohanram, General Manager, Precision Machinery Division and Mr S Karunakaran, Joint General Manager (Engg.), R & D Centre (Metal Cutting), deserve to be complimented for their continuing efforts in spreading and popularising the culture of mechatronics.

The coverage of this book demonstrates the keen desire of the authors to systematically link the concepts to the processes and present the entire subject in a logical sequence. I congratulate the authors for bringing out their experience and expertise in the form of this book for the benefit of practising engineers and students.

I understand that many technical institutions, polytechnics and universities have introduced mechatronics in their curriculum. I am sure that the industrial expertise and wealth of knowledge presented in this book will fulfil the long-felt need of the students and engineers in understanding and application of this concept.

**N Ramanuja**
*Chairman & Managing Director*
*HMT Limited*

# Preface

Mechatronics, a concept of Japanese origin, was first introduced during the early 1970's to qualify the dual alliance of electronics and computer technology to practical control applications in mechanical systems. The Japanese were quick to exploit the advantages of this integrated technology through their consumer products while others are yet adopting this concept to realise its benefits.

Mechatronics aims at lending better value to products and systems. The combination of mechanical, electronics and information engineering is optimised to take full advantage of the ability of microelectronics to reduce the demand on mechanical systems of a product. High technology CNC machines, robots and automated manufacturing systems are truly mechatronic products. Needless to mention that the future manufacturing systems will all be compelled to adopt more and more of mechatronic systems—automation, CAD/CAM, flexible manufacturing system (FMS), and computer integrated manufacturing (CIM)—crossing the traditional boundaries that exist in engineering. The synergy derived from this fusion will lower the cost of production and improve quality and reduce lead time.

It is in this context that HMT started training and retraining its workforce in the area of mechatronics. Owing to persisting demands from customer industries, this training was extended to our customer engineers also. I am happy to mention that this training programme has helped our customers in proper selection, utilisation and maintenance of advanced equipment in production.

The programme material prepared for this purpose has been revised and enriched in content, and is being brought out in the present form by Tata McGraw-Hill Publishing Company. In fact, our association with the publishers dates back to early 1980s when they first published our book *Production Technology*, which is recognised as one of the best reference books for students and practising engineers. I thank the publishers for their efforts.

*Mechatronics* is an exhaustive coverage of the various aspects from design to testing of high technology CNC machines. The various mechatronic elements and systems used in building a CNC machine are explained in detail. The contributors to this book are our R&D engineers with vast experience in the areas of CNC machines and advanced manufacturing technology and I congratulate them for their contributions.

In coordinating the work of the authors and editing the text to a uniform standard of excellence, Mr S Karunakaran, Joint General Manager (Engineering) and Mr N S Dasharathy, Assistant General Manager (Technical Publications) of R&D Centre (Metal Cutting) have done a commendable job. I am also grateful to the innumerable individuals and institutions for providing us with photographs and the necessary permissions to reproduce copyrighted materials. My sincere appreciation goes to all the executives, engineers and personnel who have contributed to this book.

In the coming years, HMT intends to bring out many such publications covering the areas of production engineering and advanced manufacturing technology.

**J P MALIK**
*Director,*
*Machine Tool Business Group*
*HMT Limited*

# The Mechatronics Team

**Chief Editor**

S Karunakaran

**Associate Editors**

N S Dasharathy

A Janaki

**Contributors**

| | |
|---|---|
| H R Anandaram | P K Ray Chaudhuri |
| A N Badhe | C Sekharan |
| H S Diwakara | A Shantharam |
| S Karunakaran | M V Suryanarayana |
| M A Khallaq | H N Uma |
| K Nalankilli | T K Venkatesh |
| E Palani | K J Vijayamma |
| S Raju Sagi | |

**CAD Drafting**

| | |
|---|---|
| G Babu | P Thippanna |
| K Nagaraj | G Udayakumar |
| G Ramachandra | |

# Acknowledgements

A project of this magnitude would have been impossible to conceive, had it not been for the many hands that joined ours.

We sincerely thank them all and the following in particular for their contribution in developing this work.
- R Raghavendra Rao, INA Bearing Co., Bangalore
- Satish Chandra, Rane (Madras) Limited, Mysore
- D S Nagabhushan, Rane (Madras) Limited, Mysore
- N Srinivasa Murthy, Motor Industries Co. Ltd., Bangalore
- L Narayana
- H S Ramachandra
- S R Ganapathy

We are also grateful to the innumerable institutions for rendering their services during the writing of this book.
- Ashok & Lal, Chennai, India
- Bureau of Indian Standards, New Delhi, India
- Cincinnati Milacron Inc., USA
- Engineering & Scientific Equipment Ltd., UK
- ETP Transmissions AB, Sweden
- FAG OEM und Handel AG, Germany
- Gerwah-Praezision, Germany
- GE Fanuc Automation North America Inc., USA
- GF Georg Fischer, Germany
- Gettys Corporation, USA
- INA Waelzlager Schaeffler KG, Germany
- Indian Machine Tool Manufacturers' Association, New Delhi, India
- Institute of Information Studies, Bangalore, India
- Dr Johannes Heidenhain, Germany
- KTM, UK
- Mac Marketing Corporation, Bangalore, India
- Ringfeder GmbH, Germany
- Siemens AG, Germany
- SKF Transol, France
- THK Co. Ltd., Japan
- Tsubakimoto Precision Products Co. Ltd., Japan

# Contents

About HMT Limited — *v*
Foreword — *vii*
Preface — *ix*
The Mechatronics Team — *xi*
Acknowledgements — *xiii*

**CHAPTER 1  Introduction** — 1

- 1.1 What is Mechatronics? *1*
- 1.2 Scope of Mechatronics *2*
- 1.3 Key Issue *3*
- 1.4 About This Book *5*

**CHAPTER 2  Introduction to Modern CNC Machines and Manufacturing Systems** — 6

- 2.1 Introduction *6*
- 2.2 Advantages of CNC Machines *6*
- 2.3 CNC Machining Centre Developments *7*
- 2.4 Turning Centre Developments *10*
- 2.5 Tool Monitoring on CNC Machines *17*
- 2.6 Other CNC Developments *20*
- 2.7 Advanced Manufacturing Systems *20*
- 2.8 Benefits of an FMS *22*
- 2.9 Trends in Adoption of FMSs *24*

## CHAPTER 3   Electronics for Mechanical Engineers   29

    3.1   Introduction   *29*
    3.2   Conductors, Insulators and Semiconductors   *30*
    3.3   Passive Components used in Electronics   *34*
    3.4   Transformers   *41*
    3.5   Semiconductors   *42*
    3.6   Transistors   *58*
    3.7   Silicon Controlled Rectifiers (SCR)   *68*
    3.8   Integrated Circuits (IC)   *71*
    3.9   Digital Circuits   *74*

## CHAPTER 4   Mechanical Systems for Electronics Engineers   79

    4.1   Basic Concepts   *79*
    4.2   Materials   *101*
    4.3   Heat Treatment   *106*
    4.4   Electroplating   *115*
    4.5   Standards   *117*

## CHAPTER 5   Design of Modern CNC Machines and Mechatronic Elements   135

    5.1   Introduction   *135*
    5.2   Machine Structure   *135*
    5.3   Guideways   *137*
    5.4   Feed Drives   *147*
    5.5   Spindle/Spindle Bearings   *169*
    5.6   Measuring Systems   *177*
    5.7   Controls, Software and User Interface   *181*
    5.8   Gauging   *182*
    5.9   Tool Monitoring System   *182*

## CHAPTER 6   Assembly Techniques   184

    6.1   Guideways   *184*
    6.2   Ballscrew and Nut   *195*
    6.3   Feedback Elements   *207*
    6.4   Spindle Bearings   *214*
    6.5   Shop Tools and Equipments for Assembly   *236*
    6.6   Hydraulics   *237*

## CHAPTER 7   *Drives and Electricals*   240

- 7.1 Drives   *240*
- 7.2 Spindle Drives   *242*
- 7.3 Feed Drives   *244*
- 7.4 DC Motors   *246*
- 7.5 Servo-principle   *252*
- 7.6 Drive Optimisation   *276*
- 7.7 Drive Protection   *281*
- 7.8 Selection Criteria for AC Drives   *281*
- 7.9 Electric Elements   *283*
- 7.10 Wiring of Electrical Cabinets   *287*
- 7.11 Power Supply for CNC Machines   *289*
- 7.12 Electrical Standard   *289*
- 7.13 Electrical Panel Cooling (Air Conditioning)   *289*

## CHAPTER 8   *CNC Systems*   308

- 8.1 Introduction   *308*
- 8.2 Configuration of the CNC System   *308*
- 8.3 Interfacing   *319*
- 8.4 Monitoring   *319*
- 8.5 Diagnostics   *322*
- 8.6 Machine Data   *322*
- 8.7 Compensations for Machine Accuracies   *323*
- 8.8 PLC Programming   *325*
- 8.9 Direct Numerical Control (DNC)   *333*

## CHAPTER 9   *Programming and Operation of CNC Machines*   338

- 9.1 Introduction to Part Programming   *338*
- 9.2 Coordinate System   *340*
- 9.3 Dimensioning   *340*
- 9.4 Axes and Motion Nomenclature   *341*
- 9.5 Structure of a Part Program   *342*
- 9.6 Word Addressed Format   *344*
- 9.7 G02/G03 Circular Interpolation   *348*
- 9.8 Tool Compensation   *351*
- 9.9 Subroutines (Macros)   *352*
- 9.10 Canned Cycles (G81-G89)   *353*
- 9.11 Mirror Image   *361*
- 9.12 Parametric Programming (User Macros) and R-Parameters   *363*
- 9.13 G96 S... Constant Cutting Speed and G97 Constant Speed   *364*

9.14 Machining Cycles  *364*
9.15 Programming Example for Machining Centre  *373*
9.16 Programming Example for Turning Centre  *376*

## CHAPTER 10  *Testing of Machine Tools*  378

10.1 Introduction  *378*
10.2 Verification of Technical Specifications  *379*
10.3 Verification of Functional Aspects  *379*
10.4 Verification During Idle Running  *380*
10.5 Verification of Machine Tool Accuracy and Workpiece Accuracy  *395*
10.6 Metal Removal Capability Test  *417*
10.7 Other Tests  *424*
10.8 Safety Aspects  *427*

## CHAPTER 11  *Industrial Design, Aesthetics and Ergonomics*  434

11.1 Introduction  *434*
11.2 Elements of Product Design  *435*
11.3 Ergonmic Factors for Advanced Manufacturing Systems  *441*

## CHAPTER 12  *Introduction to Computers and CAD/CAM*  448

12.1 Introduction to Computers  *448*
12.2 CAD/CAM Systems  *450*

*Further Reading*  454

*Index*  455

CHAPTER **1**

# Introduction

## 1.1 WHAT IS MECHATRONICS?

Mechatronics is a term coined by the Japanese to describe the integration of mechanical and electronic engineering. More specifically, it refers to a multidisciplinary approach to product and manufacturing system design. It represents the next generation machines, robots and smart mechanisms for carrying out work in a variety of environments—predominantly factory automation, office automation and home automation—as shown in Fig. 1.1.

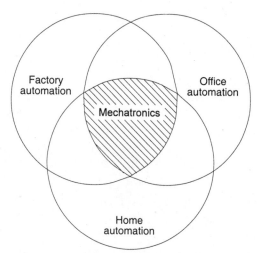

**Fig. 1.1** Domains of mechatronics

As a discipline, mechatronics encompasses electronics enhancing mechanics (to provide high levels of precision and reliability) and electronics replacing mechanics (to provide new functions and capabilities). Some examples where mechanics has been enhanced by electronics are numerically controlled machine tools which cut metal automatically, industrial robots and automatic bank tellers. The products where electronics replaces mechanics include digital watches, calculators, etc.

However, the products that really blur the distinction between electronics and mechanics are the machines and robots driven by numerical control. Japan is the first country in the world to have mastered the NC machines technology, and as a result the Japanese machine tool industry has flourished. This is because the Japanese have mastered mechatronics, the fusion of precision mechanics and electronics in design, engineering and manufacture, which is popularly depicted by Japanese as given in Fig. 1.2.

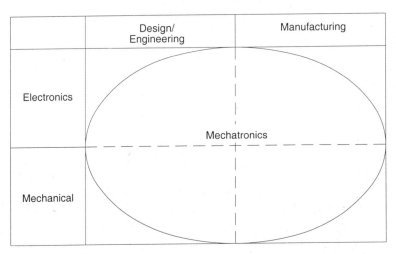

Fig. 1.2 Concept of mechatronics

## 1.2 SCOPE OF MECHATRONICS

Until the 1970s, most industrial products—machine tools, manufacturing equipment, consumer products, home appliances, etc. were largely mechanical systems with very limited electrical or electronic content. However, since the seventies there has been a dramatic change in the technology of these products, mainly, an increasing content of electric and electronic systems integrated with the mechanical parts of the products, viz., mechatronics. Examples of products which have already moved to mechatronic technology from simple mechanical products are:

- Machine tools incorporating computer numerical control (CNC), electric servo-drives, electronic measuring systems, precision mechanical parts such as ball-screws, antifriction guideways, etc.
- Electronic watches incorporating fine mechanical parts and sophisticated electronic circuits.
- Electronic consumer products—washing machines, electronic cooking appliances, etc.; office equipment—fax, plain paper copiers, etc. and entertainment products—VCRs and videos, etc.

In the last twenty years, the production technology has seen the introduction of high precision measuring instruments such as electronic gauges and measuring instruments, inprocess gauges and quality control instruments, laser measuring systems, etc. to ensure high dimensional accuracies, as well as increased productivity on the shopfloor.

In the domain of factory automation, mechatronics has had far reaching effects in manufacturing and will gain even more importance in future. Major constituents of factory automation include CNC machines, robots, automation systems, and computer integration of all functions of manufacturing. Basically these advanced manufacturing solutions consist of mechatronic systems. Low-volume, more variety, higher levels of flexibility, reduced lead time in manufacture, and automation in manufacturing and assembly are likely to be the future needs of customers.

Proper application, utilisation and maintenance of these high technology products and systems is an important aspect that enhances the productivity and quality of products manufactured by the customers. To ensure correct selection of equipment, an accurate estimation of the techno-economics of various manufacturing systems, developments in the high technology machines and equipment are studied in detail. Also, proper maintenance of various mechatronic elements/assembly, diagnostics, etc. can increase the life of the various mechatronic elements, which in turn will enhance the life of the product/system. Such inputs in mechatronics can be best given by the manufacturers of hi-tech machines and manufacturing systems. In the long run, this will result in a true understanding of the needs for design, manufacture and supply of such systems by the manufacturer for the effective utilisation by the customers. These needs are bound to increase in future. In fact, the machine tool manufacturers are now being called upon to offer a total manufacturing for solution in production, by the customer, rather than supply of just the stand-alone machines. This trend is already evident in many of the advanced countries and Indian industries are not going to be an exception.

Evidently, the design and manufacture of future products will involve a combination of precision mechanical and electronic systems and mechatronics will form the core of all activities in products and production technology.

In order to produce quality products of international standards, the industries must excel in this new technology—mechatronics. This calls for training and re-orienting the workforce at all levels in the design and manufacture of mechatronics-based products and equipment. Technicians and engineers have to be given the basic knowledge and practical training in precision mechanics, electronics and a combination of the two disciplines, viz., mechatronics.

## 1.3 KEY ISSUE

Mechatronics is not the only means to achieve progress in the company. The key issue is, how to become competitively attractive in the light of the fact that for all products and services there is competition. To survive in the long run, international competitiveness is also essential; not necessarily, that we trade internationally but certainly to combat the threat of global competition at home and abroad.

Four factors are to be considered in the *value portfolio* of a company and the way in which it faces its market place. These are:

- Cost
- Quality
- Product specifications
- Delivery

## Cost

*Cost* reduction must be the first priority since price is the most important consideration for the buyer. There is an urgent need to minimise the following.
- Work-in-process inventory
- Inventory turnover
- Manufacturing costs
- Rejections/reworks
- Assembly costs

## Quality

*Quality* has many definitions:
- Quality is the fitness for use,
- Quality is meeting the customers' requirement,
- Quality is right the first time, etc.

All these definitions mean the same and are not in conflict as long as it is understood in the company that quality is everyone's business. We need to look at quality as a competitive weapon to become world players. That is, quality has to be thought of as a strategic weapon; a matter of daily attention.

## Product Specification

To exploit *product specification* as a value portfolio, basically we need to focus on the designs. To apply design as a leverage, product design should be able to meet the customers' requirements and be easily produced with high quality. Two simple design concepts which enhance design quality, both for the consumer products and the industrial products are:
- Minimising the number of parts
- Use of a modular design

It is also important that the design be closely integrated with the rest of the organisation. The product designers have a vital role to play because the designs possess enormous power to prevent or cause problems later.

## Delivery

The value portfolio *delivery* can be achieved only if the suppliers, carriers and the customers come to a mutual understanding that they are partners in profit. Suppliers have to be treated as long-term partners and supporting their development will enhance quality and reduce costs. Similarly, the carriers also have an important role to play for the manufacturing industries to adopt the just-in-time (JIT) concepts. Low cost, high-service, total transportation are the goals being set by some of the transportation companies abroad, after many manufacturing industries started adopting the JIT concepts.

## 1.4 ABOUT THIS BOOK

This book has been conceived to provide technical information on the most important mechatronic products—the CNC machine tools and manufacturing systems. In the wake of factory automation, and the concepts appearing on the industrial manufacturing scene, it becomes imperative to understand the CNC machine tools—their design, working, use and maintenance in the day-to-day operations—which are the basic building blocks in such systems. For maintaining such high cost equipment and reducing the down time in production, there is a need to understand the product technology, design and assembly, and trouble shooting. Towards this need, through this book, an attempt has been made to provide the technical details of CNC machines, the various working principles of mechatronic elements used and their maintenance. Information on drives, CNC systems, programming and testing, and evaluation are also included to orient the production engineers in production, maintenance and servicing to utilise the information in their day-to-day working. Various standards, testing details and other technical details on aesthetics/ergonomics are also included to provide comprehensive information on typical CNC products.

CHAPTER **2**

# Introduction to Modern CNC Machines and Manufacturing Systems

## 2.1 INTRODUCTION

Development of computer numerically controlled (CNC) machines is an outstanding contribution to the manufacturing industries. It has made possible the automation of the machining processes with flexibility to handle small to medium batch quantities in part production.

Initially, the CNC technology was applied on basic metal cutting machines like lathes, milling machines, etc. Later, to increase the flexibility of the machines in handling a variety of components and to finish them in a single set-up on the same machine, CNC machines capable of performing multiple operations were developed. To start with, this concept was applied to develop a CNC machining centre for machining prismatic components combining operations like milling, drilling, boring and tapping. Further, the concept of multi-operations was also extended for machining cylindrical components which led to the development of turning centres.

## 2.2 ADVANTAGES OF CNC MACHINES

CNC machines offer the following advantages in manufacturing.

- Higher flexibility
- Increased productivity
- Consistent quality
- Reduced scrap rate
- Reliable operation
- Reduced non-productive time
- Reduced manpower
- Shorter cycle time
- Higher accuracy
- Reduced lead time
- Just-in-time (JIT) manufacture
- Automatic material handling

- Lesser floor space
- Increased operational safety
- Machining of advanced materials

## 2.3  CNC MACHINING CENTRE DEVELOPMENTS

Over the last three decades, developments in CNC machining centres have been taking place worldwide. In India, these developments have been realised with a lag of five to ten years. Initially, the development of machining centres started with vertical spindle configuration comprising three basic servo axes—two for the table movement and one for the spindle head—vertical machining centre (VMC) (Fig. 2.1). The same configuration with vertical spindle head replaced by an indexable tool turret—turret machining centre (TMC) (Fig. 2.2)—was developed which enabled, automatic selection of the required tool by indexing the turret without manual intervention during the programmed cycle of operation. Both VMC and TMC facilitate machining of plate type of components or one side of a cubical component.

Though TMC eliminated manual tool change within the capacity of the tool turret, its advantage was restricted to the manufacture of components requiring only a limited number of tools. This limitation has been overcome with the introduction of a large tool magazine and an auto tool changer (ATC) for automatic transfer of tools from the magazine to the spindle and vice versa (Fig. 2.3).

To facilitate machining on four sides of a cubical or prismatic component in a single set-up, the concept of horizontal machining centre (HMC) with a magazine containing more tools of the order of 32 and above and an index/rotary table to orient the component for machining on different faces (Fig. 2.4) was developed. Later, the idle time of

**Fig. 2.1**  Vertical machining centre without ATC

**8** *Mechatronics*

**Fig. 2.2** Turret machining centre

machine due to setting of the job on machine table has been eliminated in both VMC and HMC by the introduction of an automatic pallet changer (APC). Here, the next workpiece could be loaded/set on the additional pallet on APC while the machine is busy cutting the previous workpiece. After the machining is complete, the pallet with the finished workpiece and the pallet on APC having a raw component could be exchanged automatically without much loss of machine time (Figs 2.5 and 2.6).

While VMC and HMC could be respectively utilised for machining only on one face and four faces of the component in a set-up, complete machining of all five exposed faces of the cubical component in a single set-up was possible with a feature to change the spindle configuration automatically from horizontal to vertical and vice versa, as the case may be within the programmed cycle. These machines are called universal machining centre (UMC). Figure 2.6 shows different concepts of spindle head changeover in UMC.

Over a period of time, to reduce the operator intervention during machining and to increase the machine utilisation, the following additional features were introduced.

(a) Tool/work monitoring
(b) Increased tool capacities on the tool magazine

**Fig. 2.3** Vertical turning centre without pallet changer

(c) Direct numerical control (DNC) to upload and download the part programmes in the foreground or in the background.

The following features on machining centres contribute to higher productivity, reduced non-cutting time and improved part accuracy.

(a) Higher spindle speeds of the order of 10,000-20,000 rpm for higher material removal rates in relatively softer materials.
(b) Very high rapid traverse rates of around 40-60 m/min for faster tool positioning and high cutting feed rates for increased metal removal rates.
(c) Faster tool change systems with a tool change time as low as 0.6 s for tool-to-tool and 3 s for chip-to-chip to reduce the non-cutting time.
(d) Digital servo-control of main spindle and standard interface for axis servo-control for improved part accuracy.
(e) Efficient thermal control on machine elements such as spindle, ballscrews and bearings for better part accuracy.

Trends in development indicate that the following advanced features will be available in the future on machining centres.

**Fig. 2.4** Horizontal machining centre with auto tool changer and without auto pallet changer

(a) Integration of high speed controllers for increased feed rates in contouring without affecting the part accuracies.
(b) Introduction of robust adaptive controls for process optimisation, improved part accuracy and increased tool life.
(c) Provision of non-contact position sensing and gauging systems for measurements on-the-fly together with necessary corrective actions.

## 2.4 TURNING CENTRE DEVELOPMENTS

Historically, the CNC technology was applied on turning machines with a conventional horizontal bed configuration having two servo axes—one for the saddle and the other for the cross slide—and with an indexable tool turret (Fig. 2.7). Over a period of time the concept of slant bed configuration was adopted for higher rigidity, better chip disposal and easy access for loading and unloading of components together with disc type turret to accommodate more number of tools (Fig. 2.8).

Though the CNC technology was applied first on turning machines, combining the machining operations like drilling and milling was introduced only in the recent past. The advantages of combining various operations on a single machine resulted in

**Fig. 2.5** Horizontal turning centre with auto pallet changer

- reduction in number of machines
- reduction in floor space
- reduced work-in-process and lesser number of operations resulting in increased productivity
- increased consistency in quality
- reduction in the cost of manufacture.

The multi-operation machining on a single machine led to the development of the following features on turning centres.

(a) Addition of C-axis for precise rotary motion as servo axis and precise rotary positioning of spindles with live tools in the turret for off-axis milling, drilling, keyway and pocket milling, etc. These machines are generally referred to as turn-mill centres (Fig. 2.9(c)).

(b) Introduction of a second turret with two additional axes to the basic machine making it a 4-axes machine with twin turrets for simultaneous machining with a reduced cycle time in machining (Fig. 2.9(d)).

(c) Addition of sub-spindle for completion of second side turning operation on the same machine by transferring the part from the main spindle to a sub-spindle (Fig. 2.9(e)). The part is picked off from the rotating main spindle by the sub-spindle rotating at synchronous speed with the main spindle traversing on an independent linear servo axis. In yet another version the part is transferred through a pick-off chuck mounted in one of the tool stations on the turret.

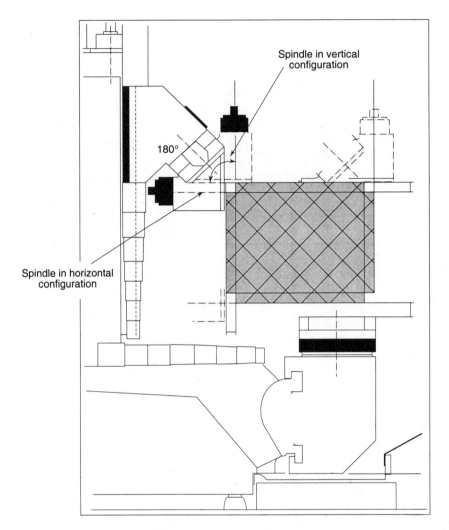

**Fig. 2.6(a)** Universal machining centre in vertical and horizontal configurations through swivelling of spindle head by 180° in 45° inclined plane

(d) Introduction of a milling head with two linear axes and a tilting axis on a basic 3-axes turning centre for powerful milling and precise keyway operations on turned parts on a single machine in the same set-up. The turning machine provided with high power independent milling head is referred to as mill-turn centre (Fig. 2.10). Thus, such machines have 6-axes and they can be further configured to a 8-axes machine with the addition of a second turret and two more independent axes for increasing the productivity of the machine.

Further developments on turning centres were in the areas of automation/unmanned production, process control, improvement of productivity features and increased cutting performance. The following are some of the developments.

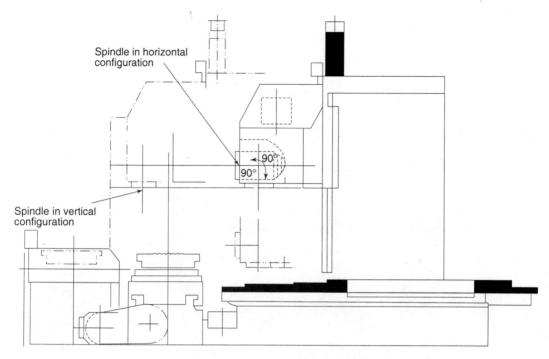

**Fig 2.6(b)**  UMC in horizontal and vertical configurations by tilting spindle head by 90°

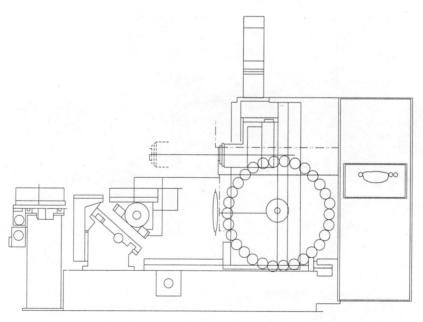

**Fig. 2.6(c)**  UMC in Horizontal spindle head configurations with pallet table mounted on 45° inclined plane indexable through 180° to facilitate 5-face machining

14  Mechatronics

Fig. 2.7  CNC lathe

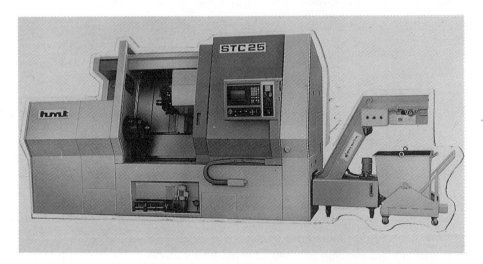

Fig. 2.8  Slant bed turning centre

(a) Provision of drum type magazine with a tool changer for automatic tool replacements (Fig. 2.11).
(b) Introduction of sensor systems for main spindle bearings to detect forced vibrations and abnormal tool conditions to equip adaptive controls and more accurate turning.

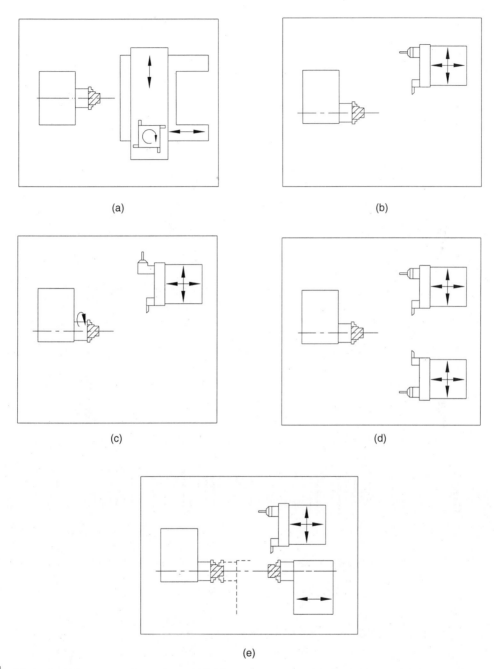

**Fig. 2.9** Different configurations of CNC turning centres (a) Two-axes and an indexable tool turret with horizontal bed (b) Two-axis, slant bed configuration with a disc type turret. (c) Turn-mill centre (d) Four-axes version for simultaneous machining with two turrets for reduction of cycle time (e) Additional sub-spindle for second-side turning operation

## 16 Mechatronics

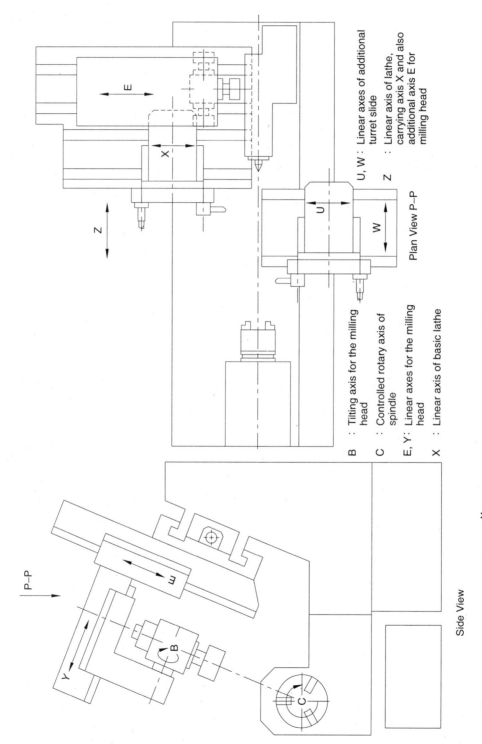

**Fig. 2.10** Eight-axes mill-turn centre

B : Tilting axis for the milling head
C : Controlled rotary axis of spindle
E, Y : Linear axes for the milling head
X : Linear axis of basic lathe
U, W : Linear axes of additional turret slide
Z : Linear axis of lathe, carrying axis X and also additional axis E for milling head

⇘ **Fig. 2.11** Lathe with drum tooling

(c) Higher spindle speeds of the order of 8000 rpm, higher slide rapid traverse rates of 20–30 m/min, and faster turret indexing as fast as 0.1 to 0.2 s for station-to-station and 1–2 s for tool change with ATC.

(d) Introduction of coolant-through-the-tool and programmable coolant jet for improving the chip breaking and flushing of chips.

(e) Provision of full digital servo control of main spindle for accurate off-axis work and better control of spindle accelerations and decelerations.

## 2.5  TOOL MONITORING ON CNC MACHINES

With the introduction of advanced features on CNC machines and with the introduction of flexible manufacturing system (FMS) and computer integrated manufacturing (CIM) tool monitoring systems to detect tool wear and breakage assumed importance as the operations on such systems are normally unmanned or minimally manned.

A large number of systems are available for automatic tool monitoring on CNC machines. In general, these systems can signal tool wear and breakage and automatically perform corrective actions or prevent damages.

The development of these systems made unmanned production possible and opened up the possibility of one operator to operate several machines at the same time. The systems available have their own operating principles and their own fields of application. However, the basic principles are as follows:

- registration of tool working hours
- measurement of output power
- measurement of cutting forces
- geometric measurements of tools or products

Generally these systems are integrated on machining centres or lathes and are used for milling, drilling and/or turning operations.

## Monitoring by Registration of Tool Working Hours (Tool Life Management)

The tool life monitoring system is used on machining centres and lathes. The tool life time (with a safety margin) is defined and input into the CNC system before hand, on the basis of experience. At the end of the tool life, the machining operation is terminated and the tool is replaced by the operator with a new tool or automatically with a sister tool available in the tool magazine.

## Monitoring by Measurement of Electric Current

By measuring the electric current drawn by the spindle or the feed motor as the case may be, the power drawn by the motor is determined. Monitoring is performed through comparative values such as fixed load limit, time-dependent current pattern, etc. These monitoring systems based on current measurement have a limited accuracy where small loads are concerned. They also have a large time constant between tool breakage and signalling, owing to many mechanical transmissions between machining station and measuring station. Therefore the principle is less suited to detect tool breakage, where a time lapse of 20–60 ms between breakage and signalling cannot fully exclude consequential damage.

## Monitoring by Force Measurements

Rapid signalling of tool breakage (in 1–2 ms) is possible by measurements of the cutting force. Sensors for force measurements, viz., piezoelectric quartz crystals or strain gauges are mounted at a location where the cutting forces are transmitted to the machine, as closely as possible to their origin either in the main spindle or in feed spindles.

## Monitoring via Geometric Measurements

Tool breakage, particularly drills, can be detected by measuring the tool length. Three dimensional measurements of tool or product can be used to determine tool wear directly or indirectly. The product geometry and tool positioning can be corrected or a new tool can be mounted.

Various tool monitoring systems as listed in Table 2.1 have been developed which signal tool wear and breakage and which switches off the machine or have a new product and tool placed automatically. The systems differ in operating principle, in applicability and reliability, in price and in features.

## TABLE 2.1
### Tool monitoring systems

| Basic principle | System | Measurement | Reference | Machine | Operation | Remarks |
|---|---|---|---|---|---|---|
| Registration of tool "working hours" | Tool life monitoring | Preventive | Experimental data | Integrated on machining centres and lathes | Universal | No separate detection of breakage |
| Measurements of electric current | Monitoring of load limits | In-process | Model machining value + limits; experimental data | Integrated on machining centres | Mainly drilling | Inaccurate for small powers |
| | Current pattern monitoring | | Fluctuating power of model machining | and lathes | Universal | Slow (reacts 20–60 ms after breakage |
| | Adaptive feed-rate control | | Constant power | | Milling turning | |
| Force measurements | Piezoelectric force transducers | In-process | Rapid force alterations | Lathes (in principle also machining centres) | Turning (universal) | Detection of breakage, chipping, collision, vibrations, rapid detection of breakage (1–2 ms) |
| | Transducers with strain gauges | | Model machining value + limits; experimental data | Machining centres and lathes (in some cases integrated) | Turning, drilling milling | Somewhat slower detection |
| Geometric measurements | Detection of breakage by tool length measurement or presence checking | Before or after (before and after) process (clamped product) | — Pressure/contact switch<br>— Photoelectric system<br>— Measuring probe<br>— Mechanical feeler | Integrated on machining centres and lathes | Drilling | |
| | Measurements of tool and product with probe | | Rated values of tool length and position; product geometry | Machining centres and lathes (in some cases integrated) | Universal | |

## 2.6 OTHER CNC DEVELOPMENTS

While the benefits of CNC have been exploited to the maximum for higher productivity and flexibility in the case of machining centres and turning centres, they have been best utilised by grinding centres for the production of precision components. This is achieved by the introduction of automatic dressing and gauging systems, in addition to combining operations like internal and face grinding operations, external and internal grinding operations, etc. Also, the productivity of CNC grinding machines has been improved through auto changing of wheels and work loading and to higher levels of technology for unattended operation.

The applications of hi-tech CNC is not only limited to basic metal cutting processes only, but has also been exploited in the development of modern sheet metalworking machines such as CNC turret punch press, CNC bending machines, CNC gear cutting machines, and other machines for non-traditional machining such as laser and electro-discharge machining.

## 2.7 ADVANCED MANUFACTURING SYSTEMS

While the various evolutionary stages in the development of CNC machines provided a means for efficient part production, over a period of time and with rapid developments in electronics, the machine tool technology has graduated from the concept of stand-alone machine tools to system-oriented manufacturing. This resulted in the introduction of flexible manufacturing cells (FMC), flexible manufacturing systems (FMS), and computer integrated manufacturing (CIM). Thus the emphasis has shifted from mechanical hardware in the case of conventional manufacturing to a combination of mechanical hardware, electronics and software which now accounts for 30-50% of the value of modern manufacturing systems (Fig. 2.12).

The development of flexible manufacturing, beyond the stand-alone CNC machines has come about in the last ten years due to an increased demand for flexibility in manufacturing. The need arose because of the shift in market demand and customer preferences, which required product variety in manufacturing. As a result, the concept of manufacturing has moved away from mass production to batch production in medium to large batches, and the FMC and FMS production systems meet these needs. Typically, an FMC/FMS is a manufacturing cell/system consisting of one or more CNC machines connected by an automated material handling system, all operated under the control of a central computer, along with other auxiliary sub-systems like component load/unload station, automatic tool handling system, tool presetter, component measuring equipment, wash station, etc. Figure 2.13 shows the major constituents of an FMS. Figure 2.14 gives the integration of various constituents of an FMS.

The characteristic features of an FMS are as follows.

(a) Solves the production problem of mid-volume and mid-variety parts for which neither the high production rate transfer lines nor the highly flexible stand-alone CNC machines are suitable.

Introduction to Modern CNC Machines and Manufacturing Systems 21

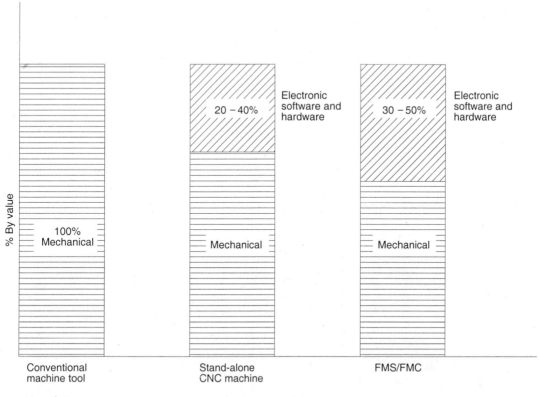

Fig. 2.12 Valuewise contents of mechanical and electronic software and hardware in different manufacturing facilities

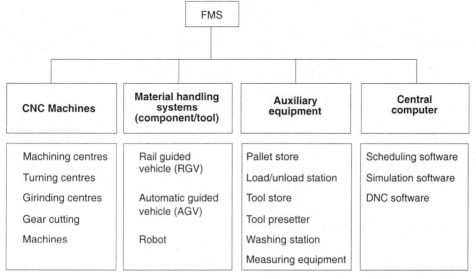

Fig. 2.13 Main constituents of an FMS

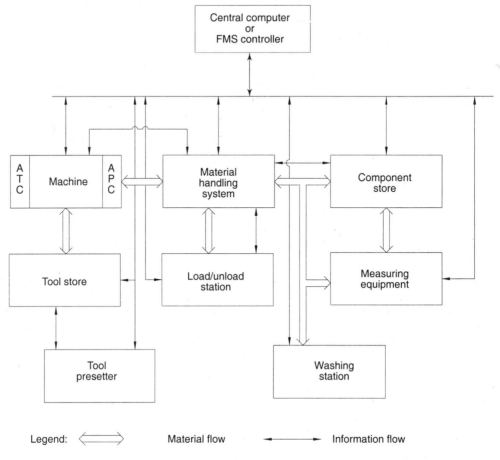

**Fig. 2.14** Integration of various constituents of an FMS

(b) Designed to process simultaneously, several types of parts in a defined mix.
(c) Equipped with sophisticated flexible machine tools which are capable of processing a sequence of different parts with negligible tool changeover time.
(d) Parts are transferred from machine to machine by a computer controlled material handling system.

### 2.8 BENEFITS OF AN FMS

Prime benefits of an FMS are listed below.

- Flexibility to change part variety
- Higher productivity
- Higher machine utilisation
- Balanced output

- Less rejections
- High product quality
- Reduced work-in-process and inventory
- Better control over production
- Just-in-time manufacturing
- Minimally manned operation
- Easier to expand

A study conducted on the utilisation of a stand-alone machining centre in Europe indicated a utilisation of 2554 hours out of the 8760 hours available, which works out to 29% (Fig. 2.15). The idle time includes sundays, holidays, night shifts, absenteeism, technical faults, organisational stoppages, workpiece and job changes. With the application of FMSs, the utilisation of CNC machining centres increased to 53% from 29%, as can be seen in Fig. 2.16.

As reported by a US company, with a sample of 20 US operating FMSs, the benefits of existing installations as in 1987 are given in Table 2.2.

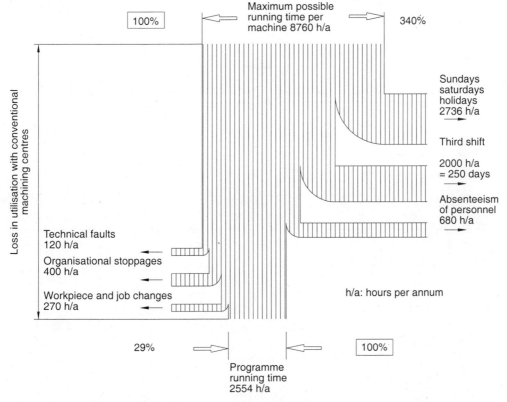

**Fig. 2.15** Utilisation of stand-alone CNC machining centre

**24** *Mechatronics*

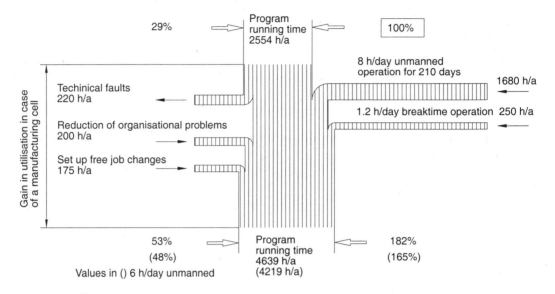

**Fig. 2.16** Improvement in utilisation of CNC machining centre in FMS

TABLE 2.2

| Criterion | Previous method | FMS |
|---|---|---|
| Number of machines | 29 | 9 |
| Direct labour | 70 | 16 |
| Machine efficiency | 20% | 70% |
| Processing time: Days | 18.6 | 4.2 |
| Number of operations | 15 | 8 |
| Floor space | 1500 cm$^2$ | 500 cm$^2$ |
| Product cost | $ 2000 | $ 1000 |

## 2.9　TRENDS IN ADOPTION OF FMSs

FMS was adopted in the US way back in 1967, but its widespread application has been rather slow. In 1981, the world population of FMSs was estimated to be only 115, with 25 FMS installations each in USA, Western Europe and 40 in Japan. But by 1986, the population had grown to over 200.

An industrywise application of FMSs is given in Fig. 2.17.

Thus, with developments over a period of time, the manufacturing industry now has a spectrum of production alternatives to choose from depending on part variety and production volume (Fig. 2.18).

The production of a limited variety of parts in high volume, which is typical of automobile industries, is served by transfer lines. When the volume is large, but the variety

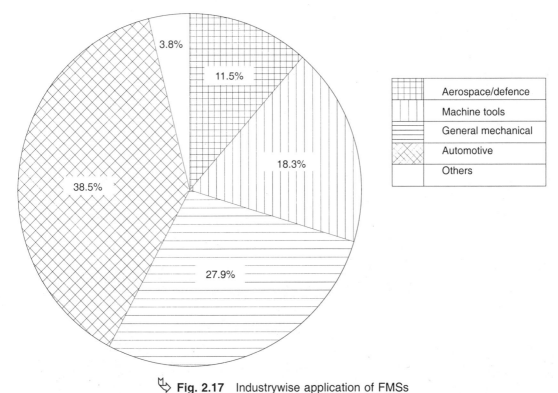

**Fig. 2.17** Industrywise application of FMSs

of parts increase, the transfer line is designed with some built-in flexibility to handle the production requirements.

Flexible manufacturing systems ideally suit the production requirement of large part variety but small-to-medium batch volume. Examples fitting this lot include machine tools, industrial machinery and general engineering industries. Industries manufacturing items such as tractors, earthmoving, agricultural machinery and defence-related components have production requirements of a fairly large volume and wider variety of parts. This need is being met by what may be termed as dedicated FMSs/FMCs where the basic production equipment is flexible, but the volume is high enough to dedicate individual machines to specific parts.

Presently, many firms abroad, desiring increased manufacturing flexibility are finding that a stepwise approach to advanced machine tool technology through a flexible cell implementation is more manageable than wholesale investment in a large FMS. Such smaller modular systems are both easier to integrate into the existing manufacturing set-up and to upgrade to a more complex configuration when needed.

## 2.9.1 Stepwise approach to FMS

A leading machine tool manufacturer abroad has outlined a stepwise adoption of a FMS for prismatic parts production as detailed in steps below and represented schematically in Fig. 2.19.

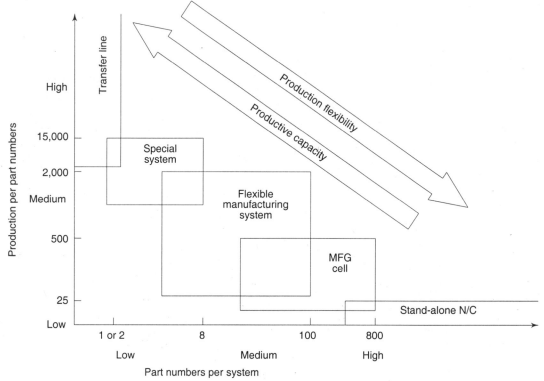

**Fig. 2.18** Spectrum of production alternatives

Step 1 : A stand-alone machine with features for unmanned operation allows the user to build up experience in palletising work and presenting a mixed batch of parts to the machine.

Step 2 : The addition of a rail guided vehicle and up to 15 pallet stands enhances performance in two ways. The range of components that can be handled without operator involvement increases and the period of unmanned operation can be extended.

Step 3 : Machines with up to 15 pallet stands each can be added. At this level, pallets are dedicated to individual machines. Load/unload is carried out at the load/unload station allotted to each machine. The rail guided vehicle responds to calls from the individual machine control systems.

Step 4 : The same system hardware parameters as in step 3 plus an independent transport control. Any pallet can now be loaded on to any machine. The transporter works according to priorities established by the operator instead of answering machine calls for the next component on a queueing basis.

Step 5 : With the addition of a host computer, the cell can be expanded to include more machines and to support amongst others the operations of inspection and wash. A host computer upgrades the system to exploit all the facets of automated manufacturing technology, including all commercial and engineering functions leading to computer integrated manufacturing.

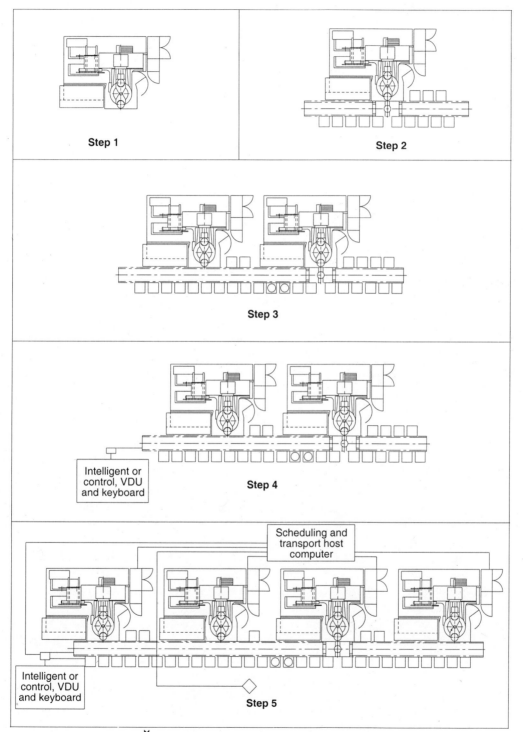

**Fig. 2.19** Stepwise approach to FMS

Such an FMS is not only adopted in advanced countries but also in the Indian industry. In the last five years some of the leading automobile, machine tool and defence sectors in the country have also installed FMSs. HMT has already developed an FMS on a pilot basis and has undertaken the supply of an FMS to the defence sector under a joint working agreement with an overseas manufacturer. Though the FMS requires higher initial investment, its benefits are substantial in the long run. Hence the adoption of FMSs in the Indian industry is likely to grow in the future as demand for flexibility and productivity increases.

CHAPTER 3

# Electronics for Mechanical Engineers

## 3.1 INTRODUCTION

According to the definition given by the Institute of Electronics and Electrical Engineers (IEEE), USA, electronics is the science and technology of passage of charged particles in gas, vacuum or in semiconductors. Electronics deals with the conduction of electricity in devices like radio tubes and solid state devices like transistors to get results not possible with ordinary electrical equipments.

Initially, communication was the main purpose of electronics, but gradually it has been adopted in other fields also.

Electricity flows in electric motors, incandescent lamps, electric furnaces and transformers through copper wires or some metallic parts. All these are electrical devices. But, during a stroke of lightning, electricity traverses through space. The high electric potential associated with lightning forces the electric current to pass through air. Similarly in vacuum tubes, electric current is made to flow in vacuum between the electrodes. When electricity flows through open space or vacuum as in the case of lightning or vacuum tubes instead of being confined to metallic conductors, it is termed as electronic. For example, the fluorescent lamp can be considered electronic as electrons flow in space from one end of the tube to the other end, whereas, the common incandescent lamp cannot be considered electronic even though it has a glass enclosure because the electric current flows through metallic filaments. An ordinary electrical equipment enters the electronic class whenever its circuit includes electronic devices such as electron tubes or solid state devices which are formed by junctions of semiconductor materials. Understanding electronics includes the understanding of ordinary electrical devices and circuits. No matter what electronic devices are used, the equipment is still electrical. In other words, all that is electronic is also electrical.

The salient features of electronic devices are:

1. Electronic devices can respond to very small control signals and produce a correspondingly large signal, i.e. electronic devices can amplify the input signals.
2. Electronic devices can respond at speeds far beyond the speeds that one comes across in mechanical and electrical devices. The speed at which some electronic devices work is unimaginable in mechanical and electrical devices.

3. Some electronic devices are photo-sensitive.
4. Electronic devices can rectify alternating current into direct current.
5. Some electronic devices can produce radiations such as X-rays.

## 3.2 CONDUCTORS, INSULATORS AND SEMICONDUCTORS

In our day-to-day activities, we frequently come across materials which allow electric current to flow through them freely. These are good conductors of electricity, e.g. metals. There are others which resist the flow of electricity through them. These materials which do not allow electric current to flow through them are called insulators, e.g. glass, wool, plastic, dry wood, etc.

In addition to conductors and insulators, there are some materials which fall in between these two extremities. They are neither good conductors of electricity nor bad conductors like insulators. This class of materials are called semiconductors, e.g. graphite, silicon, germanium, etc.

To understand why some materials behave as conductors, insulators or semiconductors, the atomic structure of the materials has to be understood.

All matter is made up of atoms consisting of a central positively charged nucleus, around which one or more negatively charged particles of negligible weight called electrons are orbiting in specific orbits. In any atom the positive charge on nucleus is equal to the total negative charge of its orbiting electrons. As such the atom as a whole is electrically neutral. Since unlike charges attract, there is a force of attraction between the nucleus and the electrons. The electrons remain bound to the atom by means of the force of attraction that exists between the nucleus and electrons. This force of attraction reduces with the increasing distance of the electrons from the nucleus. Hence, the electrons in the outermost orbit are more loosely bound to the nucleus than those in the inner orbits. The electrons in the outer most orbit are called valence electrons and the outermost orbit is called the valence orbit. For electric current to flow through a material, it should have free electrons in its atoms which can constitute a flow of current under the influence of an electric field.

The electrons can be dislodged from their orbits by imparting them a specific amount of energy in the form of heat, voltage, magnetic field or light. Once the electrons get dislodged from the orbits, they are free to form a current under the influence of an electric field. The amount of energy required to dislodge electrons from the orbits once again depends on the material the number of free electrons present in the atom of the material and the amount of energy required to dislodge the valence electrons from their orbits and decide whether it is a good conductor, insulator or a semiconductor of electricity. In case of metals, this energy requirement is very low and the thermal energy imparted to the electrons at room temperature is enough to free a sufficiently large number of electrons from their orbits. In fact, in a metal there are so many free electrons which cannot be considered as belonging to any particular atom and form what is known as electron gas. Hence metals behave as good conductors of electricity. On the other hand, in insulators this energy requirement is very high and normally, cannot be acquired through applied thermal energy or electric field and as such there are very few free electrons in insulators; hence they are bad conductors of electricity. In case of semiconductors, this energy requirement is in between that of metals and insulators. Hence they behave as semiconductors of electricity.

No understanding of the basics of electronics will be complete without the knowledge of certain basic concepts and components used in electronics. A brief description of such concepts and devices is given below.

## *Voltage*

The electric voltage is the applied electric level or potential. As an anology the electric level can be compared with the water level. Water flows from a higher level to a lower level when connected through a pipe. Similarly, electric current flows from higher electric level (voltage or potential) to lower electric level when connected through a conductor. Hence level and potential are analogous. Sea level is considered as the zero level or datum for specifying levels, whereas earth potential is considered as datum for specifying electric potential. In a circuit, positive and negative terminals of the battery are usually regarded as the points of highest and lowest potential.

## *Potential*

A body raised above the ground level has a certain amount of potential energy stored in it, which is equal to the amount of work done in raising it to that height. If this body is left free to fall, it falls because of the gravitational attraction of earth and proceeds from a point of higher potential energy to one of lower energy. If we imagine an isolated positive charge $q$ placed in air, like earth's gravitational field, it has its own electrostatic field which theoretically extends to infinity.

In Fig. 3.1, if a charge $X$ is very far from $q$, say at infinity, then force on it due to charge $q$ is zero. As $X$ is brought nearer to $q$, a force of repulsion acts on it, as similar charges repel each other. Hence, work or energy is required to bring it to a point $A$ in the electric field. Thus charge $X$ has some electric potential energy at point $A$ (Fig. 3.1).

In a gravitational field, sea level is usually taken as zero potential. In electric field, infinity is chosen as the theoretical place of zero potential, although in practice, earth's potential is chosen as zero potential. Since earth is a large conductor, its potential remains unaltered (constant) due to gaining or losing of charges. The voltage or electric potential at a point in an electric field can be defined as the work done in bringing one coulomb of positive charge to that point from infinity against the electric field.

The potential difference between two points in an electric field is the work done in taking one coulomb of positive charge from one point to the other.

The voltage is measured in volt (V). If in shifting one coulomb of positive charge from infinity to a certain point in the electric field, the work done is one joule, then the potential at that point is one volt.

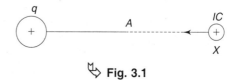

Fig. 3.1

## Current

An electric current may be defined as the movement of electric charge along a definite path. The magnitude of current is measured by the rate of flow of charge. In other words magnitude of electric current is the amount of flow of electric charge per second through any conductor. If $q$ coulombs of charge pass through a given point in $t$ s then the current is given by $q/t$ coulombs/sec. The unit of current is ampere (A).

One ampere is the current corresponding to a flow of one coulomb of charge per second through a conductor. As one coulomb of charge corresponds to the combined charge on $6.24 \times 10^{18}$ electrons, one ampere of current is the flow of $6.24 \times 10^{18}$ electrons per second through a conductor.

A current in a conductor produces the following characteristic effects:

1. Heating effect.
2. Magnetic field.
3. Chemical transformation when passed through certain solutions.

Some practical applications of the heating effect are incandescent lamps, electric heaters, electric cookers, electric furnaces, etc. Magnetic effect is used in electromagnets and chemical effect is used in electroplating.

## Resistance

Resistance may be defined as the property of a material due to which it opposes the flow of electricity through it. Metals are good conductors of electricity. They offer very low resistance to the flow of current, whereas, there are other substances which offer a relatively greater opposition to the passage of electricity through them. They are said to be relatively poor conductors of electricity. Carbon is one such material. The electrical resistance is similar to friction in mechanics.

The practical unit of resistance is ohm ($\Omega$). A conductor is said to have a resistance of one ohm if it permits one ampere of current to flow through it when one volt potential is applied across its terminals.

The resistance offered by a conductor is dependent on the following factors.

(a) It varies directly as its length
(b) It varies inversely as the cross-section of the conductor
(c) It depends on the nature of the material
(d) It also depends on the temperature of the conductor. If the effect of temperature can be neglected then,

$$R = \rho \frac{l}{a}$$

where $R$ = resistance, $l$ = length of the conductor, $a$ = area of cross-section.

The constant $\rho$ depends on the nature of the material of the conductor and is known as specific resistance or resistivity. Specific resistance is defined as the resistance between the opposite faces of a cm cube of that material.

*Ohm's law* Ohm's law states the relation between voltage, resistance and current for a given conductor. It can be stated as

$$V = IR$$

where, $V$ = voltage applied, $I$ = current, and $R$ = resistance.

## Electric Power

Electric power is the rate of expenditure of electric energy. The electric energy spent per second is called power. The electric power is given by,

$$P = VI$$

where, $P$ = power, $V$ = voltage applied, and $I$ = current.
Applying Ohm's law we have

$$P = I^2 R = \frac{V^2}{R}$$

Electric power is measured in terms of a unit called watt (W). One watt is the power corresponding to the energy spent per second by one ampere current, when the applied voltage is one volt. Alternatively, if the electric energy spent per second is equal to one joule then the power dissipated is equal to one watt.

## Direct Current (dc)

Direct current is a unidirectional current which does not change appreciably in value like the one supplied by a battery. The dc voltage is a unidirectional voltage as that of a battery.

## Alternating Current or Voltage (ac)

In the alternating current or voltage, the circuit direction of current reverses at regularly recurring intervals of time. The shape of the curve obtained by plotting the instantaneous values of voltage or current against time is called its wave form or wave shape. An alternating current or voltage may not always take the form of a symmetrical wave. Figure 3.2 illustrates some forms of ac waves.

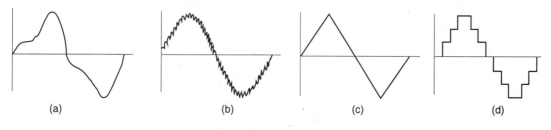

**Fig. 3.2** AC waveforms

The sine wave is most frequently encountered in electronics. A sine wave form is as shown in Fig. 3.3.

In this wave form, the voltage or current rises from zero to a positive peak, falls back to zero, then rises to a negative peak and again comes back to zero and this pattern repeats regularly. One complete set of positive and negative values of alternating quantity is known as a cycle. Figure 3.3 shows one cycle of a sine wave. A cycle is sometimes specified in terms of angular measure. A cycle is said to spread over 360° or two $\pi$ radians.

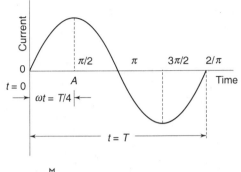

Fig 3.3 Sine wave

The number of such cycles per second is called frequency and the time taken for one cycle is called a period.

Domestic supply is a sinusoid having a frequency of 50 cycles/sec. Frequency is expressed in hertz (Hz). A frequency of one hertz equals one cycle/sec.

In an ac wave, the amplitude varies at every instant of time and hence a sort of effective value has to be defined for calculation purposes. The ac value is expressed in terms of a value called root mean square (rms) value.

The rms value of an alternating current is given by that steady (dc) current which when flowing through a given circuit, for a given time, produces the same heat as that produced by the alternating current when flowing through the same circuit for the same time. It is given by the square root of the average value of the squares of the instantaneous values taken over one cycle.

## 3.3 PASSIVE COMPONENTS USED IN ELECTRONICS

### Resistor

Resistor is a component which offers a specific amount of resistance to the flow of current. Usually resistors are made up of a film of carbon or a coil of nichrome wire. Resistors are used in electronics for limiting current, biasing active components and as voltage dividers. These resistors are available in some standard values and power ratings as well as tolerance ranges.

Figure 3.4 shows the symbolic representation of a resistor in a circuit.

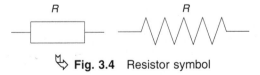

Fig. 3.4 Resistor symbol

**Resistance in series** When resistors are connected end-to-end as in Fig. 3.5, they are said to be connected in series. The equivalent resistance or total resistance of this combination is equal to the sum of the individual resistances so connected, since the same current flows through all resistors. In other words, the resistances connected in series can be replaced by a single resistance whose value is equal to the sum total of the individual resistances in series.

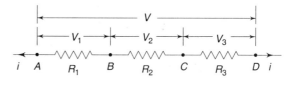

**Fig. 3.5** Resistors in series

***Resistance in parallel*** Resistors are said to be connected in parallel if two or more resistors are connected across each other as shown in Fig. 3.6.

Any number of resistors in parallel can be replaced by a single resistance of value such that its reciprocal is equal to the sum of the reciprocals of all the branch resistances. In Fig. 3.6, if the equivalent resistance is $R$, then it is given by:

$$\frac{1}{R} = \frac{1}{R_1} + \frac{1}{R_2} + \frac{1}{R_3}$$

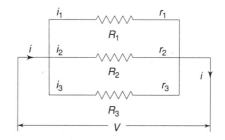

**Fig. 3.6** Resistors in parallel

## Potential Divider

A potential divider is used for obtaining a variable voltage from a constant voltage supply. Figure 3.7 shows a simple potential divider.

The whole of the supply voltage is dropped across the resistance $AB$ and by changing the position of the sliding contact $C$ over the resistance $AB$, any voltage from zero to the supply voltage can be obtained across the load.

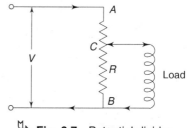

**Fig. 3.7** Potential divider

## Capacitor (Condenser)

A capacitor essentially consists of two conducting surfaces separated by a layer of insulating medium called a dielectric. A dielectric is an insulation medium through which an electrostatic field can pass. The main purpose of the capacitors is to store electric energy. Figure 3.8 shows the construction of a capacitor and its symbolic representation.

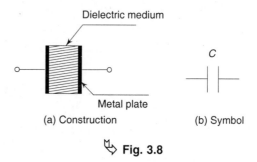

**Fig. 3.8**

In Fig. 3.9, a capacitor is connected in series with a resistance $R$ and a switch $S$ to a battery of voltage $V$. When the switch $S$ is closed, there is a momentary flow of electrons from the plate connected to positive terminal of the battery to the other plate of the capacitor. As electrons are withdrawn from the plate connected to positive terminal of the battery, it becomes positively charged and these electrons collect on the other plate and hence it becomes negatively charged. Thus a potential difference is established between the plates of the capacitor. The momentary flow of electrons gives rise to a charging current. This charging current is maximum at the instant when the switch $S$ is closed and plates are uncharged, but it then decreases and finally ceases when the potential difference across the plates becomes equal to the battery voltage $V$. This charging of capacitor can be compared with the charging of water tank $T$ from a reservoir $R$ as shown in Fig. 3.11. On opening the valve $V$, the water rushes through the valve to tank $T$. Initially the rush would be high and then the flow decreases and would cease once the levels in $R$ and $T$ become equal. The curve in Fig. 3.9 shows the charging current and voltage across the capacitor with respect to time.

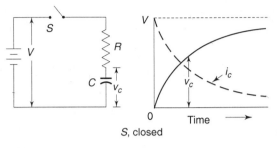

**Fig. 3.9** Charging of $G$ after $S$ is closed

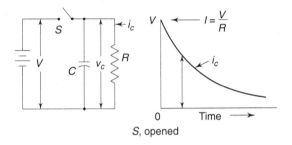

**Fig. 3.10** Discharging of $C$ after $S$ is opened

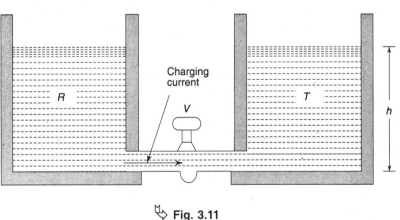

**Fig. 3.11**

Once the capacitor is charged, it can act like a temporary voltage source. If switch $S$ is opened the charged capacitor can source a temporary current as shown in Fig. 3.10. The discharging current is maximum at the instant of opening of switch $S$ and decreases exponentially with time as illustrated by current vs. time curve in Fig. 3.10.

The property of the capacitor by which it stores charge on its plates is called capacitance. The charge $Q$ that accumulates on the plates of the capacitor is proportional to the applied voltage $V$.

Since, $Q \propto V$, $Q = CV$

where $C$ is the proportionality constant. The proportionality constant C is the capacitance of the given capacitor. Capacitance of a capacitor is defined as the amount of charge required to create a unit potential difference between its plates.

We have $\quad C = \dfrac{Q}{V}$

where $Q$ = amount of charge on the plates of capacitor in coulomb, $C$ = capacitance, and $V$ = potential difference between the plates in volts.

The unit of capacitance is farad (F). One farad is defined as the capacitance of a capacitor which requires a charge of one coulomb to establish a potential difference of one volt between its plates. As farad is a very big unit of capacitance, much smaller units like microfarad (μf) and picofarad (pf) are generally used.

As illustrated in Fig. 3.9, the charging current is drawn from the battery momentarily and the current stops flowing once the capacitor is charged to the battery voltage. Hence we can say that a capacitor blocks the passage of dc. The charging time of the capacitor is dependent on the product of resistance $R$ and capacitance $C$. The product $RC$ is referred to as time constant.

When a dc voltage is applied to a capacitor, it momentarily acts like a short circuit and once the capacitor is charged, it acts like an open circuit.

When a capacitor is connected to alternating voltage (ac), the continuous change and periodical reversal of the applied voltage causes continuous change in state charge of the capacitor with a continuously changing current. If ac voltage is applied to a pure capacitive circuit, the capacitor will always be in a discharged state and hence potential difference between its plates will be zero. In other words, it acts like a short circuit for ac voltages.

As the capacitor blocks dc, this property of the capacitor is used to block dc component from an ac signal which is superimposed with dc and to transmit only ac signal to the output. As capacitors act like short circuit for ac signals, they are used to filter ac voltage from dc voltage. They are used as reservoir capacitors to smoothen pulsating dc.

When ac voltage is applied to a capacitor, the resulting ac current is as shown in Fig. 3.12. In a pure capacitive circuit the current leads the voltage by 90° when ac voltage is applied to the circuit.

### Inductor

A wire wound in the form of a coil forms an inductor. The property of an inductor is that it always tries to maintain a steady flow of current through it and opposes any change in current.

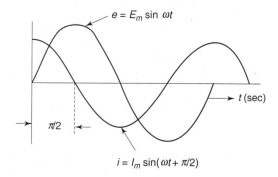

**Fig. 3.12** AC voltage and current in a capacitor

We know that whenever a current flows through a conductor, it produces a magnetic field around it in a plane perpendicular to the conductor as shown in Fig. 3.13. Conversely, when a wire is passed through a magnetic field an electromagnetic force (emf) is induced in the conductor. The polarity of this emf is dependent on the direction in which the conductor is moved.

Imagine a coil of wire similar to the one shown in Fig. 3.14, connected to a battery through a rheostat. Whenever an effort is made to increase the current through it, it is opposed by the instantaneous production of a counter emf. Energy required to overcome this opposition is supplied by the battery. This energy is stored in the inductor in the form of additional magnetic flux produced.

At this stage, if an effort is made to decrease the current through the coil, the current decrease is delayed due to the production of a self-induced emf. The property of the inductor due to which it opposes any increase or decrease of current (current growth or fall) by production of a counter emf is known as self inductance. This is because, whenever a current is passed through a coil of wire, the current through it produces a magnetic field with its axis down the centre of the coil, as shown in Fig. 3.15. The turns of the coil intersect this growing magnetic field and hence an emf is induced in the coil. The polarity of this induced emf is always opposite to the polarity of the applied voltage. This induced emf is referred to as back emf or counter emf. The back emf always tries to oppose the current growth or fall through the inductor and is proportional to the rate at which the current changes. This property is analogous to inertia in a material body. We know by experience that initially, it is difficult to set a heavy body into motion, but once in motion, it is equally difficult to stop it. Similarly, in a coil it is initially difficult to establish a current through it, but once it is established, it is equally difficult to withdraw it. Hence inductance is sometimes called electrical inertia.

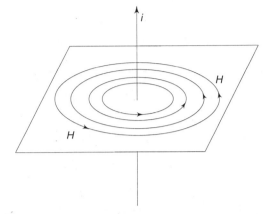

**Fig. 3.13** Magnetic field produced by current

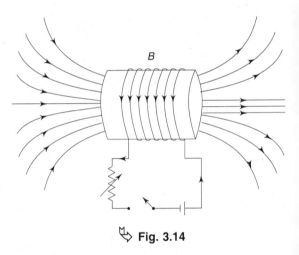

**Fig. 3.14**

The back emf $e$ is given by

$$e \propto \frac{di}{dt}$$

where $i$ = current through inductor.

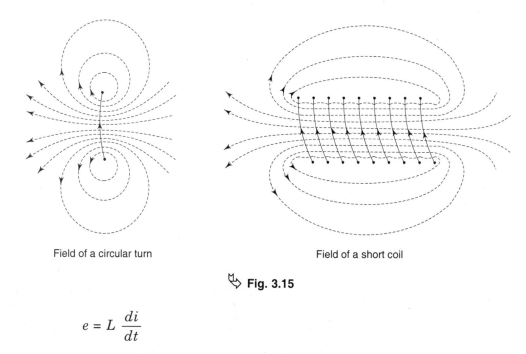

Field of a circular turn    Field of a short coil

Fig. 3.15

$$e = L \frac{di}{dt}$$

where the proportionality constant $L$ is called self inductance of the coil.

The inductance is measured in terms of a unit called henry. One henry is the self inductance of an inductor which produces a back emf of one volt when the current through it changes at a rate of one ampere per second. Since henry is a very large unit of inductance, smaller units like millihenry ($10^{-3}$ henry) and microhenry ($10^{-6}$ henry) are used.

Whenever an ac voltage is applied to an inductor, the current through it is changing at every instant and this current change is opposed by the inductor by production of counter emf. The effect of this is that the current through the inductor lags behind the applied ac voltage by 90° in a pure inductive circuit as illustrated by the curves in Fig. 3.16. However, in practice it is difficult to have a pure inductive circuit and hence the phase lag is a little less than 90°. The inductors are widely used in smoothing and sensing circuits.

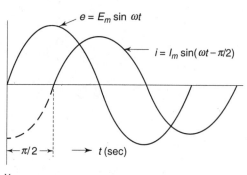

Fig. 3.16  AC voltage and current inductor

### 3.3.1 Combinations of Resistance, Capacitance and Inductance

In practice it is impossible to get pure resistors, inductors and capacitors.

Any ac circuit can be regarded as consisting of separate elements of resistance, inductance and capacitance connected together in different ways. The Ohm's law in its dc form cannot be applied to these circuits because of current and voltage phase relationship. Ohm's law, $V = IR$ still holds good but in an ac circuit $R$ should be substituted by impedance ($z$), which may be defined as ac resistance. The ratio of voltage to current gives resistance in a dc circuit but in an ac circuit it gives impedance.

When circuits consist of several elements in series or parallel, the component current and voltage is combined vectorially to find the resultant. It is useful to consider various combinations of these elements and calculate the resulting impedance. Table 3.1 shows the various formulae that can be used to calculate the impedance. The impedance of a capacitor is known as capacitive reactance and the impedance due to inductance ($L$) is known as inductive reactance. The inductive reactance $X_L$ of an inductance $L$ is given by

$$X_L = \omega L$$

where, $\omega = 2\pi f$, $f$ being the frequency of applied ac. The capacitive reactance $X_c$ of a capacitance is given by

$$X_c = \frac{1}{\omega C}$$

where, $\omega = 2\pi f$. Inductive reactances are positive while capacitive reactances are negative.

**TABLE 3.1 Impedance**

| Circuit | | Impedance |
|---|---|---|
| R (resistor) | (a) | $R$ |
| L (inductor) | (b) | $j\omega L$ |
| C (capacitor) | (c) | $-\dfrac{j}{\omega C}$ |
| R, L in series | (d) | $R + j\omega L$ |
| R, C in series | (e) | $R - j\dfrac{1}{C\omega}$ |
| C parallel with R | (f) | $\dfrac{R}{1 + j\omega CR}$ |
| R, L, C in series | (g) | $R + j\left(\omega L - \dfrac{1}{\omega C}\right)$ |
| R, L in series, parallel with C | (h) | $\dfrac{1}{\dfrac{R - j\omega L}{R^2 + \omega^2 L^2} + j\omega C}$ |

## 3.4 TRANSFORMERS

A transformer is a device by means of which electric power in one circuit is transformed to electric power of the same frequency in another circuit. It essentially consists of two or more inductive windings wound on the same core, which could be air for high frequency or some magnetic material for low frequency. It can raise or lower the voltage in a circuit but with a corresponding decrease or increase in current. In its simple form it consists of two inductive coils which are electrically separate but magnetically linked by winding them on the same core as shown in Fig. 3.17.

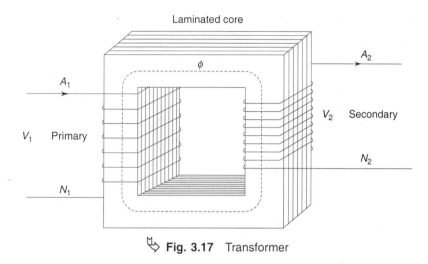

**Fig. 3.17** Transformer

One of the coils is connected to a source of alternating voltage which sets up an alternating magnetic flux, most of which is linked with the other coil. The alternating magnetic field produces an alternating voltage of the same frequency in the other coil by way of magnetic induction. If the second coil circuit is closed, a current flows in it and hence electric energy is transferred from the first to the second coil. The first coil which is connected to ac mains is called the primary winding and the other from which energy is drawn out is called secondary winding. In brief, a transformer is a device that

(a) transfers electric power from one circuit to another
(b) does so without changing the frequency
(c) does this by electromagnetic induction.

In any transformer

(a) the voltage transfer is directly proportional to the ratio of the number of turns of the primary to the secondary winding.
(b) the current transfer is inversely proportional to the ratio of the number of turns of the primary to the secondary winding.
(c) there are losses due to resistance of the windings and energising current taken by the core called copper loss and core loss of the transformer respectively.

In a transformer (Fig. 3.17), if $V_1$ is the primary voltage, $V_2$ is the secondary voltage and $N_1$ and $N_2$ are the number of turns in the primary and secondary windings respectively, then the relation between voltages and number of turns is given by

$$\frac{V_1}{N_1} = \frac{V_2}{N_2}$$

If a 10 volt potential is applied to the primary winding and there are 10 turns in the primary and 5 turns in the secondary, then 5 V potential appears on secondary winding; conversely, if 5V is applied to the secondary then 10 V will appear on the primary winding.

If $A_1$ is the current in the primary and $A_2$ is the current in the secondary then,

$$V_1 A_1 = V_2 A_2$$

If, $V_1 = 10$ V, $A_1 = 1$ A, and $V_2 = 5$ V, then 2 A current can be drawn from the secondary.

If the secondary voltage is higher than the primary (secondary has more number of turns compared to the primary) then the transformer is said to be a step-up transformer. On the other hand if the secondary voltage is lower than the primary (secondary has less number of turns compared to the primary) then it is said to be a step-down transformer.

## 3.5 SEMICONDUCTORS

### 3.5.1 Intrinsic Semiconductors

A semiconductor in its pure form is called intrinsic semiconductor. Silicon and germanium are the two most important and commonly used semiconductors in electronic devices. Figure 3.18 shows the structure of a silicon and germanium atom.

Both silicon and germanium are tetravalent, having four electrons in their outermost orbits. Both these materials are crystalline in nature. The crystal structure of these materials comprises a regular repetition in three dimensions of a unit cell having the form of a tetrahedron with an atom at each vertex. Figure 3.19 shows the crystal structure of silicon illustrated symbolically in two dimensions. Each atom in a silicon crystal contributes four valence electrons. Each silicon atom shares one of its valence electrons with the neighbouring four silicon atoms. Due to the sharing of valence electrons between atoms, a bond is established between them. This bond is called a covalent bond. The fact that the valence electrons bind one atom to the next also

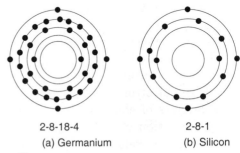

**Fig. 3.18** Atomic structure (a) Germanium (b) Silicon

2-8-18-4 (a) Germanium

2-8-1 (b) Silicon

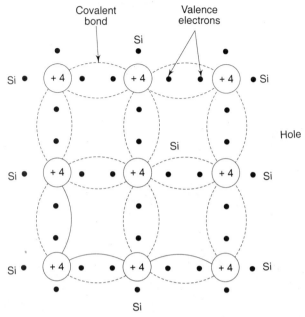

**Fig. 3.19** Crystal structure of silicon illustrated symbolically in two dimensions

results in the valence electrons being tightly bound to the nucleus. Hence, despite the availability of four valence electrons, the crystal has no free electrons at very low temperatures. At very low temperatures (say 0° K) the crystal behaves like an insulator as there are no free electrons available in the crystal. However, as the temperature increases and reaches the room temperature, some of the covalent bonds are broken because of the thermal energy imparted to the electrons. Due to this breaking of covalent bonds some of the electrons get liberated and the conductivity of the crystal improves. Figure 3.20 shows the silicon crystal structure with an electron dislodged from the covalent bond.

The liberated electron leaves a vacancy in the crystal. The black dot in Fig. 3.20 represents the liberated electron and the hollow circle represents the vacancy left by it in the crystal. The vacancy that is created in the crystal by the free electron is called a hole. The free electrons can act as current carriers. In addition to electrons, holes also serve as carriers of electricity. The formation of a hole and the mechanism by which it aids in propagation of current is illustrated in Fig. 3.21. The hole has an affinity for electrons and it is relatively easier for a valence electron in the neighbouring atom to leave its covalent bond and occupy the vacancy (hole). An electron moving from a bond to fill a hole leaves a hole in its initial position. Hence, the hole effectively moves in the direction opposite to that of the electrons. Hence holes can be considered as positive charges aiding propagation of current just like electrons. Here we have a mechanism for the conduction of electricity which does not involve free electrons. This phenomenon is

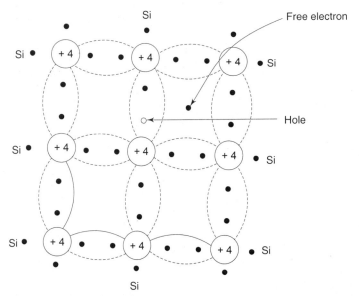

**Fig. 3.20** Silicon crystal with a broken covalent bond

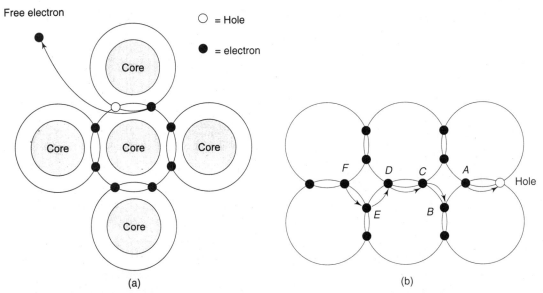

**Fig. 3.21** Formation of hole and propagation of current (a) Production of a hole (b) Hole movement

schematically illustrated in Fig. 3.22, where a circle with a dot represents a filled covalent bond and an empty circle represents a hole. Figure 3.22 shows a hole in position 6. Now if an electron from position 7 moves into the hole at position 6, then it leaves a hole in position 7. From these figures it is evident that the hole moves in the opposite direction of electrons. As far as the electric current is concerned, the hole behaves like a positive charge equal in magnitude to the charge on electron.

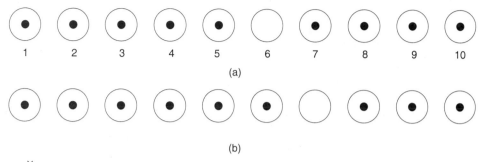

**Fig. 3.22** The mechanism by which a hole contributes to the conductivity

Due to the thermal production of free electrons and holes, pure silicon and germanium crystals act like semiconductors at room temperature. The conduction in an intrinsic semiconductor at absolute zero temperature and at higher temperatures is illustrated in Fig. 3.23.

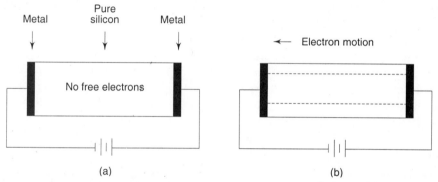

**Fig. 3.23** Conduction at (a) absolute zero temperature and (b) higher temperatures

### 3.5.2 Extrinsic Semiconductor

Even though pure silicon and germanium have some free electrons and holes at normal temperature, the number of free electrons and holes is still very low and of no use practically. In order to improve the conductivity of these materials some impurities are intentionally added to these materials. This addition of impurities to an intrinsic semiconductor is called doping. Such semiconductor materials in which impurities are added to improve conductivity is called extrinsic semiconductor (impure semiconductor).

Usually either a pentavalent impurity such as phosphorus, antimony or a trivalent impurity such as indium, gallium or boron is added to pure semiconductors.

Figure 3.24 shows the crystal structure of the silicon crystal in which a pentavalent impurity such as antimony (Sb) is added. The antimony atom replaces a silicon atom in the crystal lattice. As usual four valence electrons of the impurity atom are shared by the adjacent four silicon atoms. The fifth electron is not shared by any atom and hence it is free. So whenever a pentavalent impurity atom is added to pure silicon crystal, a free electron is donated and hence pentavalent impurities are called donors.

The semiconductor which is formed by doping with a pentavalent impurity is called an *N*-type semiconductor, since it has more electrons as current carriers and electrons are negatively charged. Here *N* stands for negative.

On the other hand if a trivalent impurity is added, it creates a vacancy at one of its covalent bonds as illustarated in Fig. 3.25. In this type of semiconductor formed by doping with trivalent impurity, holes are in majority and hence this type of semiconductor is called a *P*-type semiconductor where *P* stands for positive. As a trivalent impurity leaves a vacancy which can accept electrons from adjacent atoms, these types of impurities are called acceptors. The fundamental difference between a metal and a semiconductor is that the current in metals is only because of the negative charge (electrons), whereas current in a semiconductor is due to both negative and positive charges (electrons and holes).

### 3.5.3 Semiconductor Devices

The first semiconductor device to be announced in 1947 was a point contact diode. A point contact diode has a metal base on which a germanium semiconductor wafer is mounted as shown in Fig. 3.26. An external lead is brought out from the metal base. A short piece of spring wire made out of tungsten alloy is mounted in such a way that it presses against the semiconductor crystal. Another external lead is brought out from the spring. The whole thing is encapsulated in a bead like glass shell. This device works like a unidirectional valve and allows current to flow in only one direction.

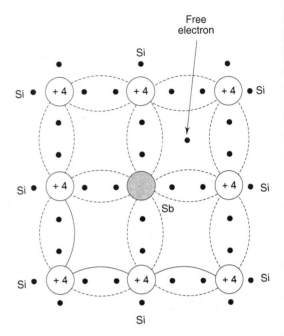

Fig. 3.24 Crystal lattice with a silicon atom displaced by a pentavalent (antimony) impurity atom

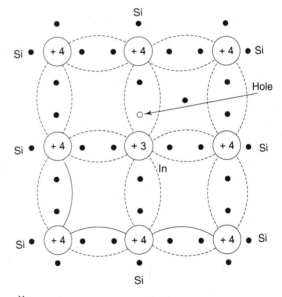

Fig. 3.25 Crystal lattice with a silicon atom displaced by an atom of a trivalent (indium) impurity

In this device the contact area between point contact and the semiconductor is very small. This makes the capacitance between the elements very small. Because of small shunt capacitance, it can work at high frequencies. This type of diode is used as detector diode in radios. The disadvantage of this diode is that the contact surface is very small and hence it cannot handle higher currents.

One year later, the junction diode was announced which had a junction of a $P$-type semiconductor material and an N-type material. This type of diode is called a $P$-$N$ junction diode. Because of a larger contact area, it can handle a much larger power as compared to point contact diode.

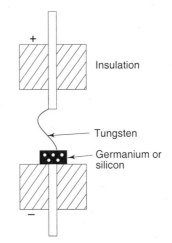

**Fig. 3.26** Point-contact semiconductor rectifier

## P-N Junction Diode

If $P$-type impurities are introduced into one side and $N$-type impurities into the other side of a single crystal of semiconductor, a $P$-$N$ junction is formed as shown in Fig. 3.27. Two leads are brought out from both $P$ and $N$ regions. The lead that is brought out from $P$ region is called an anode and the lead brought out from $N$ region is called the cathode. This device acts like a unidirectional valve and allows current to flow in one direction only. Figure 3.27 also shows the schematic representation of a junction diode. The arrow in the symbol denotes that the direction of easy flow of conventional current is from the anode to the cathode.

If the $P$-$N$ junction diode is connected in a circuit with its anode connected to positive terminal of the battery and cathode connected to negative terminal of the battery as shown in Fig. 3.28, the diode offers a low resistance path to flow of current and hence a current flows through the diode. When a diode is so connected it is said to be forward biased.

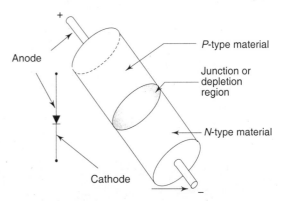

**Fig. 3.27** P-N Junction rectifier or diode

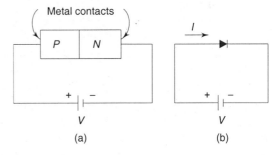

**Fig. 3.28** (a) A P-N Junction biased in the forward direction (b) The rectifier symbol is used for the P-N diode

On the other hand if the polarity of the battery is reversed, i.e. *P*-type region is connected to negative terminal of the battery and *N*-type region is connected to positive terminal of the battery, as shown in Fig. 3.29, the diode is said to be reverse biased. In this type of connection no current flows through the diode.

**Diode action** Figure 3.30 illustrates the diode action. Figures 3.30(a) and (b) show the current carriers and immobile atoms in *P*- and *N*-type semiconductors. As shown in Fig. 3.30(a), a *P*-type semiconductor has

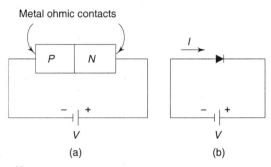

**Fig. 3.29** (a) A *P-N* junction biased in the reverse direction (b) The rectifier symbol is used for the *P-N* diode

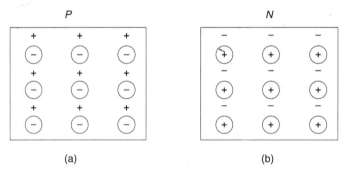

**Fig. 3.30** Current carriers and immobile atoms (a) *P*-type (b) *N*-type

more holes represented by a + sign. The immobile acceptor in *P*-type semiconductor are ions represented by circles with a − sign. Similarly, the *N*-region has a greater number of electrons which are represented by a − sign. The immobile donor ions in *N*-type semiconductor are represented by circles with a + sign. Figure 3.31(a) shows *P-N* junction at the instant of formation of the junction and Fig. 3.31(b) shows the *P-N* junction on the lapse of time on the formation of the junction. At the instant of formation of the *P-N* junction, *P*-type material has holes as the majority current carriers (indicated by a + sign) and *N*-type material has electrons as majority carriers of current (indicated by a − sign). Some of the electrons from *N*-type material and holes from *P*-type material cross the junction due to thermal diffusion. The electrons and holes that cross the junction recombine, making a small region around the *P-N* junction void of any charge. This region or layer is called depletion region or layer. The thickness of this depletion region is of the order of 5 microns ($5 \times 10^{-6}$ m). Once the depletion region is formed, the *P-N* junction can be compared to a charged capacitor as positive charges are on one side of the depletion region and negative charges are on the other side of the depletion region, and a potential difference exists between them. This potential difference acts like a potential barrier which prevents further diffusion and recombination of charges. When the

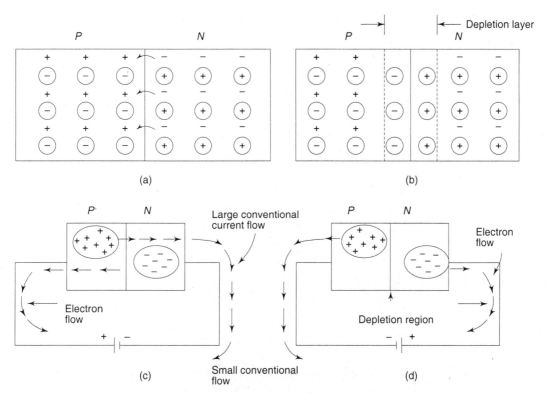

**Fig. 3.31** Movement of electrons and positive charges (a) At the instant of formation (b) Depletion layer (c) Forward bias (d) Reverse bias

diode is forward biased, the positive terminal of the battery connected to $P$-region repels the holes present in $P$-region and hence they cross the junction and come to $N$-region. When they are in $N$-region, they are attracted by the negative terminal of the battery resulting in a current. Similarly, the electrons present in $N$-type region cross the junction and come to $P$-region from where they are collected by the positive terminal of the battery connected to $P$-region. In other words, when the diode is forward biased the battery potential overcomes the barrier potential. The negative terminal supplies fresh electrons which cross the junction and are collected by positive terminal of the battery, resulting in a continuous current across the junction.

When the diode is reverse biased, majority of the current carriers present in $P$- and $N$-type region are attracted by the terminals of the battery and both $P$-type and $N$-type regions become void of any charges. In other words, the battery potential aids the barrier potential and hence no current carriers cross the junction. Hence, normally zero current flows across the junction. Actually, a very small current does flow in the diode because of thermal production of a small number of electrons and hole pairs in the semiconductor crystal. These electrons and holes are forced across the junction and are collected by the connected battery, resulting in a very small reverse current. The forward and reverse characteristic of a diode is illustrated in Fig. 3.32. Here the diode

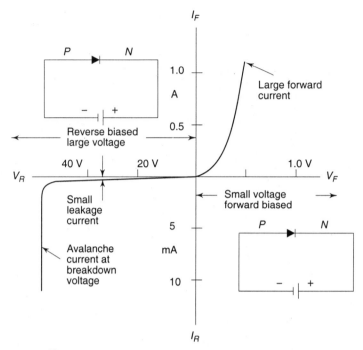

**Fig. 3.32** Diode current versus applied voltage

currents for various applied voltages are plotted against the applied voltage for both forward and reverse bias.

As can be seen from the VI characteristic of diode for forward bias, the diode current increases steeply for every increase in applied voltage after an initial small voltage. This small voltage, called the cut in voltage, depends on the semiconductor material. For germanium it is of the order of 0.2 V and for silicon it is around 0.6 V. However, the current is very small in the reverse direction till breakdown voltage. After this critical voltage, large reverse current flows through the diode and it may get damaged if the current is not limited by an external resistor. Beyond breakdown voltage, the diode is said to be operating in breakdown region.

### Diode as Rectifier

Power supply is the most basic requirement of any electronic circuit and practically all electronic systems require one or more power supplies for working, the simplest and the most effective being a battery pack. While it is simple and has low output impedance, it suffers from a limited useful life. Most of the common power supplies convert the mains ac supply into required dc supply and also comprise a rectifier circuit which converts ac voltage to dc voltage. As diode conducts in one direction only, it can be used as a rectifier to convert ac to dc.

Figure 3.33(a) shows a half-wave rectifier circuit having a single diode and Fig. 3.33(b) shows the input waveform. During the positive half cycle of line voltage, the diode is

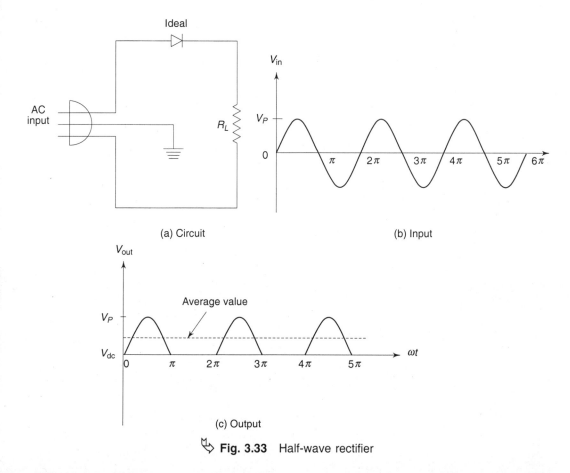

**Fig. 3.33** Half-wave rectifier

forward biased. During the negative half cycle it is reverse biased. As the diode conducts when it is forward biased, the positive half cycle is transmitted to the output and it appears across the load $R_L$. The negative half cycle does not appear across the load resistor $R_L$ as diode does not conduct during this half cycle of the input signal. Hence the output of the circuit across $R_L$ will be a pulsating dc as shown in Fig. 3.33(c). The average value of this half wave signal is $V_{dc}$ as shown in Fig. 3.33(c) and is given by

$$V_{dc} = \frac{V_P}{\pi}$$

where, $V_P$ is the peak value of input signal.

**Centre tap rectifier** Figure 3.34(a) shows a centre tap rectifier which uses a centre tap transformer and two diodes. In this circuit, during positive half cycle the upper diode $D_1$ is forward biased and the lower diode $D_2$ is reverse biased and hence there is current through $D_1$, load resistor and upper half winding of the transformer. The circle on diode $D_2$ in Fig. 3.34(c) indicates that it is not conducting. During the negative half cycle of

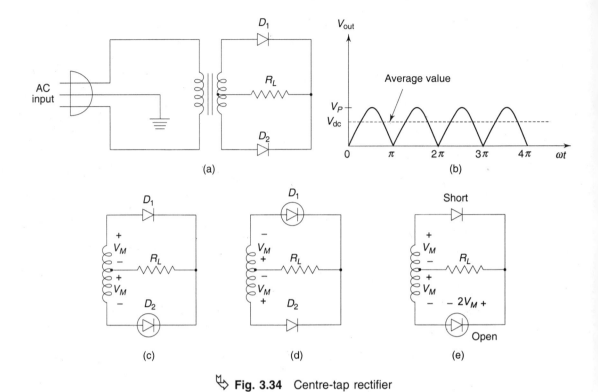

**Fig. 3.34** Centre-tap rectifier

input the diode $D_2$ conducts and current is passed through the lower diode, load resistor $R_L$ and lower half winding of the transformer. Note that during both half cycles the current through the load resistor is in the same direction. Hence the output across load resistor $R_L$ is of the form shown in Fig. 3.34(b). As the output appears during both half cycles of the input wave, this circuit is called a full wave rectifier. The average or dc voltage of a full wave rectifier is twice that of a half wave circuit and is given by

$$V_{dc} = 2\,\frac{V_P}{\pi}$$

where, $V_P$ is the peak input voltage.

**Bridge rectifier** Four diodes can be connected in the form of a bridge as shown in Fig. 3.35(a) to get a bridge rectifier, which is also a full wave rectifier. During positive half cycles, the diodes $D_2$ and $D_3$ conduct and during negative half cycles, diodes $D_1$ and $D_4$ conduct. In either case the current through load resistor $R_L$ is in the same direction, resulting in full wave rectification of the input. Figure 3.35(b) shows the time output wave form and Figs. 3.35(c) and (d) show the rectifier during positive and negative half cycles respectively.

The bridge rectifier is the most widely used rectifier. Its main advantage is that it eliminates the requirement of a centre tap transformer. But the rectifier creates a prob-

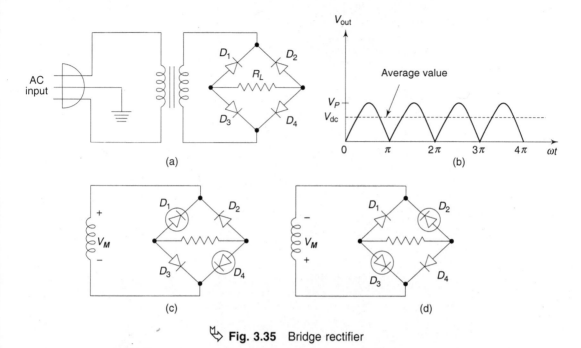

**Fig. 3.35** Bridge rectifier

lem when the secondary voltage is low—the two diode drop (1.2 V) becomes significant in this case.

### 3.5.4 Filtering of Rectified Output

The output from the rectifier is pulsating dc and the uses of this pulsating dc are limited to charging batteries running dc motors and a few other applications. What we really need is a dc voltage that is constant in value similar to the voltage from a battery. To convert the rectified output into constant dc voltage, the ac variations in the output have to be filtered or smoothened.

A capacitor filter is usually used to filter out the pulsations in the rectified output. Figure 3.36(a) shows a capacitor input filter. A capacitor is connected across the output. Figures 3.36(b) and (d) show the output voltage for half wave and full wave circuit with a capacitor filter connected.

During the first quarter half cycle of the input voltage, the diode is forward biased and is conducting. During this period the capacitor gets charged to peak voltage $V_P$. Just past positive peak, the diode is reverse biased as the capacitor voltage is higher than the input voltage. The charged capacitor now acts like a battery supplying current through load resistor $R_L$. Because of this discharge the capacitor loses a small part of its charge, and the voltage across the capacitor decreases slightly. During the next positive input peak, the diode again turns on and recharges the capacitor to peak value $V_P$. This repeats during every cycle. Hence the signal at the output is almost a constant voltage. The only deviation from a pure dc voltage is a small ripple caused by the charging and

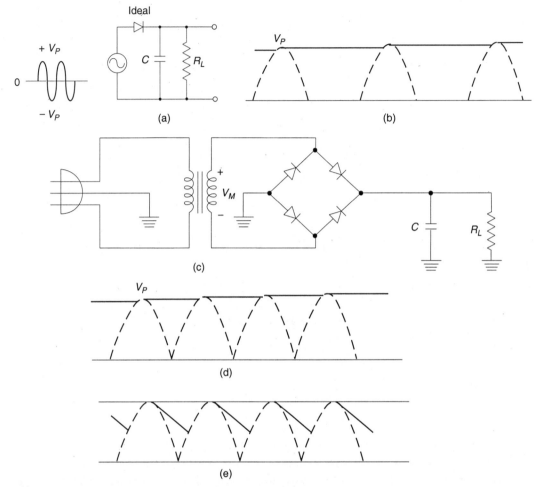

**Fig. 3.36** Capacitor-input filter (a) Half-wave circuit (b) Half-wave output (c) Full-wave bridge (d) Full-wave output (e) Output with heavy load

discharging of the capacitor. The discharging of the capacitor is dependent on the load. Heavier the load, higher will be the ripple at the output.

A full wave rectifier working into a capacitor filter produces even better results as the capacitor is charged twice as often as a half wave rectifier.

### 3.5.5 Special Diodes

*Zener Diode*

In a normal diode, the reverse breakdown occurs at a very high voltage, beyond which the current increases steeply for every increase in applied voltage. By varying the doping level, diodes with lower breakdown voltages, i.e. from 2 to 200 V can be manufactured. These diodes are operated in breakdown region by reverse biasing them. These diodes are called zener diodes and breakdown occurs at a constant reverse voltage for a given zener diode. The voltage at which breakdown occurs is known as zener voltage. These diodes can be used as constant voltage sources by applying reverse voltages that exceed the zener breakdown voltage. Zener diode is the back bone of voltage regulators. Figure 3.37 shows the schematic symbol and diode curve of a zener diode.

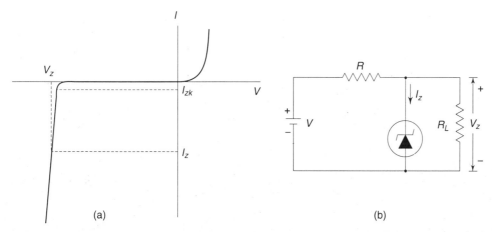

**Fig. 3.37** Schematic symbol and diode curve of a zener diode (a) The volt-ampere characteristic of an avalanche, or zener diode (b) A circuit in which such a diode is used to regulate the voltage across $R_L$ against changes due to variations in load current and supply voltage

The breakdown in zener diode occurs because of avalanche and zener effects. When the applied reverse voltage reaches the breakdown value, the minority carriers are accelerated to reach velocities high enough to dislodge valence electrons from outer orbits. The newly liberated electrons can then gain enough velocities to free further electrons. This results in sufficient free electrons to cause breakdown of the diodes. This effect is called avalanche effect. Avalanche effect is predominant in diodes with breakdown voltages higher than 6 V. When a diode is heavily doped, the depletion layer becomes very narrow. Because of this, the electric field across the depletion layer becomes very intense. When the electric field becomes intense enough it can pull electrons out of the valence orbits. The creation of free electrons in this way is called zener effect. This effect is predominant in diodes with breakdown voltages less than 4 V. Originally it was thought that the zener effect was the only breakdown mechanism in breakdown diodes. For this reason the name zener diode came into widespread use before avalanche effect was discovered.

## Tunnel Diode

By increasing the doping level of a backward diode (zener diode) we can get breakdown at zero volts. This happens when the concentration of impurities is increased to more than 1000 times the amount used in ordinary breakdown diodes. The diode starts conducting at zero volts in both positive and negative directions. The forward curve of such a diode is altered as shown in Fig. 3.38. Such a diode is known as tunnel diode (also known as esaki diode).

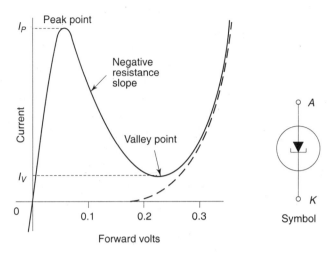

**Fig. 3.38** Behaviour of a tunnel diode

In the forward direction the current increases and reaches a peak value $I_P$ when the diode voltage equals $V_P$. Then the current starts decreasing with the increase in voltage and reaches a valley current $I_V$ when voltage equals $V_V$. Then onwards, the current starts increasing as in any ordinary diode. The typical value of $V_P$ and $V_V$ is 0.1 V and 0.5 V respectively. The diode curve between $V_P$ and $V_V$ exhibits negative resistance characteristic as current decreases with increase in voltage. In other words, in the negative resistance region the tunnel diode produces power instead of absorbing it. This characteristic of tunnel diode makes it very useful in very high frequency oscillators.

### *Light Emitting Diodes (LED)*

In a forward biased diode, free electrons cross the junction and recombine with holes. We can produce free electrons by imparting energy to them. Similarly whenever an electron recombines with a hole, it radiates energy. In a rectifier diode this energy goes off as heat. But in a light emitting diode (LED) this energy is radiated as light. By using elements such as gallium, arsenic and phosphorus LEDs that can radiate light—red, green, yellow, orange or infrared—can be manufactured. LEDs producing visible radiation are often used in instrument displays, digital readouts, digital clocks, calculators, etc. LEDs producing infrared radiation are used in burglar alarm systems and other areas requiring invisible radiations. The advantages of LEDs over incandescent lamps are long life, low voltage, fast on/off switching. The symbolic representation of an LED is shown in Fig. 3.39.

### *Photodiode*

A reverse biased diode has a small current because of its thermally produced minority carriers. When the *P-N* junction of the diode is housed in a glass package, strong light hitting the junction increases the reverse current. A photodiode is a normal diode optimised for its sensitivity to light and is housed in a glass package. The glass window

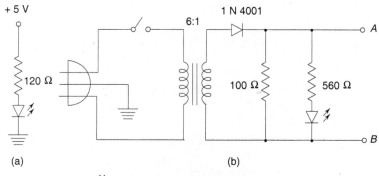

(a)  (b)

**Fig. 3.39** LED indicator

lets the incoming light to pass through the package and hit the *P-N* junction. This light produces additional holes and electrons, giving rise to a higher reverse current. Photodiodes are always used in reverse biased condition. The stronger the light, lower will be the back resistance of the diode and lower will be the output voltage across the diode. Figure 3.40 shows the schematic symbol of a photodiode. The arrows in the symbol represent the incoming light.

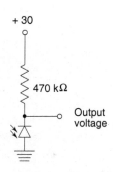

**Fig. 3.40** Photodiode

## Opto Isolator

An opto isolator combines an LED and a photodiode in a single package as shown in Fig. 3.41. The LED supply voltage forces a current $I_{IN}$ through LED. The light from LED hits the photodiode and sets up a current $I_{OUT}$. If LED current $I_{IN}$ varies, $I_{OUT}$ also varies. If LED current has an ac variation, the photodiode current will have an ac variation of same frequency. There is an electrical isolation between input LED circuit and output photodiode circuit. The resistance between input and output is greater than $10^{12}$ ohms.

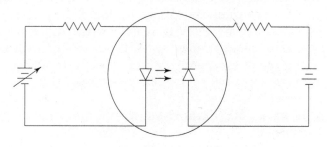

**Fig. 3.41** Optical isolator

## 3.6 TRANSISTORS

A junction transistor consists of either a *N*-type semiconductor sandwiched between two *P*-type semiconductors or a *P*-type semiconductor sandwiched between two *N*-type semiconductors. In the former case the transistor is referred to as a *PNP* transistor and in the latter case as an *NPN* transistor. The semiconductor sandwich is very thin and lightly doped. The semiconductor junctions so formed are housed in a hermatically sealed case of either plastic or metal. Three leads are brought out from the three semiconductor regions. The sandwiched semiconductor region is called base. The other two regions are called emitter and collector. The collector region is larger in size and more heavily doped. Figure 3.42 shows *NPN* and *PNP* transistors and their symbolic representation. The arrow on the emitter lead in the symbol of transistor indicates the low resistance direction to conventional current flow.

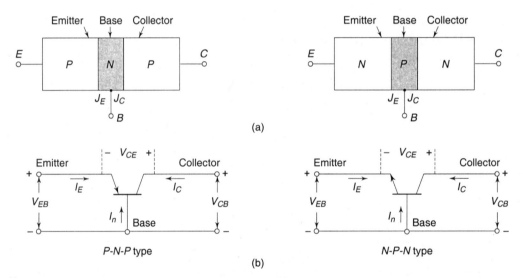

**Fig. 3.42** (a) A *PNP* and an *NPN* transistor. The emitter (collector) junction is $J_E$ ($J_c$)
(b) Circuit representation of the two transistor types

The transistor can be considered as two semiconductor diodes connected back-to-back in a crystal. The working of the transistor can be explained by making use of Fig. 3.43(a), (b) and (c). Figure 3.43(a) shows an *NPN* transistor with both emitter and collector diodes reverse biased and Fig. 3.43(b) shows an *NPN* transistor with both the diodes forward biased. By applying the diode theory, we can expect that as both diodes are reverse biased in the former case, negligible current flows through them and in the latter case there will be a large current in either of the diodes. The explanation given is practically true. Figure 3.43(c) shows an NPN transistor with emitter diode forward biased and collector diode reverse biased.

As in the previous case, if we apply diode theory in this case also, there should be a large current in the emitter diode and negligible current in the collector diode. But this

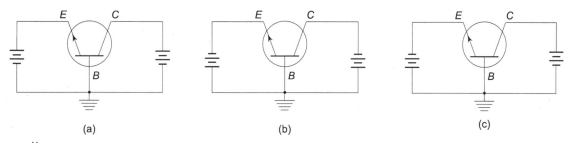

**Fig. 3.43** Biasing (a) Both diodes off (b) Both diodes on (c) Emitter forward biased and collector reverse biased

is not what happens in practice. Practically there will be a large current in emitter diode and an equally large current in collector diode also. This is referred to as transistor action. The transistor action can be explained by using Figs 3.44(a) and (b).

Figure 3.44(a) shows an *NPN* transistor with emitter junction forward biased and collector junction reverse biased. The electrons are the majority current carriers in *N*-type semiconductors. The electrons present in collector region are drawn by the positive terminal of the collector battery $V_{CB}$ as shown in Fig. 3.44(a). The electrons present in emitter region are repelled by the negative terminal of the battery $V_{EB}$ connected to emitter. As the base region is very thin and lightly doped, the electrons moving away from the emitter region have enough life time to cross the base region and enter the collector region. Once they are in the collector region, they are collected by the positive terminal of the battery $V_{CB}$. Now the negative terminal of the battery $V_{EB}$ supplies fresh electrons to the emitter region and these electrons are once again collected by the collector region as shown in Fig. 3.44(b). This results in a continuous large current in both emitter and collector diodes. Actually, a small portion of the electrons moving from emitter, across the base junction, recombine with a small number of holes present in the base region, resulting in a small base current. If we neglect this base current, then collector current is almost equal to emitter current. The forward biased emitter diode can be considered as a small resistance and the reverse biased collector diode can be considered as a very high resistance equivalent to back resistance of a diode. Since collector current is almost equal to emitter current, it is clear that current in the low resistance input is

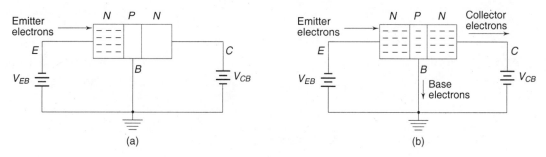

**Fig. 3.44** Emitter electrons captured by collector

## 3.6.1 Transistor as an Amplifier

Figure 3.45 explains how a transistor can amplify input signal. As shown in Fig. 3.45, $V_i$ is the input signal and $V_0$ is the output signal. The emitter diode is forward biased and collector diode is reverse biased. Hence forward biased emitter can be represented as a resistance $r_{eb}$ having a low resistance value equivalent to the forward resistance of a diode. Similarly a reverse biased collector diode can be replaced by another resistance $r_{cb}$ having a very high value. The collector current $I_c$ is about 95% of emitter current, $I_e$ and the base current $I_B$ is about 5% of $I_c$. The input voltage $V_i$ is the voltage across the input resistance $r_{eb}$.

Hence, $\qquad V_i = I_e \, r_{eb}$

Similarly output voltage is the voltage across output resistance $r_{cb}$.

Hence $\qquad V_o = I_c \, r_{cb} = 0.95 \, I_e \, r_{cb}$

The voltage gain is given by

$$\alpha = \frac{V_o}{V_i} = \frac{0.95 \, I_c \, r_{cb}}{I_e \, r_{eb}} = 0.95 \, \frac{r_{cb}}{r_{eb}} \text{ as } I_c \simeq I_e$$

As $r_{cb}$ is very high when compared to $r_{eb}$, the voltage gain is very high. Typically $r_{cb}$ is about 500 k$\Omega$ and $r_{eb}$ is about 50 $\Omega$. Hence the voltage amplification is given by

$$\alpha = \frac{0.95 \times 500 \times 10^3}{50} = 9500$$

As can be seen from these calculations the transistor can amplify the input voltage by 9500 times. But here we have assumed an infinite load resistance which will not be the

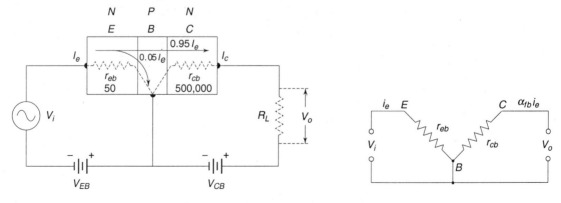

**Fig. 3.45** Transistor as amplifier

case in practice. The load resistance lies between 10,000 and 25,000 Ω. Hence, the voltage amplification lies between 200 and 500.

## 3.6.2 Transistor Configurations

A transistor can be used in three basic configurations, viz. common base configuration, common emitter configuration and common collector configuration.

### Common Base Configuration

The circuit shown in Fig. 3.46(a) is commonly referred to as common base configuration as the base is common to both input and output circuits.

Figure 3.46(b) shows the characteristics of collector current vs collector voltage. As can be seen from the curves for a given emitter current, the collector current is constant and is independent of collector voltage. Collector current increases with emitter current alone.

Common base configuration works like a controlled current source.

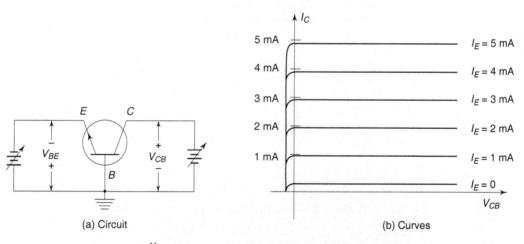

(a) Circuit  (b) Curves

**Fig. 3.46** Common base configurations

### Common Emitter Configuration

The common emitter configuration is shown in Fig. 3.47(a). In this circuit, emitter is common to both input and output.

This is the most widely used configuration. Figure 3.47(b) shows the collector current curves drawn for various base currents $I_B$. As seen from the curves, the collector current is not much dependent on collector voltage $V_{CE}$. It is fairly constant over a wide range of $V_{CE}$. It may be observed that base current is much smaller than collector current. The collector curve can be divided into three regions.

When $V_{CE}$ approaches zero, the collector current drops off. This region is known as saturation region. In this region $V_{CE}$ is typically only a few tenths of a volt. In saturation region the collector diode is coming out of reverse bias.

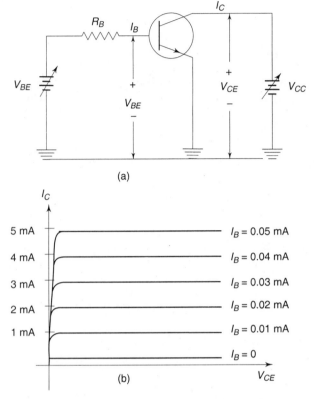

**Fig. 3.47** Common emitter configuration (a) Circuit (b) Transistor curves

If $V_{CE}$ is too large the collector diode will breakdown. The region beyond this is known as breakdown region. The active region is where the collector current is almost constant. This is the region between saturation and breakdown region.

### Common Collector Configuration

Figure 3.48 shows the third basic transistor circuit, namely, common collector configuration. No load resistance remains in the collector circuit, instead the output is obtained across $R_L$ located in emitter circuit. The battery $V_{CE}$ provides reverse bias to collector. The positive potential applied to the base causes a current $I_B$ to flow. The resulting collector current $I_c$ flows out of the emitter lead through $R_L$ until it nearly equals the voltage $V_B$, the base voltage. As the voltage across $R_L$ approaches the base potential $V_B$, the forward bias on the

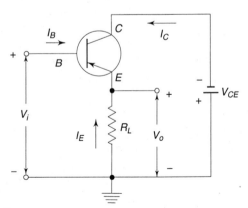

**Fig. 3.48** The transistor common-collector configuration

emitter is decreased so that the emitter current is just enough to keep the drop across $R_L$ equal to base potential $V_B$. As the output voltage follows the change in input, this configuration is also called emitter follower. This circuit is useful for controlling a low impedance output circuit by means of a high impedance input circuit. As the voltage gain is equal to unity it is used as buffer circuit for impedance matching.

### 3.6.3 Transistor as Switch

When transistor is connected in common emitter configuration as in Fig. 3.47(a), the collector current is much larger than the base current and typical current amplification is more than 100. As the base current goes on increasing, the collector current and the drop across the collector resistor also increases. When the drop across collector resistor almost equals the supply voltage, the drop $V_{CE}$ across the transistor is only a fraction of a volt, so small that the transistor seems to act like a closed contact. On the other hand when the base current is very small, the collector current is also small, resulting in the whole supply voltage being dropped across the transistor. $V_{CE}$ in this case equals the supply voltage and transistor acts like an open circuit. Hence the transistor works like a switch depending on the base current.

### 3.6.4 Unijunction Transistor (UJT)

A unijunction transistor (UJT) shown in Fig. 3.49 consists of a small bar or crystal of $N$-type silicon semiconductor, which is lightly doped and two metallic leads are brought out from the ends of this bar. The resistance between the ends of this bar is about 10 k$\Omega$. The two leads from the ends of this bar are known as base 1 and base 2. A piece of $P$-

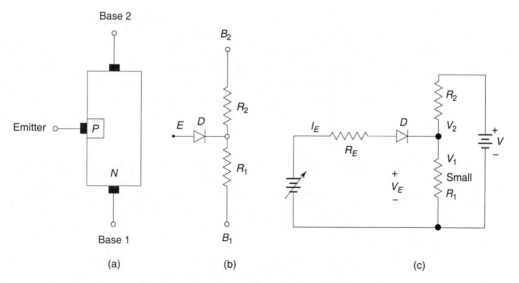

**Fig. 3.49** Unijunction transistor (a) Structure (b) Equivalent circuit (c) Producing stand off voltage

type material is formed near the middle of the bar and is known as emitter. Thus there is a single *P-N* junction in a unijunction transistor. The emitter is heavily doped. Internally UJT acts like a voltage divider consisting of two series resistors $R_1$ and $R_2$, while *P-N* junction acts like a diode $D$ driving the junction of $R_1$ and $R_2$ as shown in Fig. 3.49(b). When the emitter diode is not conducting, the interbase resistance $R_{BB}$ is a sum of $R_1$ and $R_2$. When the interbase voltage is applied between the two bases as shown in Fig. 3.49(c), the voltage across $R_1$ is given by

$$V_1 = \frac{R_1}{(R_1 + R_2)} V = \frac{R_1}{R_{BB}} V = \eta V$$

The quantity $\eta$ is called intrinsic stand-off ratio which is nothing but the voltage divider factor of UJT. The value of $\eta$ lies between 0.5 and 0.8.

The voltage $V_1$ is called the intrinsic stand off voltage as it keeps the emitter diode reverse biased for all emitter voltages less than $V_1$. We have to apply slightly more voltage than $V_1$ for the emitter diode to conduct. If in Fig. 3.49(c) the emitter voltage is less than $V_1$, the diode does not conduct. When the emitter supply is slightly more than $V_1$, the diode turns on. Since *P*-region is heavily doped as compared to *N*-region, holes are injected from *P*-region to the lower half of VJT. As *N*-region is lightly doped the holes injected have a longer lifetime. The flooding of the lower half of UJT with holes drastically lowers the resistance $R_1$. Because of this the drop across $R_1$ reduces to a low value and the emitter current increases.

### Application of UJT as Relaxation Oscillator

UJT is used as a relaxation oscillator to produce trigger pulses for firing SCRs. Figure 3.50 shows a UJT relaxation oscillator. The capacitor charges towards $V_{cc}$ through the resistor $R_1$. As soon as the capacitor is charged to a voltage slightly higher than the intrinsic stand-off voltage, the UJT is turned on and the capacitor discharges through the resistor between base 1 and the ground. As the capacitor gets discharged, the UJT is turned off and once again capacitor starts charging. Because of this turning on and off of UJT, sharp pulses appear at base 1 and base 2 as shown in Fig. 3.50. By changing the time constant $R_c$, the time between the pulses can be changed.

### 3.6.5 Field Effect Transistors (FET)

The field effect transistor is a semiconductor device which operates on the principle of control of current by an electric field. There are two types of field effect transistors, the junction field effect transistor (JFET or simply FET) and insulated gate field effect transistor or metal oxide semiconductor field effect transistor (MOSFET).

### Junction Field Effect Transistor

The structure of *N*-channel FET is shown in Fig. 3.51. It has an *N*-type semiconductor bar with two ohmic contacts at the two ends. One of these ohmic contacts is known as source and the other as drain. The semiconductor bar between the source and the drain

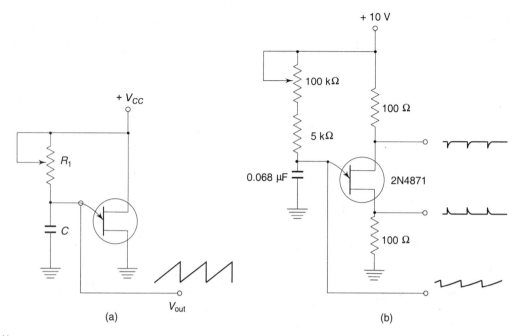

**Fig. 3.50** Relaxation oscillators (a) Sawtooth generator (b) Trigger and sawtooth generator

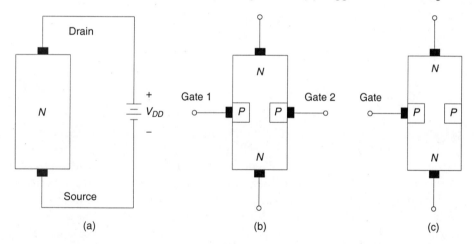

**Fig. 3.51** Structure of N-channel FET (a) Channel (b) Dual-gate JFET (c) Single-gate JFET

is known as channel. In a *P*-channel, FET the semiconductor bar is of *P*-type material instead of *N*-type material. Two *P*-type semiconductor regions are embedded in the sides of the *N*-type channel as shown in Fig. 3.51(b). Each of these *P*-type regions is called a gate. When separate leads are connected to each gate, the device is known as dual gate FET. In a single gate FET, these gates are internally shorted and a single lead is brought out.

Figure 3.52(a) shows the normal polarities for biasing a *N*-channel FET. The gate is always reverse biased by applying a negative voltage between gate and source. Since the gate is reverse biased, a very small current flows in the gate lead.

The name field effect is related to the depletion layers around each *P-N* junction. Figure 3.52(b) shows the depletion layers. Free electrons moving between the source and drain must pass through the narrow channel between the depletion layers. The size of the depletion layers determines the width of the conducting channel. The more negative the gate voltage is, narrower would be the conducting channel as the depletion layers get closer to each other. In other words the gate voltage controls the current between the source and the drain. The more negative is the gate voltage, the smaller will be the current.

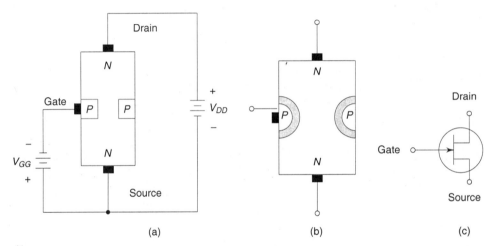

**Fig. 3.52** (a) Normal JFET bias voltages (b) Depletion layers (c) *N*-Channel JFET symbol

The key difference between a transistor and FET is that the base of transistor is forward biased, whereas the gate of FET is reverse biased. The gate voltage of FET controls the current, i.e., the FET is a voltage controlled device, whereas the transistor is a current controlled device where base current controls the collector current. As the gate of FET is always reverse biased, the input resistance is very high as compared to that of transistor. The disadvantage of FET is that it is less sensitive to changes in input voltage than a transistor. Transistor is a bipolar device as the current is due to both holes and electrons, whereas FET is a unipolar device as only the majority carrier contributes to the current.

### *Insulated Gate FET (MOSFET)*

The junction FET uses the gate that has a direct electrical contact with the channel but a reverse bias is used to prevent a flow of gate current, so that the gate acts as if it is insulated from the channel. The disadvantage here is that only negative voltages can be applied to the gate. Instead of this insulating action of reverse bias, the insulated gate

FET uses a thin layer of insulating silicon dioxide formed between the gate and the channel. There is no electrical contact between the gate and the channel and hence the disadvantages of leakage current and the need for reverse bias can be overcome.

Figure 3.53 shows the construction of an insulated gate FET. A channel of $N$ material is diffused into the top block of a $P$-type semiconductor with enlarged ends. The enlarged ends of the $N$-type semiconductor forms the source and the drain of the FET. A thin layer of silicon oxide is grown over the channel. A metallic gate is deposited over this insulating oxide film. As the gate is insulated from channel by a layer of metal oxide this FET is called MOSFET (metal oxide semiconductor FET).

As in JFET, here also the drain current is controlled by the application of a control voltage to the gate.

Figure 3.54 shows the schematic symbol for $N$-channel and $P$-channel MOSFET.

### 3.6.6 Performance Curves

The curves in Fig. 3.55 show the drain current curves plotted for various gate source

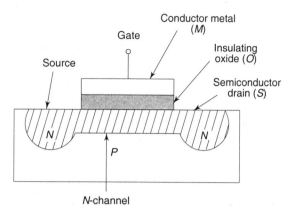

**Fig. 3.53** Insulated-gate FET; the MOSFET

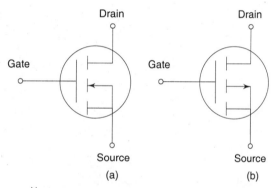

**Fig. 3.54** MOSFET symbols (a) $N$-channel (b) $P$-channel

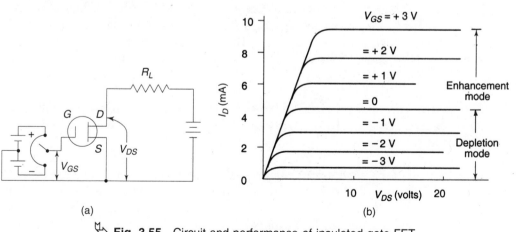

**Fig. 3.55** Circuit and performance of insulated-gate FET

voltages for a *N*-channel FET. As can be seen from the curves, the drain current is almost independent of drain voltage. The drain current is mainly dependent on gate voltage. As the gate is insulated from channel, both positive and negative control voltages can be applied to the gate. This gives rise to two modes of operation. For an *N*-channel MOSFET the negative control voltages at the gate result in reduction of drain current as shown by the performance curves in Fig. 3.55. This mode of operation where the drain current decreases with control voltage is known as depletion mode. For positive control voltages the drain current increases for *N*-channel MOSFET and this mode of operation is known as enhancement mode.

## 3.7 SILICON CONTROLLED RECTIFIERS (SCR)

The silicon controlled rectifier (SCR) is a *P-N-P-N* device having four layers of *P-* and *N-*type silicon as shown in Fig. 3.56(a). The three terminals are called the anode, cathode and gate.

The SCR works like a controlled rectifier where conduction of the rectifier can be controlled by a triggering signal at the gate terminal. SCR is fired or turned on by a pulse of current flowing through its gate. The SCR remains open till a trigger hits the gate. Then the SCR latches and remains closed even if the triggering pulse at the gate disappears. The only way to turn off the SCR is by decreasing the anode current to nearly zero.

The SCR can be visualised as two separate transistors—*N-P-N* and the *P-N-P*—connected as shown in Fig. 3.56(b).

In the absence of any voltage at the gate terminal, the base current of both transistors $T_1$ and $T_2$ is zero and both the transistors are in off condition. Hence the device does not conduct even if a positive voltage is applied between anode and cathode. If a trigger pulse is applied to the gate at this stage, it will result in a small current at the base of the transistor $T_2$. This base current is amplified and a higher current appears at the collector of $T_2$. This current acts as input to transistor $T_1$ which is again amplified at the collector of $T_1$ and this current flows through the base of $T_2$. Now $T_2$ has a higher base current which again is amplified by $T_2$. This phenomenon repeats itself resulting in a regenerative action which goes on increasing the collector current of both transistors till reach saturation. Now the SCR is in on state. It may be noted that once a gate pulse is applied, it will take the SCR to conduction and it has no effect afterwards and can be removed. Once the SCR starts conducting, it can be switched off only by reducing the voltage across it to zero or by reducing the current below the holding current. When an ac voltage is applied across SCR, the SCR gets turned off during negative half cycle as it is reverse biased.

Figure 3.57 shows the input gate pulses and the output when ac voltage is applied across the SCR. As can be seen, output is present when the gate pulse coincides with the positive going half cycle of the input wave. The device will then be switched on and remains on until the voltage across it falls to zero. However, during negative half cycle there is no output even when the gate pulse is present. Three conditions of gate pulse are illustrated in Fig. 3.57.

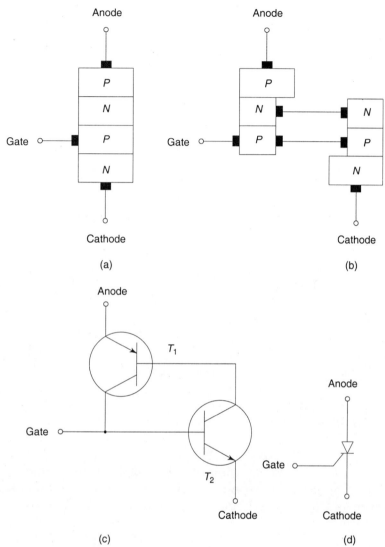

**Fig. 3.56** Silicon controlled rectifier (a) Structure (b) Equivalent structure (c) Equivalent circuit (d) Schematic symbol

(a) *Pulse 1*: The gate pulse coincides with the start of positive half cycle. Output is available during the complete duration of positive half cycle.
(b) *Pulse 2*: The gate pulse appears during negative half cycle. No output is present as the device is reverse biased.
(c) *Pulse 3*: The gate pulse appears some time after the start of positive half cycle. Output is available from the instant the gate pulse hits the gate till the end of that positive half cycle.

Thus, it can be seen that the average output from the SCR can be varied by altering either the frequency, phase or width of the gate pulse. As the speed of the dc motors can

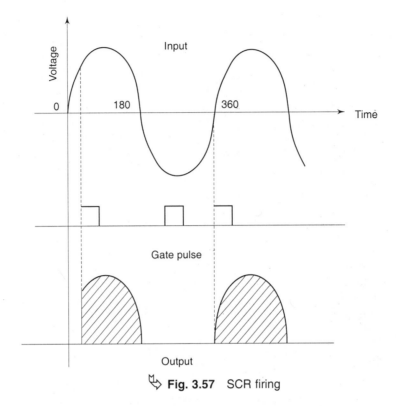

**Fig. 3.57** SCR firing

be varied by varying the voltage applied at the armature, this method of variation of average output voltage is very useful for speed control of dc motors.

## Triac

Triac is a combination of two SCRs, connected back-to-back in a single package. It can block the voltage of either polarity, but starts current flow in either direction with a current pulse in or out of its single gate. Figure 3.58 shows the structure, schematic

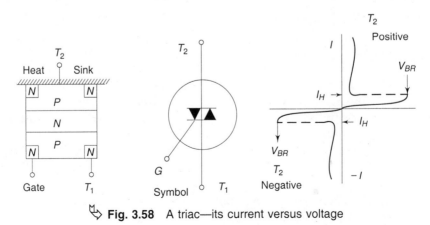

**Fig. 3.58** A triac—its current versus voltage

symbol and current versus voltage curve. Either $T_1$ or $T_2$ can be the anode. When $T_2$ is positive, the triac curve resembles that of an SCR and a current pulse into the gate will turn on the device. When $T_2$ is negative, the curve will be a mirror image of the former and a current pulse out of the gate will conduct the device. While most SCRs can be turned off by voltage reversal during negative half cycle of ac supply (up to 30 kHz), such rapid voltage reversals across triac do not turn it off, since the triac merely conducts in the opposite direction. But when the frequency of input ac is less than 60 Hz, it is low enough to permit turn off of the triac during zero crossing of input wave when no gate pulse is present. Triac is usually used in dimmerstats.

## 3.8 INTEGRATED CIRCUITS (IC)

Circuits which use separate circuit elements—resistors, capacitors, inductors, transistors, diodes, etc.—connected together are called discrete circuits. Printed circuit boards provide ways to interconnect hundreds of discrete devices on a single board. But, in these assemblies, the space consumed is very large. Integrated circuits (IC) permit the presence of hundreds of transistors, diodes and resistors to be formed and connected together to realise a complete complex circuit within the size of a small pill. All these circuit elements are made at the same time on the same tiny wafer of silicon. In a monolithic integrated circuit, hundreds of diodes, transistors and resistors are formed or grown by the same processes at the same time by chemically treating certain parts of the surface of a tiny piece of silicon. The main advantages of ICs are:

- Save space (small size)
- Provide improved reliability (lesser number of soldered joints)
- Improve performance
- Matched devices (as all transistors are manufactured simultaneously by same processes)
- Low cost (due to mass production)

Depending on the scale of integration, the ICs are classified into medium scale integration (MSI), large scale integration (LSI) and very large scale integration (VLSI) chips.

The basic structure of an IC (Fig. 3.59) consists of four distinct layers. The bottom layer (1) is $P$-type silicon called substrate, over which the IC is built. The second layer (2) is a thin $N$-type layer which is grown on the substrate. All components are built within this layer by a series of diffusion steps. The most complicated fabricated component is a transistor and all other elements are constructed with one or more of its similar processes required for transistors. The third layer (3) is a thin layer of silicon oxide ($SiO_2$). It acts as a barrier to protect portions of the wafer against impurity penetration. The fourth layer (4) is an aluminium layer which provides necessary interconnection between components. Figure 3.60 illustrates the various steps required for the manufacture of a monolithic integrated chip. These steps are:

(i) Epitaxial growth
(ii) Isolation diffusion
(iii) Base diffusion

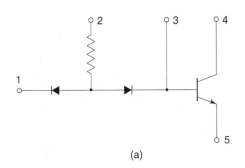

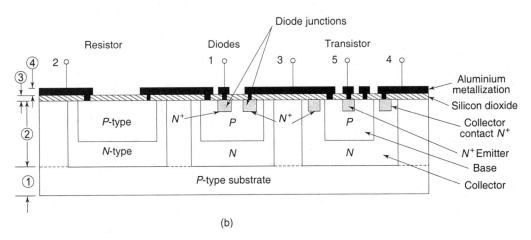

**Fig. 3.59** (a) A circuit containing a resistor, two diodes, and a transistor (b) Cross-sectional view of the circuit in (a) when transformed into a monolithic form (not drawn to scale). The four layers are (1) substrate, (2) N-type crystal containing the integrated circuit, (3) silicon dioxide, and (4) aluminium metallization

(iv) Emitter diffusion
(v) Metallisation

During epitaxial growth, a layer of $N$-type semiconductor is grown over a $P$-type substrate. All components are formed inside this epitaxial layer.

During isolation diffusion, isolated regions of $N$-type semiconductor are formed inside the epitaxial layer by diffusing $P$-type impurity at selected places. A photolithographic technique and chemical etching process is used to make openings in $SiO_2$ layer through which $P$-type impurity is diffused.

During base diffusion, $P$-type impurity is diffused into the crystal to form the base of transistor.

During emitter diffusion $N$-type impurity is diffused to form the emitter of the transistor.

During metalisation, a thin film of aluminium is grown over the entire wafer by vacuum evaporation. The aluminium on undesired areas is removed by the photoetching process. This metalisation provides interconnection between various components of the IC.

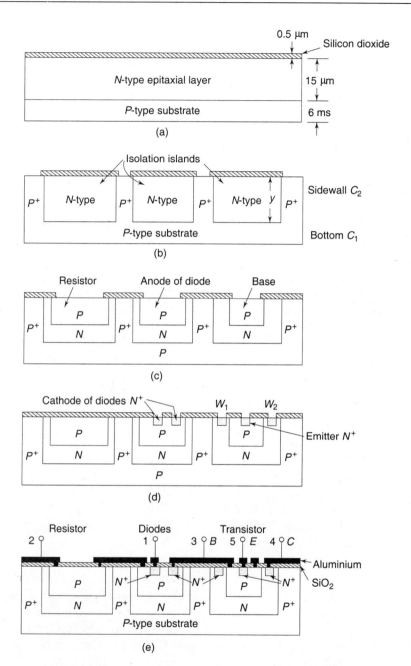

**Fig. 3.60** The steps involved in fabricating a monolithic circuit (not drawn to scale)
(a) Epitaxial growth (b) Isolation diffusion (c) Base diffusion
(d) Emitter diffusion
(e) Aluminium metalisation

## 3.9 DIGITAL CIRCUITS

The devices used in a digital system function in a binary manner. These devices exist in two possible states—ON and OFF. A switch can either be ON or OFF and no in between state is possible. A transistor is used either in Cut-Off or in saturation and not in the active region. A node may be at a high voltage, say $4 \pm 1$ V or $0.2 \pm 0.2$ V, but no other values are allowed. These two quantised states are designated by 1 or 0, high or low, true or false, ON or OFF. The most widely used designation is 1 or 0.

### 3.9.1 Logic gates

There are some simple building blocks of a digital system which control their output based on the conditions of the inputs and these are called logic gates. These gates are actually decision making circuits which reproduce some mental processes. Logic is an accurate statement of facts. A logical statement states the logical relation of inputs with the output. A gate is a logic circuit with one or more input signals but only one output signal. These signals are digital in nature and can exist in either high or low state only. There are many instances in control and computational circuitry where a certain operation must take place only when a certain set of conditions occur at the input. Logic gates are used to realise such logics. There are basically three types of gates—OR gate, AND gate and NOT gate—and all other gates are a combination of these three.

#### OR Gate

The operation of OR gate can be explained by a simple circuit with two switches $A$ and $B$ connected in parallel and a lamp connected in series with these switches to the supply as shown in Fig. 3.61.

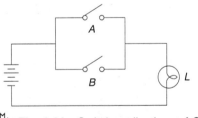

**Fig. 3.61** Switch realisations of OR

In this figure, the lamp is ON whenever any one or both the switches are in ON state. The lamp is OFF when both the switches are in OFF state. The same principle can be applied for more number of switches connected in parallel with $A$ and $B$ and we can say the lamp is ON if one or more switches are ON and lamp is OFF when all the switches are OFF. Similarly, OR gate has two or more inputs and a single output and operates according to the principle: *The output of an OR gate assumes '1' state if one or more of the inputs assume '1' state.*

Figure 3.62 shows the circuit of a three-input OR gate. Here when any one of the inputs $A$, $B$ and $C$ are in 'high' state, the diode connected to that input conducts and the current through the resistor $R$ will result in a drop across this resistor and the output goes 'high'. When all the inputs are in 'low' state, all diodes are OFF and the output is low.

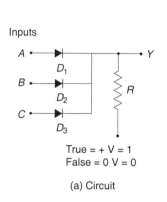

| Inputs | | | Output |
|---|---|---|---|
| A | B | C | D |
| 0 | 0 | 0 | 0 |
| 1 | 0 | 0 | 1 |
| 0 | 1 | 0 | 1 |
| 1 | 1 | 0 | 1 |
| 0 | 0 | 1 | 1 |
| 1 | 0 | 1 | 1 |
| 0 | 1 | 1 | 1 |
| 1 | 1 | 1 | 1 |

True = + V = 1
False = 0 V = 0

(a) Circuit  (b) Truth table  (c) Logic OR symbol

**Fig. 3.62** Diode OR gate

A table listing all the input possibilities and their corresponding output is called a truth table. The truth table for a three-input OR gate is shown in Fig. 3.62(b) and Fig. 3.62(c) shows the schematic symbol of a three-input OR gate.

Logically, the OR operation is expressed by $Y = A + B + C$, where $Y$ is the output, $A$, $B$ and $C$ are the inputs and '+' indicates logical OR operation. Such logical expressions are called Boolean expressions.

## AND Gate

A logical AND gate has two or more inputs and a single output, and works according to the principle: *The output of an AND gate assumes '1' state if and only if all the inputs assume '1' state.*

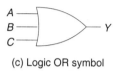

**Fig. 3.63** Switch realisations of AND

The working of AND gate can be explained by means of a circuit having two switches $A$ and $B$ connected in series with a lamp to the supply as shown in Fig. 3.63. In this circuit the lamp is ON only if both the switches are in closed (ON) state. The lamp is OFF if any or both the switches are in open (OFF) state. The output of AND gate is 'high' only when all the inputs are in 'high' state. Figure 3.64(a) shows three-input diode AND gate. Here, when all inputs are in 'low' state, all the input diodes conduct and only the drop across the diode appears at the output. When any of the inputs is low, the diode connected to that input conducts, forcing the output to go 'low'. When all the inputs are high, all diodes are in OFF state and the supply voltage appears at the output.

The AND operation is expressed by $Y = A \cdot B$, where $Y$ is the output, $A$ and $B$ are inputs and '·' denotes logical AND operation.

## NOT Gate

NOT gate has one input and one output. It is simply an inverter which inverts the input. The output of a NOT gate is high when the input is low and output is low when

# 76  Mechatronics

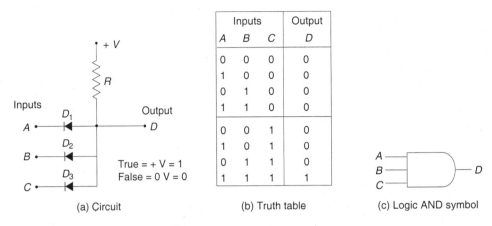

**Fig. 3.64**  Diode AND gate

input is high. Figure 3.65 shows the circuit, symbol and truth table of NOT gate. In the circuit shown in Fig. 3.65, whenever the input is '0' the transistor is in cut-off state as no base current flows and the output assumes '1' state as whole supply voltage gets dropped across the transistor. When the input becomes '1' the transistor switches on and voltage drop across transistor reduces to a low voltage. Hence output assumes '0' state.

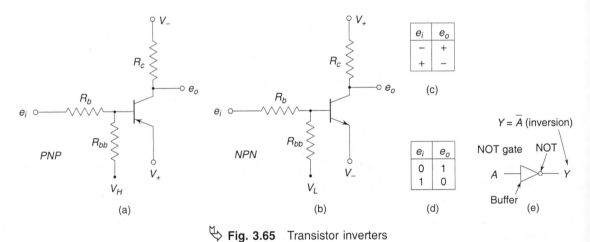

**Fig. 3.65**  Transistor inverters

## NOR Gate

An OR gate followed by a NOT gate forms a NOR gate (NOT OR gate). The circuit, truth table and symbol for two input NOR gate is shown in Fig. 3.66 (a), (b) and (c).

## NAND Gate

An AND gate followed by a NOT gate forms a NAND gate (NOT AND gate). The circuit, truth table and symbol for two input NAND is shown in Fig. 3.67 (a), (b) and (c).

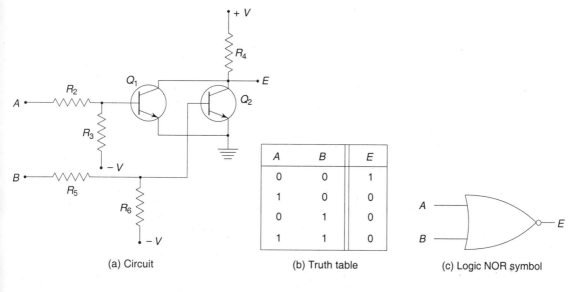

**Fig. 3.66** NOR gate

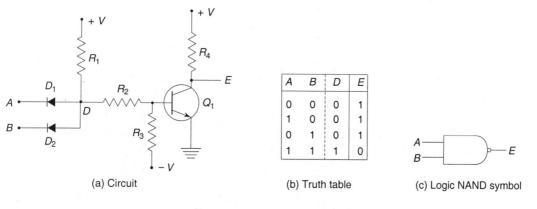

**Fig. 3.67** NAND gate

### Exclusively OR Gate (EXOR Gate)

This gate works according to the principle: *The output of EXOR gate assumes '1' state if and only if one input assumes '1' state.* EXOR gate has two inputs and one output. The symbol and truth table for EXOR gate is shown in Fig. 3.68 (a) and (b).

The Boolean expression for EXOR function is $Y = A \oplus B$, where $Y$ is the output, $A$ and $B$ the inputs and '$\oplus$' denotes EXOR operation.

78　Mechatronics

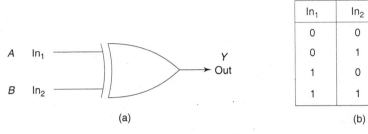

**Fig. 3.68** Exclusive OR logic symbol and truth table

CHAPTER **4**

# Mechanical Systems for Electronics Engineers

## 4.1 BASIC CONCEPTS

### 4.1.1 Force

Force is that physical quantity which causes or tends to cause a change in the state of rest or of motion of a body. The line of action of a force is a line drawn through the point of application of the force and along the direction in which the force acts. If a change in motion is prevented, force will cause a deformation or change in shape of the body to which it is applied.

In statics, it is often convenient to consider the effect of a force which acts on a rigid body that is negligibly small—the point object. The perfectly rigid body which suffers no deformation under the action of any force does not exist. Nevertheless, it is a very useful idea through which many practical problems can be simplified.

A force, when applied to a rigid body has the same effect at all points on its line of action. Hence, for many purposes it does not matter where the point of application is, as long as the line of action is known. The effect of a force at a point which is not on its line of action depends on the position of that point relative to the line of action. A force is completely defined by its

(a) Magnitude
(b) Point of application
(c) Direction (line of action)

A body is said to be in equilibrium under the action of a system of forces if all forces acting on it are in balance. A body cannot remain in equilibrium under the action of a single force. In statics, all forces come in pairs—an action and a reaction. Nevertheless, while considering the equilibrium of a system of forces, it is usually convenient to consider each force as existing separately. A body is in equilibrium under the action of two forces provided that:

(i) The forces are of equal magnitude
(ii) The forces have the same line of action but act in opposite directions.

### 4.1.2 Moment

The action of a force on a rigid body tends to:
(a) Move the body
(b) Rotate the body

The turning effect or the tendency of a force to cause a rotation about any point $O$ is measured by the product of the force ($F$) and the perpendicular distance ($x$) of the line of action of the force from the point $O$ (Fig. 4.1).

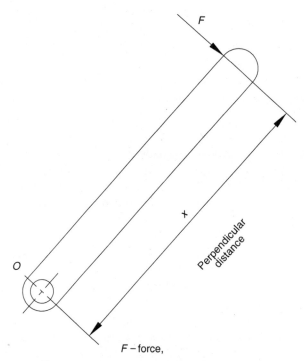

**Fig. 4.1** Action of force on rigid body

This turning effect is called a moment of the force $F$ about $O$ and the distance $x$ is called the moment arm. A moment always refers to the turning effect at a particular point.

When more than one force act on a body, the total turning effect about any given point due to all the forces is the sum of all the separate turning effects. But, since forces exert a turning effect either clockwise or anticlockwise about a point, a sign convention has to be established to account for the two turning directions.

If a body is at rest under the action of a number of coplanar forces, there must be a balance of moments, otherwise the unbalanced resultant moment will cause a rotation of the body.

## 4.1.3 Couple

A couple is formed by two equal parallel forces which are not collinear and act in opposite directions. The moment of a couple is the product of one of the forces and the perpendicular distance between the lines of action of the forces (Fig. 4.2). To specify a couple completely, we must know its moment, its direction of rotation (T) and the plane in which it acts. An important point to be noted is that a couple acting on a body can only be balanced by another couple; a single force can neither replace nor balance a couple. The moment of a couple is sometimes called torque. The term torque is usually used when dealing with the turning moment exerted on the shaft.

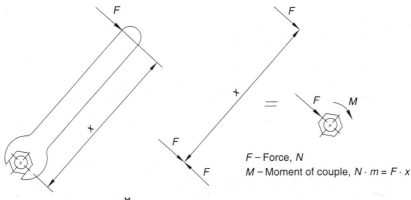

$F$ – Force, N
$M$ – Moment of couple, $N \cdot m = F \cdot x$

**Fig. 4.2** Moment of a couple

## 4.1.4 Friction

Figure 4.3 shows a body of weight $W$ resting on a surface that exerts a normal force $N$ on the body. To move the body, a force $F$ tangential to the surface has to be overcome by applying external force $P$. This force $F$ is known as the force of friction. If the magnitude of the external force $P$ is increased (Fig. 4.4), the frictional force also increases until its magnitude reaches a certain maximum value, $F_m$. If $P$ is further increased, the force of friction cannot balance it anymore and the body starts moving. The ratio of this maximum frictional force $F_m$ to the normal reaction $R$ is a constant, which depends only on the nature of the pair of surfaces in contact. Thus, $F_m/R = \mu_s$, where $\mu_s$ is called the coefficient of static friction. When sliding is just about to start, the pulling force $P = \mu_s R$. The direction of the frictional force is always opposite to that of the resultant motion. As soon as the body starts moving, the magnitude of the frictional force drops from $F_m$ to $F_k$. This frictional force $F_k$ is given by $F_k = \mu_k R$, where $\mu_k$ is called the *coefficient of sliding or kinetic friction* and is usually less than the coefficient of static friction. The sliding friction depends on the pressure between the two surfaces, on the nature and roughness of the surfaces, but is independent of the area of contact and the speed of slide.

When $P$ remains equal to $\mu_k R$ during sliding of the body, the body moves with a steady speed. When $P > \mu_k R$, the speed of the body is accelerated.

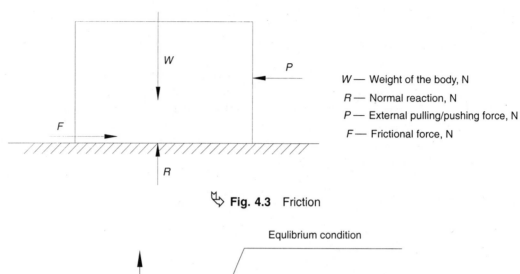

**Fig. 4.3** Friction

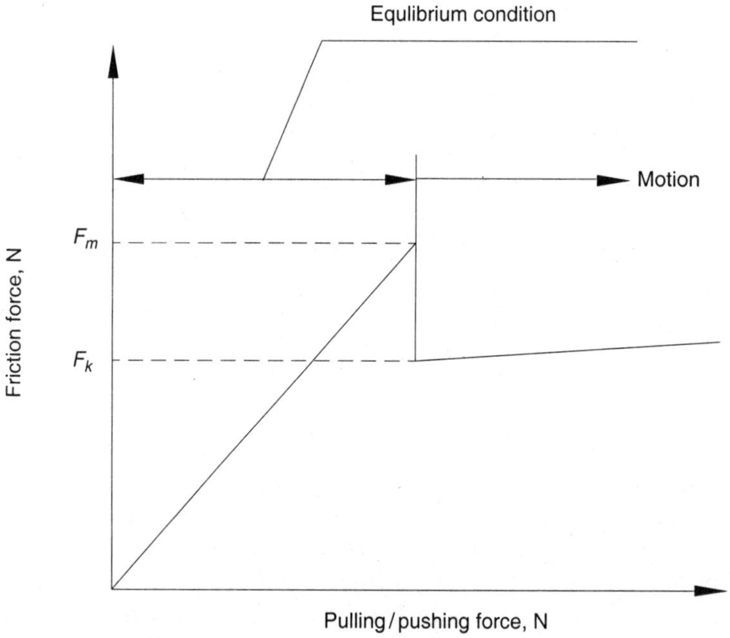

**Fig. 4.4** Frictional Force

### Wheel Friction and Rolling Friction

For a wheel to roll on the ground without slipping, there must be a force of friction on contact between the wheel and the track (Fig. 4.5). If the wheel is driven by a torque $T$, the frictional force exerts an opposing moment to resist the slip. If the wheel radius is $r$, then the maximum magnitude of the torque $T$ which can be applied to the wheel without any slipping, is equal to the product of force of the friction and the wheel radius. In addition to this, during rolling movement, there is a resistance to the rolling motion of the wheel (Fig. 4.6) due to deformation of the wheel and the track under load. This

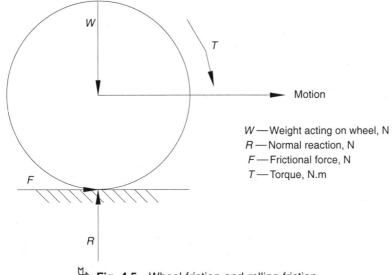

**Fig. 4.5** Wheel friction and rolling friction

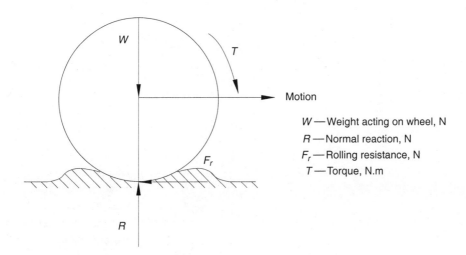

**Fig. 4.6** Rolling resistance

resistance is called the *rolling resistance*. It acts to oppose the linear motion of the axle. The ratio of maximum rolling resistance to normal reaction between the wheel and the track is known as the coefficient of rolling resistance.

### Bearing Friction

Journal bearings are used to provide lateral support to rotating shafts and axles. If the journal bearing is fully lubricated, the frictional resistance depends upon the speed of rotation, the clearance between the axle and the bearing, and the viscosity of the lubricant. The clearance between the shaft and the bearing is normally around 20 to 100 microns, depending on the bearing size. Hence, it may be assumed that the centre of a shaft and a bearing coincide approximately. When stationary, the shaft and the bearing make contact at the lowest point $A$ (Fig. 4.7). When rotating under a load, the point of contact climbs to the point $B$ at which slip commences. If N is the normal angle of friction, at this point, $F$ the frictional force is equal to the tangential component of load $W$, i.e.,

$$F = W \sin \theta, \text{ N}$$

If the shaft diameter is $d$ mm, then the frictional torque is given by

$$T_F = \frac{F_d}{2000} \text{ N.m}$$

This frictional torque $T_F$ results in a loss of power in the bearing.

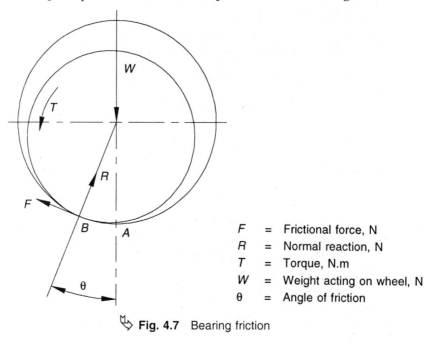

$F$ = Frictional force, N
$R$ = Normal reaction, N
$T$ = Torque, N.m
$W$ = Weight acting on wheel, N
$\theta$ = Angle of friction

**Fig. 4.7** Bearing friction

### 4.1.5 Inertia

If a force acting on a body is balanced by an equal and opposite force, then the two forces are said to be in equilibrium. Since the forces are in balance, the body will not move or if it is already moving, it will not accelerate. On the other hand, an unbalanced force will cause the body to increase its velocity. If the acceleration of a body of mass $m$ due to a force $F$ is $a$, then according to the second law of Newton

$$F = ma$$

In the above equation, since the force is required to accelerate the body, the quantity $m$ is a measure of its reluctance to accelerate. This reluctance is called the inertia of the body.

Inertia is a characteristic property of all material bodies; it is the inherent property of matter by which it continues to be in a state of rest or uniform motion in a straight line. If the mass is at rest, it requires a force to give it a motion. Greater the mass, higher is the force required. If the mass is already moving, it requires a force to change its velocity or direction, again the force required being proportional to the mass.

### 4.1.6 Stress and Strain

When an external force ($P$) acts on a body, an internal resistance ($R$) is set up within the body to balance the external load (Fig. 4.8). The resistance carried per unit area is called the *stress*, i.e.,

$$\text{Stress} = \frac{\text{Force}}{\text{Area}}$$

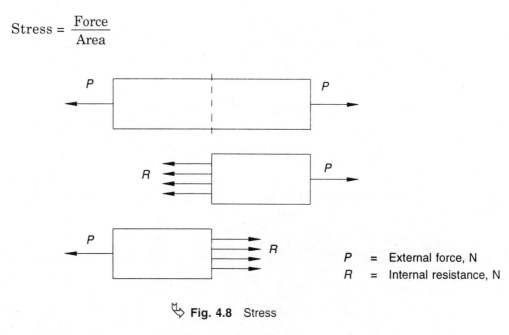

**Fig. 4.8** Stress

When the applied force tends to compress the material or crush it, the material is said to be in compression and the stress is referred to as compressive stress. When the force tends to expand the material or tear it apart, the material is said to be in tension and the stress is referred to as *tensile stress*. When the force tends to cause the particles of the material to slide over one another, the material is said to be in shear and the stress is referred to as shear stress (Fig. 4.9).

In case of a tensile or a compressive force, the area carrying the force is the area of cross-section in the plane of the material perpendicular to the direction of the force. In the case of a shear force, the area carrying the force is the area to be sheared in the direction of the line of action of the force.

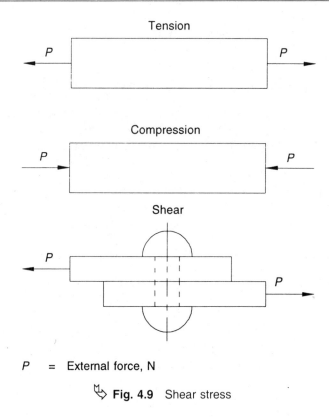

P = External force, N

**Fig. 4.9** Shear stress

Strain is the change in dimension that takes place in a material due to an externally applied force. Linear strain is the ratio of change in length with respect to its original length when a tensile or compressive force is applied (Fig. 4.10). Thus,

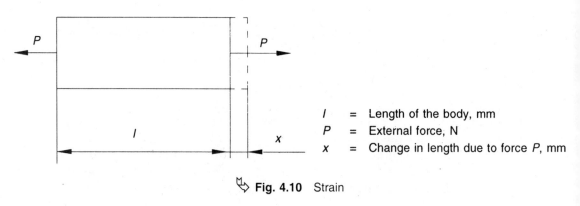

$l$ = Length of the body, mm
$P$ = External force, N
$x$ = Change in length due to force $P$, mm

**Fig. 4.10** Strain

$$\text{Linear strain} = \frac{\text{Change in length}}{\text{Original length}} = \frac{x}{l}$$

Shear strain is measured by the angular distortion caused by an external force (Fig. 4.11).

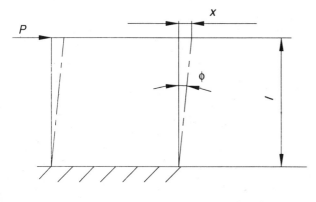

|   |   |                                        |
|---|---|----------------------------------------|
| l | = | Length of the body, mm                 |
| P | = | External force, N                      |
| x | = | Change in length due to force P, mm    |

**Fig. 4.11** Shear strain

$$\text{Shear strain} = \frac{x}{l} = \tan \phi$$

Consider the typical deflection–load (force) curve for a ductile material in a tensile test carried to destruction as shown in Fig. 4.12.

The strength of a material is expressed as the stress required to cause it to fracture. The maximum force required to break a material, divided by the original cross-sectional area at the point of fracture is called the ultimate strength. It is called the (ultimate) tensile strength if the material is in tension.

All materials are elastic to a certain extent. In the elastic region, the material will stretch if a tensile force is applied to it and returns to its original length on the removal

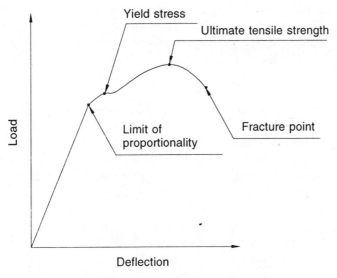

**Fig. 4.12** Deflection-load curve

of the force. There is a limit to this elastic property in every material, known as the elastic limit and if this is exceeded by increasing the applied force, the material after being stretched or compressed will not return to its original form but will remain permanently set in its deformed condition. The stress corresponding to this condition is known as the yield stress. It is found that when a material is loaded within its elastic limit, the stress is proportional to the strain produced (the linear part of the deflection-load curve). Mathematically,

$$\frac{\text{Stress}}{\text{Strain}} = \text{Constant}$$

This constant is called the *modulus of elasticity* or *Young's modulus* for tensile and compressive stress and is denoted by E.

Similarly, when a shear stress is divided by a shear strain, the constant obtained is termed as the *modulus of rigidity* and is represented by G. The values of E and G for steel and cast iron are as given below:

For steel:

$$E = 2.1 \times 10^5 \text{ N/mm}^2 \quad \text{and}$$
$$G = 8.3 \times 10^4 \text{ N/mm}^2$$

For gray cast iron:

$$E = 8.5 \times 10^4 \text{ N/mm}^2 \quad \text{and}$$
$$G = 3.85 \times 10^4 \text{ N/mm}^2$$

It is obvious that the stress allowed in any component of a machine under working condition must be less than the stress which would cause permanent deformation. A safe working stress is chosen with regard to the conditions under which the material is to work, taking into account whether the stress is due to a constant load; whether it varies or is subjected to reversal of stress and the rapidity of change of the stress. The ratio of the yield stress to the allowable stress is termed as the factor of safety.

## Bending of Beams

The term beam is given to a single rigid length of material that can support a system of external forces at right angles to its axis. The beam may be loaded either by concentrated loads or distributed loads or both (Fig. 4.13).

A beam which is fixed at one end, leaving the other end free is called a *cantilever beam*. A beam with its ends resting freely on supports is called a *simply supported beam* and a beam with both ends fixed is called a *fixed beam*.

The loads on a beam tend to shear the beam and bend it as well. These effects on the beam are measured by the shearing force and the bending moment respectively. The shearing force at any section of a beam is the algebraic sum of all the external forces acting perpendicular to the beam on one side of that section. The bending moment at any section of a beam is the algebraic sum of the moments of the forces on one side of

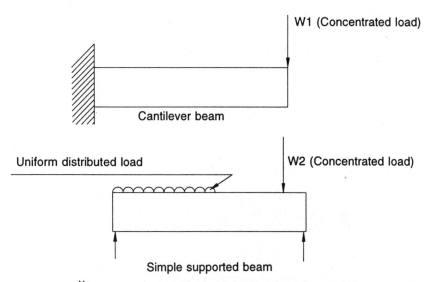

**Fig. 4.13** Application of load in bending of a beam

that section. Bending moments are simple moments of forces and are termed so because of their tendency to bend the beam.

A beam is in equilibrium if it is at rest and has no tendency to move. The conditions for beam equilibrium are given below.

(a) The sum of all the vertical forces must be zero
(b) The sum of all the horizontal forces must be zero
(c) The sum of all the moments of forces must be zero

The effect of bending of a beam is to cause tensile and compressive stresses in it. In a beam which sags as in Fig. 4.14, tensile stresses are set up in the lower half of the section and compressive stresses in the upper half. The axis having stress value as zero is referred to as the *neutral axis*. For any other layer, the stress is proportional to its distance from neutral axis layer. The stress at any distance $y$ (mm) from the neutral axis (Fig. 4.15) is given by

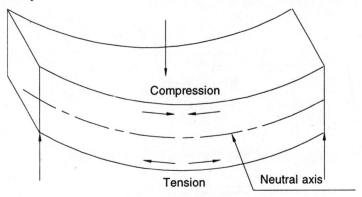

**Fig. 4.14** Bending of beams

$$\text{Stress} = E\,\frac{y}{R}$$

where, $E$ = modulus of elasticity, N/mm$^2$
$R$ = radius of curvature at the neutral axis, mm.

The deflection of a beam at any point at distance $x$ mm as in Fig. 4.15 is calculated by

$$\frac{d^2y}{dx^2} = \frac{M}{EI}$$

where, $M$ = bending moment at a distance $x$, Nm (as shown in Fig. 4.15) and
$I$ = area moment of inertia, mm$^4$

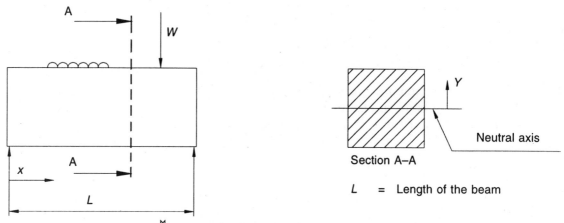

**Fig. 4.15** Stress during bending of beams

## 4.1.7 Torsion

The moment of a force applied to a shaft, which tends to twist or turn it is termed as *twisting moment* or a *turning moment* or a *torque*.

Torsion is a state of being twisted, therefore, a shaft transmitting the torque is said to be in torsion. When a shaft is twisted, each lamination of its cross-section (thin circular discs of the shaft) rotates slightly relative to the next lamination. Thus, the shaft suffers a shear stress. This shear stress at any point in a cross-section at a distance $r$, as shown in Fig. 4.16, from the axis of the shaft is given by:

$$\tau = G\,\theta\,\frac{r}{L}$$

When a torque is applied to the shaft, it develops an internal resistance and the resulting deflection can then be calculated (Fig. 4.16) as:

$$\theta = \frac{TL}{GJ}\,1000$$

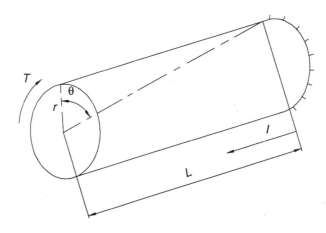

*l*   = Length of shaft under twist, mm
*r*   = Radius of point at which shear stress is calculated, mm
*L*  = Length of shaft, mm
*T*  = Applied torque, Nm
*θ*  = Angle of twist, radian

**Fig. 4.16** Torsion

where,   $\theta$ = angle of twist, radian
$T$ = applied torque, Nm
$J$ = polar moment of inertia, mm$^4$
$\tau$ = shear stress at outer surface, N/mm$^2$
$L$ = length of shaft under twist, mm
$G$ = modulus of rigidity, N/mm$^2$
$r$ = radius of point at which shear stress is calculated, mm

## 4.1.8 Theory of Metalcutting

Metalcutting is a machining process in which a finished part of the desired size, shape and accuracy is obtained by removing excess material in the form of chips. Basically, it is the result of a relative movement between the cutting tool and the material which has to be machined. The *cutting movement*, i.e. the relative movement between the cutting tool and the workpiece material results in the removal of an amount of metal corresponding to the depth of the cut. The *feed movement* brings new material in front of the cutting edge after a cut has been completed.

    In some cutting processes, the cutting and feed movements occur simultaneously and without interruption, like in turning and milling operation. Whereas in planing and slotting, the operation is interrupted at the end of the cutting stroke and the feed movement is given before the start of the next cutting stroke. The cutting tool consists of two intersecting surfaces to form the cutting edge (Fig. 4.17). The surface along which the chip flows is known as the *rake surface*. The surface which is relieved to clear the newly machined surface is known as the flank. The cutting face or the rake face makes an angle $\gamma$ with the normal to the finished surface and is called the *rake angle*. The angle $\alpha$ between the flank and the new workpiece is known as the clearance or relief angle.

Most of the cutting operations carried out either with a single point or a multi-point tool relate to the same basic principle. The multi-point tool can be regarded as a group of several single point tools joined together and constrained to move in such a way that each individual tool contributes to the process of metal removal.

### Chip Formation

The portion of the material that has been cut away from the work material by a cutting tool is called the chip. When a force is applied to the cutting tool, the material of the workpiece ahead of the cutting edge gets deformed, owing to the shearing action. Continued application of the force causes the material to rupture and it either reaches the point of plastic flow, passing across the face of the tool, or it fractures and breaks away in the form of a chip. During the cutting process, the resistance caused by deforming the workpiece material and frictional forces act in the form of a cutting force on the tool (action) and on the workpiece (reaction). The various parts of the machine tool (structure, slide, workpiece and tool carriers) should be able to carry these cutting loads and the driving elements should be able to transmit the corresponding forces and torques at the required velocities.

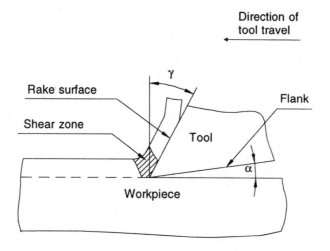

$\gamma$ = Rake angle
$\alpha$ = Relief angle

**Fig. 4.17** Cutting edge

## Cutting Force

The cutting forces acting on the tool point in a turning operation are shown in Fig. 4.18. Here, $P_z$ is the main or tangential force which determines the power requirement during cutting and acts in the direction of the cutting speed, $P_x$ is the axial force which acts in the direction of the feed movement and $P_y$ is the radial force which acts in the direction of the depth of the cut.

The ratio of $P_z$, $P_x$, and $P_y$ varies with the geometry of the tool and to some extent the feed rate. For a 60° approach angle, they are approximately in the ratio of 3:1:1.2.

The cutting forces acting on the cutting tool in milling operation are shown in Fig. 4.19. For different milling processes, the following equations are generally used to find out the values of $P_x$ and $P_y$.

1. *Symmetrical Face Milling*

   $P_x = 0.5\ P_z - 0.55\ P_z$

   $P_y = 0.25\ P_z - 0.35\ P_z$

2. *Asymmetrical Face Milling*

   $P_x = 0.5\ P_z - 0.55\ P_z$

   $P_y = 0.3\ P_z - 0.4\ P_z$

3. *End Milling (30° Helical flute cutter)*

   $P_x = 0.15\ P_z - 0.25\ P_z$

   $P_y = 0.45\ P_z - 0.55\ P_z$

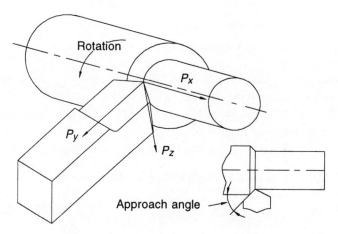

$P_x$ = Axial cutting force, N
$P_y$ = Radial cutting force, N
$P_z$ = Main/tangential cutting force, N

**Fig. 4.18** Cutting forces acting on cutting tool in turning operation

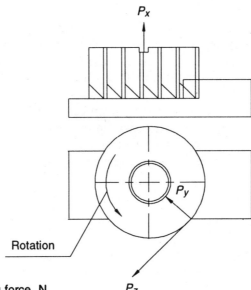

$P_x$ = Axial cutting force, N
$P_y$ = Radial cutting force, N
$P_z$ = Main/tangential cutting force, N

**Fig. 4.19** Cutting forces acting on cutting tool in milling operation

## Drive Requirements for Cutting Movement

The quantity of material removed per unit time is given by

$$Q = s\,t\,v \ \text{cm}^3/\text{min}$$

where, $s$ = feed, mm/rev of workpiece/cutter
$t$ = depth of cut, mm
$v$ = cutting speed, m/min = $\pi Dn/1000$
$D$ = diameter of workpiece/cutter, mm
$n$ = number of revolutions per minute, rpm of workpiece/cutter

For a constant material removal rate,

$$Q = s\,t\,v = \text{constant}$$

Power required for the cutting movement is given by

$$\text{Power} = s\,t\,v\,\frac{K_s}{6120}, \ \text{kW}$$

where, $K_s$ is the specific cutting force per unit chip cross-section (kgf/mm²) and is often specified as a material parameter. This specific cutting force depends upon the chip cross-section. It decreases with increasing chip cross-section. But for constant material removal application, the chip cross-section is constant. Therefore, the specific cutting force is also constant. Hence, the basic requirement for the cutting movement for a constant material removal rate is a constant power. Since, in the cutting processes like turning, milling, grinding, etc., the cutting movement is given to the spindle, the spindle drive should be able to deliver constant power over the entire speed range. A typical spindle drive characteristic is shown in Fig. 4.20.

## Requirements of Feed Drive

The basic requirement of a feed drive is to overcome the machining forces and the frictional forces. For most of the machining processes, the cutting forces in the radial and axial directions are proportional to the tangential force $F_z$, which is constant for a given material removal rate. Similarly, for practical purposes, the frictional forces are also independent of the feed drive speed. Thus, the total resistive force to be overcome by the feed drive is independent of the speed. This requires a constant torque over the entire speed range of feed drive. The typical speed-torque characteristics of feed drive are given in Fig. 4.21.

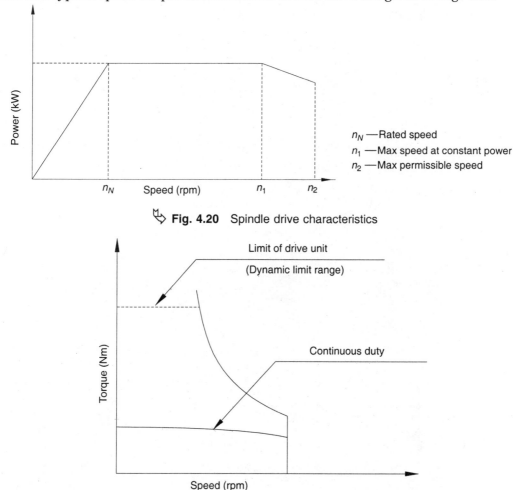

**Fig. 4.20** Spindle drive characteristics

$n_N$ — Rated speed
$n_1$ — Max speed at constant power
$n_2$ — Max permissible speed

**Fig. 4.21** Speed-torque characteristics of a feed drive

## Calculation of Drive Requirements on Feed Motor Shaft

The complete feed drive of computer numericallly controlled (CNC) machine slide consists of the following [Figs 4.22(a) and 4.22(b)].

1. The ac/dc servomotor
2. Drive unit
3. Mechanical transmission elements
4. Feedback element

For the selection of an optimum servomotor and the drive, it is necessary to find the drive requirements on the motor shaft. These are

(a) Load Torque
(b) Load Inertia
(c) Maximum Speed

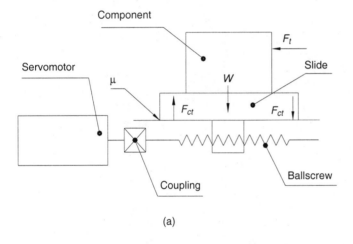

(a)

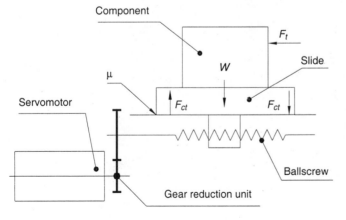

$F_t$ – Cutting force, N
$F_{ct}$ – Reaction at the bearings due to the moment of cutting force, N
$\mu$ – Coefficient of friction, N
$W$ – Weight of the sliding mass, N

(b)

**Fig. 4.22** Components of a feed drive of a CNC machine slide

Generally, the following procedure is followed to find the load torque, load inertia and maximum speed requirements on the motor shaft.

### Load Torque Calculation

The load torque required at the ballscrew or pinion shaft is calculated by the following equation:

$$T_L = \frac{F\,l}{2\pi\eta} + T_c, \quad \text{Nm} \ldots \text{ in case of ballscrew drive}$$

$$T_L = \frac{F\,r}{\eta} + T_c, \quad \text{Nm} \ldots \text{ in case of rack and pinion drive}$$

Where, $F$ = the force required to make the slide move in axial direction, N
$\eta$ = the efficiency of the drive system
$l$ = the lead of the ballscrew, m
$r$ = the pitch radius of the pinion, m
$T_c$ = the friction torque of the ballscrew nut section and/or the support bearing section converted to ballscrew or pinion shaft, Nm

The force $F$ depends on the table weight, coefficient of friction at the surface, the position of the cutting force applied, the axis configuration (horizontal or vertical) and on whether any counterbalance is used in the vertical axis. In case of a horizontal configuration, the value of $F$
*during non-cutting time is,*

$$F_{NC} = \mu W + F_g, \quad \text{N}$$

*during cutting time is,*

$$F_C = F_t + F_g + \mu(W + F_{ct}), \quad \text{N}$$

where, $W$ = weight of the sliding mass, N
$\mu$ = coefficient of friction
$F_g$ = drag force within the guideways due to preload and seals, N
$F_t$ = cutting force, N
$F_{ct}$ = the reaction at the bearings due to the moment of cutting force, N

In case, the load torque is of a periodic nature and the load cycle can be constructed, the load torque requirement is calculated on the basis of the root mean square value of the torque during one cycle.

When the ballscrew/pinion is directly connected to motor shaft, the load torque required at the motor shaft is the same as that on ballscrew/pinion shaft. In case any gear reduction unit is used as in Fig. 4.22(b), the torque required on motor shaft is calculated by

$$T_M = \frac{T_L}{i}, \quad \text{Nm}$$

where, $i$ = reduction ratio of gearbox

## Load Inertia Calculation

During the positioning of the slide, the moving masses have to be accelerated and decelerated with the maximum acceleration value possible. In order to compute on the inertia torque, one must know the total inertia of the system as referred to the motor shaft. The following basic formulae are used to calculate the inertia.

1. *Inertia of Cylindrical Bodies*

   Inertia of a cylindrical body rotating around its own axis is calculated as follows.

   For solid bodies, $\quad J = \dfrac{MD^2}{8}$, kgm$^2$ $\quad$ [Fig. 4.23(a)]

   For hollow bodies, $\quad J = \dfrac{M(D^2 + d^2)}{8}$, kgm$^2$ $\quad$ [Fig. 4.23(b)]

   Where, $M$ = mass of the body, kg
   $D$ = outside diameter of body, m
   $d$ = inside diameter of body, m

   The inertia of the ballscrew, gear, timing pulleys, spacers, etc. is calculated using the above formulae.

2. *Inertia of a Mass Moving along the Linear Axis*

   The inertia of linearly moving mass is calculated as follows

   $$J = M \left( \dfrac{l}{2\pi} \right)^2, \text{ kgm}^2 \; ... \text{ for ballscrew drive}$$

   $$J = M r^2, \text{ kgm}^2 \; ... \text{ for rack and pinion drive}$$

   where, $M$ = sliding mass, kg
   $r$ = pitch radius of pinion, m
   $l$ = lead of ballscrew, m

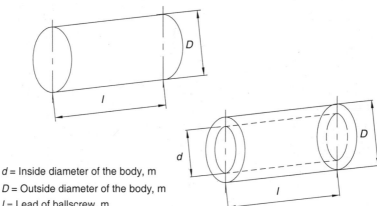

$d$ = Inside diameter of the body, m
$D$ = Outside diameter of the body, m
$l$ = Lead of ballscrew, m

**Fig. 4.23** Inertia of a cylindrical body rotating around its own axis (a) Solid body (b) Hollow body

The inertia of sliding table, and workpiece is calculated using the above formulae.

3. *Inertia of a Cylindrical Body Rotating about an Axis Other than Its Own Axis*
   In this case, the inertia of the cylindrical body is given by
   $$J = J_a + MR^2, \text{ kgm}^2 \quad \text{(Fig. 4.24)}$$
   where, $J_a$ = inertia of the body about its own axis, kgm$^2$
   $R$ = distance between the cylindrical body's own axis and the axis about which it is rotating, m
   $M$ = mass of the body, kg

   This equation is used to calculate the inertia of gears and pulleys of large diameter in which holes are made on a certain pitch circle to reduce the inertia and weight.

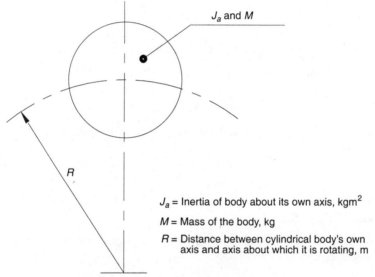

$J_a$ = Inertia of body about its own axis, kgm$^2$
$M$ = Mass of the body, kg
$R$ = Distance between cylindrical body's own axis and axis about which it is rotating, m

**Fig. 4.24** Inertia of cylindrical body rotating about an axis other than its own axis

4. *Inertia of Driven Shaft Referred to Driver Shaft When Gear Reduction Unit is Used*
   The inertia of driven shaft referred to driver shaft (Fig. 4.25) is
   $$J = \frac{J_0}{i^2}, \text{ kgm}^2 \quad \text{(Fig. 4.25)}$$
   where, $J_0$ = inertia of driven shaft, kgm$^2$
   $i$ = reduction ratio of the gear reduction unit
   Thus if, $J_s$ = inertia of sliding mass, kgm$^2$
   $J_c$ = inertia of coupling, kgm$^2$
   $J_b$ = inertia of ballscrew, kgm$^2$
   $J_{G1}$ = inertia of gear on ballscrew shaft, kgm$^2$

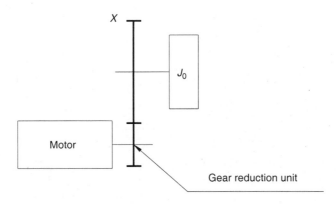

$J_0$ = Inertia of driven shaft, kgm$^2$

**Fig. 4.25** Inertia of driven shaft referred to driver shaft

$J_{G2}$ = inertia of gear on motor shaft, kgm$^2$
$i$ = reduction ratio of gear reduction unit

the total load inertia of the system referred to motor shaft in case as shown in Fig. 4.22(a) is

$$J_L = J_s + J_c + J_b, \text{ kgm}^2$$

The total load inertia of the system referred to motor shaft in case as in Fig. 4.22(b) is

$$J_L = \frac{1}{i^2}[J_s + J_b + J_{G1}] + J_{G2}, \text{ kgm}^2$$

## Maximum Speed Calculation

If $v$ is the rapid traverse rate of the slide, then maximum speed required on ballscrew or pinion shaft is given by:

$$n = \frac{v}{l} \text{ rpm in case of ballscrew drive}$$

$$n = \frac{v}{2\pi r}, \text{ rpm in case of rack-pinion drive}$$

where,  $v$ = rapid traverse rate of slide, mm/min
$l$ = lead of ballscrew, mm
$r$ = pitch radius of pinion, mm

In case, motor is directly connected to the ballscrew or pinion, the speed required on the motor is the same as that required on the ballscrew or the pinion shaft. If a gear reduction unit is used between the motor shaft and the ballscrew or the pinion shaft, then the speed of motor is calculated by:

$$n_m = n\, i, \text{ rpm}$$

where,  $i$ = reduction ratio of gear reduction unit.

## 4.2 MATERIALS

### 4.2.1 Introduction

The modern machine tools are designed to operate at higher speeds and feeds. They possess improved accuracy, higher rigidity and reduced noise levels.

The cost of raw material input is very high—of the order of 40% for general purpose machine tools. This calls for optimising the design of machine elements, selecting the right type of materials, judiciously imparting effective fabrication and treatment method.

### 4.2.2 General Principles in the Selection of Materials for Machine Tools

*Functional Requirements*

The functional requirements must be met in terms of various properties. For example, in selecting material for the main spindle of a machine tool, the modulus of elasticity and the surface hardness required for the spindle nose, bore and the locations of the bearings are important properties which need to be considered. Generally, low nickel-chromium alloy case carburised steel such as 15CrNi6 (as per DIN 17210) is selected, which meets the functional requirements.

*Ease of Fabrication*

The process of fabrication should be such that the part or component should be easy to make. If it is required in batch quantity, casting process is adopted. For example, machine tool elements or parts such as bed, headstock, etc. required in batch quantity are made out of a casting process in the foundry. If the requirement is one or two numbers, a welding process is used to fabricate the part.

*Machinability*

This is another important parameter to be considered for selecting the raw material of machine tool components as extensive machining is involved. Construction steels such as medium carbon steel (C45 as per DIN 17200) and low alloy steel (15CrNi6 as per DIN 17210, 36CrNiMo4 as per DIN 17200 and 34CrAlMo5 as per DIN 17211) are chosen for many of the parts which have good machinability. In case of castings, grey cast iron is selected.

*Cost*

Since the raw material cost plays a significant role in the overall cost of the machine tool, it becomes an important factor to be considered.

*Availability*

The chosen material must be easily available so that the cost and the delivery time are kept low. In fact, all the raw materials required for machine tools are easily available in India.

### 4.2.3 Materials Used and Application

#### General Classification of Materials

#### (a) Ferrous materials

*Castings*
- grey cast iron, spheroidal graphite cast iron (SG iron), malleable cast iron and steel castings.

These are used for production of grey cast iron, SG roin, malleable cast iron and steel castings of machine tools.

*Wrought steels*
- hot rolled and forged constructional steels such as medium carbon and low alloy steels.
- cold drawn constructional steels such as free cutting steel.
- cold work tool steels.
- high speed steels (HSS).

#### (b) Nonferrous metals and alloys
- castings of copper base alloys, and aluminium base alloys.
- wrought products of copper base alloys and aluminium base alloy forgings.

#### (c) Miscellaneous materials
- friction reducing materials like turcite, SKC3, etc., plastics, rubber products, paints, lubricants, cutting oils, etc.

The application of various materials along with requirements are given in Table 4.1. The guidelines for achievable mechanical properties and surface hardness after heat treatment of constructional steels are given in Table 4.2. The guidelines for hardness of tool and high speed steels in as provided and heat treated condition is given in Table 4.3. The guidelines for achievable surface hardness and case depth for machine tool guideways along with relevant heat treatment is given in Table 4.4.

TABLE 4.1

*Application and requirements of important materials used for machine tools*

| Material | Requirement | | | Application |
|---|---|---|---|---|
| | Ultimate tensile strength (N/mm$^2$) | Hardness as supplied (HB) | Chemical composition (%) | |
| Medium carbon steel C45 (as per DIN 17200) | 600–750 Normalised | 170–210 | Carbon : 0.42–0.50 Silicon : 0.15–0.35 Manganese : 0.5–0.8 | Parts such as gears, pinions, bolts, nuts, axles, shafts and lead screws. Can be flame/induction hardened to 45–50 HRC |

*(Contd)*

Table 4.1 (Contd)

| Material | Strength | Hardness | Composition (%) | Applications |
|---|---|---|---|---|
| High carbon steel C60 (as per DIN 17200) | 700–900 Normalised | 220–240 | Carbon : 0.57–0.65<br>Silicon : 0.15–0.35<br>Manganese : 0.6–0.9 | Moderately stressed parts such as tie bars of die casting machines and main spindle of multispindle automats. Can be flame/induction hardened to 55–60 HRC. |
| Nickel chromium alloy carburising steel 15CrNi6 (as per DIN 17210) | 900–1200 Hardened and tempered | 170–220 | Carbon : 0.14-0.19<br>Chromium : 1.4-1.7<br>Nickel : 1.4-1.7 | Parts subjected to high wear and bending stresses such as gears and main spindle used after case hardening to 55–60 HRC |
| Medium carbon low alloy through hardening steel 36CrNiMo4 (as per DIN 17200) | 750–1300 Hardened and tempered | 170–230 | Carbon : 0.32-0.40<br>Nickel : 0.9-1.2<br>Chromium : 0.9-1.2<br>Molybdenum : 0.15-0.30 | Components requiring hardness, toughness and resistance to shock loads: gears, shafts, drilling spindles, mandrels, clutch parts, etc. Can be flame/induction hardened to 54–60 HRC |
| Chromium vanadium spring steel 50CrV4 (as per DIN 17200) | 800–1300 Hardened and tempered | 200–235 | Carbon : 0.47-0.55<br>Chromium : 0.9-1.2<br>Vanadium : 0.1-02 | Ideal spring material with high ratio of yield point to tensile strength. Collets, feed fingers, disc springs, etc. |
| Nitriding steel 34CrAlMo5 (as per DIN 17211) | 800–1000 Hardened and tempered | 200–250 | Carbon : 0.3-0.37<br>Aluminium : 0.8-1.2<br>Chromium : 1.0-1.3<br>Molybdenum : 0.15-0.25 | A nitriding steel permitting an extremely hard and wear resistant case. Main spindles of grinding machines, racks, and shift bushes. Can be nitrided to 950–1100 HV |
| Free cutting steel 45S20 (as per DIN 1651) | 660–810 (Cold drawn) | 240 max. | Carbon : 0.42-0.50<br>Manganese : 0.50-0.9<br>Sulphur : 0.15-0.25 | Bright drawn free cutting steel for parts requiring extensive machining. Screws, nuts, bolts, etc. can be flame/ induction hardened to 45–55 HRC. |
| Cf53 (as per DIN 17212) | 600–750 Normalized | | Carbon : 0.50-0.57<br>Manganese : 0.40-0.70<br>Silicon : 0.15-0.35 | Ballscrew material |
| 100Cr6 (as per DIN 17230) | — | | Carbon : 0.9-1.05<br>Chromium : 1.35-1.65 | Nut material |
| Grey cast iron GG15 (as per DIN 1691) | 150 | 150–170 | Left to manufacturer | For components having simple functions such as covers and counter weights. Flame/induction hardening not recommended. |
| Grey cast iron GG20 (as per DIN 1691) | 200 | 170–210 | Left to manufacturer | For components subjected to wear, e.g., base plates, box tables, etc. Flame/induction hardening not recommended. |

(Contd)

Table 4.1  (Contd)

| Material | | | Composition | | Remarks |
|---|---|---|---|---|---|
| Grey cast iron GG25 (as per DIN 1691) | 250 | 180–220 | Left to manufacturer | | For components subjected to wear. Pattern of graphite distribution and size is controlled. Headstock, feed boxes, etc. Is having pearlitic matrix. Suitable for flame/ induction hardening. Achievable surface hardness 45-55 HRC |
| Grey cast iron GG30 (as per DIN 1691) | 300 | 190–230 | Left to manufacturer | | For machine tool beds and components such as sleeves, arm, etc. requiring good wear resistance and rigidity. Is having pearlitic matrix. Suitable for flame/induction hardening. Achievable surface hardness is 45-55 HRC |
| Spheroidal graphite iron GGG60 (as per DIN 1693) | 600 | 190–270 | Left to manufacturer | | For components requiring good strengths and wear resistance such as turret heads, cams, etc. Can be flame/induction hardened to 55-60 HRC |
| Spheroidal graphite iron GGG50 (as per DIN 1693) | 500 | 170–240 | Left to manufacturer | | General purpose SG iron possessing moderate ductility. Index gears, brake drums, swinging levers, etc. Can be flame/induction hardened to 45-55 HRC |
| Spheroidal graphite iron GGG40 (as per DIN 1693) | 400 | 200 max | Left to manufacturer | | Possess high resistance to shock loads, high ductility and good machinability. Levers, shift forks, connecting rods, etc. Not recommended for flame/ induction hardening. |
| G-CuSn12 (as per DIN 1705) | 260 | 80 min | Copper<br>Tin<br>Lead | : 84–88.5<br>: 11–13<br>: 1.0 | For components requiring high wear resistance under high load, e.g., feed screw nuts, split nuts, etc. |
| G-CuSn5ZnPb (as per DIN 1705) | 220 | 70-85 | Copper<br>Tin<br>Lead<br>Zinc | : 84–86<br>: 4–6<br>: 4–6<br>: 4–6 | For bearings requiring medium strengths and moderate anti-seizing properties. Bearings, bearing bushes, flanges. |
| G-CuAl9Ni (as per DIN 1714) | 500 | 110 min | Aluminium<br>Nickel<br>Iron<br>Manganese | : 8.5–11<br>: 1.5–4.0<br>: 1.0–3.0<br>: 2.5 max | For components requiring good strength, wear resistance and resistance to shock loads. To be used with steel components of |

(Contd)

Table 4.1 (Contd)

| Material | | | Composition | | Applications |
|---|---|---|---|---|---|
| | | | Copper | : 82 min | 30 HRC min. Worm wheels, shift forks, jibs, racks, etc. |
| G-CuPb15Sn (as per DIN 1716) | 180 | 60 min | Copper<br>Tin<br>Lead | : 75-79<br>: 7-9<br>: 13-17 | For bearing applications requiring lower strengths, but better antiseizing properties than G-CuSn5ZnPb. Grinding machine bearings, bearing bushes and thrust collars. |
| Aluminium | Not critical | 50-70 | Commercial aluminium | | Components having minor functions like covers, foot rests, wheel guards, etc. |
| G-AlSi11 (as per DIN 1725) | 150-200 | 45-65 | Silicon<br>Aluminium | : 10-11.8<br>: Rest | Silicon ensures fluidity. Used for pulleys, intricate covers, rocker arms, etc. |
| 64423 (as per IS 734) | 330 | 75-90 | Copper<br>Magnesium<br>Silicon<br>Aluminium | : 0.5-1.0<br>: 0.5-1.3<br>: 0.7-1.3<br>: Rest | Forging type of aluminium alloy for pressure tight applications. Control block applications. Control block valve housings, control housing of surface grinding machines. |
| CuZn36Pb3 (as per DIN 17660) | 330 | 90-120 | Copper<br>Lead<br>Zinc | : 60-62<br>: 2.5-3.5<br>: Rest | Possess excellent machinability. For high speed screw cutting on turning work on automatic lathes. Bushes, nuts, spacers and rings. |

*Note*: Relevant standard(s) to be referred for detailed chemical composition and other properties

## TABLE 4.2

**Achievable mechanical properties and surface hardness after heat treatment**

| Material | Mechanical properties in hardened and tempered condition | | Treatment | Hardness value | Case depth |
|---|---|---|---|---|---|
| | Thickness/ Diameter mm | Tensile strength $N/mm^2$ | | HRC | mm |
| 15CrNi6 | 30<br>63 | 90–120<br>80–110 | Carburising and hardening | 58–62 | For gears: 0.2–0.3 module, For shafts, etc. L : 0.5<br>M : 0.5–1.0<br>H : 1.0–1.5 |
| C45 | 16–40<br>40—100 | 67–82<br>63–78 | Flame/ induction hardening | L : 45–50<br>M : 50–55<br>H : 55–60 | 1.0–1.5 |

(Contd)

Table 4.2  (Contd)

| Material | | | | | |
|---|---|---|---|---|---|
| 45S20 | 30 | 70–85 | Flame/induction hardening | L : 45–50<br>M : 50–55 | 1.0–1.5 |
| | 60 | 60–75 | | | |
| C60 | 16 | 85–100 | Flame/induction hardening | 55–60 | 1.0–1.5 |
| | 16–40 | 80–95 | | | |
| | 40–100 | 75–90 | | | |
| 36CrNi Mo4 | 16 | 110–130 | Flame/induction hardening | 54–60 | 1.0–1.5 |
| | 16–40 | 100–120 | | | |
| | 40–100 | 90–105 | | | |
| 34CrAl Mo5 | up to 70 | 80–100 | Nitriding | 950–1100 | HV 0.3–0.5 |

L: Light duty; M: Medium duty; H: Heavy duty; HB: Hardness Brinell; HV: Hardness Vickers; HRC: Hardness Rockwell

TABLE

*Guidelines for hardness of tool and high speed steels in as supplied condition and heat treated condition*

| Material as per Din 17350 | Supply condition–Annealed hardness (max) (HB) | Hardness in heat treated condition* (HRC) min |
|---|---|---|
| 105WCr6 | 230 | 59 |
| 100Cr6 | 225 | 60 |
| X210Cr12 | 250 | 60 |
| S6–5-2-5 | 240-300 | 64 |
| S6-5-2 | 240-300 | 64 |
| S10-4-3-10 | 240-300 | 66 |

\* Refers to cutting end only

*Note*: The hardness values given are for general guidance only. In special cases, depending on designers' requirement and material/heat treatment limitations, other values may be specified.

Steels, as per DIN standards, and their approximate equivalent national standards are given in Annexure 4.1. Cast iron, as per DIN standards, and their approximate equivalent national standards are given in Annexure 4.2.

## 4.3  HEAT TREATMENT

### 4.3.1  Introduction

Heat treatment is an integral part of metallurgy, which deals with thermal treatment of metals and alloys. It is a process of heating, soaking and cooling of a metal or an alloy in a solid state to achieve the desired mechanical properties. This implies that there must be a heating source and a cooling media for completion of heat treatment cycle. Heating source may be conduction, convection, or radiation type and cooling (quenching) media

## TABLE 4.4
### Guidelines for heat treatment of machine tool guideways

| Material | Surface condition | | Condition of material | | Final heat treatment | Hardness in final condition | Case depth after finish grinding (mm) |
|---|---|---|---|---|---|---|---|
| | Before heat treatment | Final condition | As supplied | Prior to final heat treatment | | | |
| C45 | Machined with grinding allowance | Ground | Normalised | — | Induction or flame hardening | 50–55 HRC | 1.0–1.5 |
| 15CrNi6 | Machined with grinding allowance | Ground | Annealed | — | Case hardening | 58–62 HRC | 0.8–1.0 |
| 34CrAlMo5 | Machined with grinding allowance | Ground | Annealed or hardened and tempered | Hardened and tempered | Gas nitriding | 900–1100 HV 10* | 0.3–0.5 |
| GG30 | Machined with grinding allowance | Ground | As cast | Preferably stress relieved | Induction or flame hardening | 45–52 HRC | 1.0–2.0 for induction hardened 2.0–3.0 for flame hardened |
| X210 Cr12 | Machined with grinding allowance | Ground | Spherodise annealed | Stress relieved | Hardening and tempering | 62–64 HRC | — |

\* 10 kg load

may be mild to drastic. The heating source and a quenching media highly influences the treated metal and hence proper selection of heat treatment process is very important to achieve the desired properties with minimum distortion.

### 4.3.2 Steel—Engineering Material

Steel is an important engineering and construction material relatively soft and ductile and can be moulded in various shapes. This versatile material can be heat treated to increase strength and toughness.

Steel is basically an alloy of carbon and iron. Alloy steels are those which contain specified percentage of other elements in their chemical composition. Commonly, steels are alloyed with nickel, chromium, molybdenum, vanadium and tungsten to impart specific properties for engineering applications. Percentage of carbon in steel controls the hardness and alloying elements increases hardenability.

The iron content in steel varies from 98 to 99.9% and pure iron undergoes structural transformation at various temperatures with different crystal structures. Hence it is essential to understand the mechanism of transformation. Allotropic transformation of pure iron is shown in Fig. 4.26.

The temperature at which this phase transformation takes place is dependent on the rate of heating or cooling under ideal conditions.

*Micro Constituents of Steel*

Ferrite and iron carbide (cementite) $Fe_3C$ are the two important constituents of steel, the amount and distribution of which primarily controls the properties of steel. The growth of ferrite and cementite in lamellar form is known as pearlite. On heating the steel above the critical temperature, the dual phase of steel (ferrite + pearlite or pearlite + cementite) transforms to austenite (single phase) which on fast cooling transforms to hard martensitic phase characterised by body centred tetragonal structure. This transformation process is called hardening.

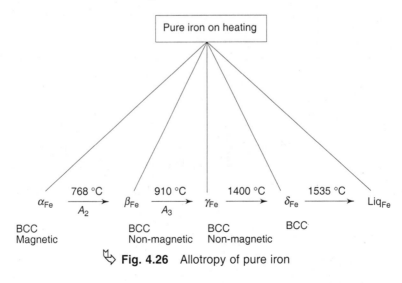

**Fig. 4.26** Allotropy of pure iron

***Ferrite*** Ferrite is a solid solution of carbon in $\alpha_{Fe}$. Maximum solubility of carbon in $\alpha_{Fe}$ is 0.04% by weight. It is a soft phase having a hardness of 60 to 100 BHN and its specific gravity is 7.86. The structure of ferrite is body centre cubic.

***Cementite*** Cementite is a compound of iron and carbon and its chemical term is iron carbide, having an approximate formula as $Fe_3C$, characterised by an orthorhombic crystal structure. It is a hard and brittle phase having a hardness of more than 600 BHN and its specific gravity is 7.86.

***Austenite*** Although austenite is not ordinarily a constituent of steel at room temperature, yet it is an important high temperature phase, the decomposition of which on cooling forms room temperature constituents. It is a solid solution of carbon in $\gamma_{Fe}$, characterised by face centred cubic structure having a hardness of 170–230 BHN. It is formed when a steel is heated to relatively high temperature above 770 °C. The limiting temperature for the formation of austenite vary with composition. Austenite can be retained at room temperature in an alloy steel only when it is cooled by suppressing the equilibrium reactions.

***Pearlite*** When plain carbon steel of approximately 0.8% carbon is cooled slowly from austenitic region, all the ferrite and cementite precipitate together in a characteristic lamellar structure known as pearlite. Its specific gravity is 7.66 and has a hardness of 160–230 BHN.

## 4.3.3 Application of Heat Treatment

Steel may be heat treated to achieve one or more of the following objectives.

(a) To increase strength and hardness of iron and steel for improved wear resistance.
(b) Relieving microstresses in the atomic lattice that increase brittleness
(c) Relieving macrostresses that cause distortion and service failure.
(d) To relieve stress induced by cold working or non-uniform cooling of hot metal.
(e) Grain size refinement of hot worked steel.
(f) To decrease hardness and to increase ductility and toughness to withstand high impact.
(g) Structural transformation for better machinability.
(h) To improve electrical properties.
(i) To change or modify magnetic properties of steel.
(j) Elimination of hydrogen gas dissolved during pickling or electroplating, which causes brittleness.

## 4.3.4 Principle of Heat Treatment

The first step in heat treatment is to austenise the steel, i.e. by heating the steel to above transformation temperature and soaking it at that temperature to homogenous austenitic phase. When steel is cooled from austenising temperature, the austenitic phase becomes unstable as soon as the temperature falls below the transformation range. As the eutectoid temperature 723 °C is crossed, the eutectoid austenite transforms into lamellar aggregate of ferrite and cementite, i.e. pearlite. The transformation of austenite to pearlite is a diffusion process and can occur best under slow cooling conditions. If the cooling rate is increased, the diffusion rate is also increased. Increased rate of cooling or continuous cooling reduces the time for transformation, and ferrite and cementite tend to become finer and finer. Very fine pearlite is referred to as sorbite and extremely fine pearlite is called as troostite. If the cooling rate is beyond critical cooling rate, the transformation of austenite to pearlite is completely suppressed and we get an altogether new product known as martensite. The transformation in steel during slow cooling is controlled by diffusion and occurs by nucleation and growth process.

The transformation products of austenite on continuous cooling are shown in Fig. 4.27.

The transformation of austenite to martensite during quenching is classified under martensitic transformation or diffusionless transformation or is sometime referred to as athermal transformation, because the transformation is not influenced by temperature. There is a whole range of temperature in which martensite is formed. The two extremes are called $M_s$ and $M_f$, where $M_s$ is the martensitic start and $M_f$ is the martensitic finish temperature. Martensitic transformation will not take place above or below the range and this range varies with the carbon content as given in Table 4.5.

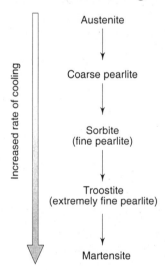

**Fig. 4.27** Transformation products of austenite on continuous cooling

TABLE 4.5

| % Carbon | $M_s$ °C | $M_f$ °C | Range ($M_s$– $M_f$ °C) |
|---|---|---|---|
| 0.2 | 460 | 270 | 190 |
| 0.4 | 380 | 160 | 220 |
| 0.6 | 330 | 40 | 290 |
| 0.8 | 260 | – 80 | 340 |

Hence heat treatment is a process of austenitic transformation into slow or fast cooling products by a diffusion or diffusionless process.

### 4.3.5 Heat Treatment Processes

#### Annealing

In general, annealing refers to any heating, soaking and cooling operation which is usually done to induce softening. Annealing may also be carried out to relieve internal stress induced by cold or hot working process.

The aim of annealing is to improve machinability, ductility and grain size refinement. Annealing is subdivided into

(a) full annealing
(b) process annealing
(c) isothermal annealing
(d) patenting

**(a) Full Annealing** Full annealing is a process of heating ferrous alloy above transformation range, soaking at that temperature for sufficient time and cooling very slowly. The transformation product is a coarse pearlite. Fully annealed steel is soft with low hardness and high ductility.

**(b) Process Annealing** Process annealing is a subcritical annealing for cold worked metals. In this process the steel is heated close to or below the lower critical temperature in the range of 550–700°C followed by desired rate of cooling. The purpose of this annealing is to soften the steel and relieve the internal stress.

**(c) Isothermal Annealing** Isothermal annealing is a process in which steel is heated above the transformation temperature to produce wholly or partly austenitic structure and it is then cooled to and held at a temperature of 660–680°C which causes transformation of austenite to relatively soft ferrite carbide aggregate.

**(d) Patenting** Patenting is a process of softening medium or high carbon steel by heating above the transformation range and cooling to 400–500 °C in air (preheater) or in molten lead or in a salt bath maintained at that temperature.

#### Normalising

Normalising is a process of heating steel to 50 °C above the upper critical temperature followed by cooling in still air. The primary purpose of normalising is to refine the grain structure prior to hardening or to reduce segregation in castings or forgings and to harden the steel slightly to improve machinability.

#### Stress Relieving

Stress relieving is a process of relieving internal stress induced by manufacturing sequence by heating the ferrous alloy at a suitable temperature of 500–650 °C and holding for sufficient time to reduce residual stress and then cooling slowly to minimise the development of new residual stress.

## Case Hardening

Case hardening is a process of hardening ferrous alloy in which the surface layer or case is made substantially harder than the interior or core. The harder surface is resistant to wear and fatigue and toughened core will resist impact and bending.

The following case hardening processes are generally used.

(a) Flame hardening
(b) Induction hardening
(c) Liquid carburising
(d) Nitriding
(e) Tufftriding

*(a) Flame Hardening* Flame hardening is a process of case hardening wherein the surface of the metal is heated very rapidly by using oxy-acetyline flame, thereby creating a thermal gradient and subsequently quenching in water or oil.

Flame hardening is planned for one or more of the following reasons.

(a) Large components impracticable or uneconomical to harden in conventional furnace, e.g. bull gears, cast iron integral bed guideways, large dies and rollers.
(b) Selective hardening of components.
(c) To control the dimensional accuracy of a large part which is difficult to control in furnace heating and quenching.
(d) To get the wear properties of an alloy steel in an inexpensive medium carbon steel.

Method of flame hardening—stationary, progressive, spinning or a combination of progressive-spinning–depends on the shape, size, composition of workpiece, area to be hardened, depth of hardness and the number of pieces to be hardened.

The success of many flame hardening applications depends largely on the skill of the operator. Large components are preheated to get the desired surface hardness and depth penetration. In flame hardening of alloy steels it is advisable to preheat the components to minimise cracking. Hardenable cast iron susceptive to cracking can be prevented by preheating. Flame hardened components are tempered to relieve the quenching stress.

Medium carbon steel with 0.4–0.5% carbon are ideal for flame hardening, but high carbon steel as high as 1.5% carbon can also be flame hardened with specific care. Normally hardening depth varies from 1 to 6 mm. Manganese bearing alloys increase the hardenability and hence free cutting steels are considered excellent for flame hardening of components requiring high wear resistance and high core strength. Alloy steels are hardened and tempered and finally flame hardened. Grey cast iron, ductile iron and pearlitic malleable iron having combined carbon of 0.4–0.8% can be flame hardened. Carburised parts of plain or alloy steels can be flame hardened to 60 ± 2 HRC.

*(b) Induction Hardening* Induction hardening is a process of case hardening of a ferrous alloy by heating it above the transformation range by means of electrical induction and cooled as desired.

In this process, the component to be surface hardened is covered by a copper conductor in the form of copper block or tube which is not in contact with the iron or steel to be hardened. A high frequency alternate current passes through the coil or block and the surface of steel is heated by induced current to cross the upper critical temperature. The current is shut off and water is sprayed through the perforation in the surrounding block, thereby hardening the surface of steel. Depending on the type of steel, surface hardness of about 60 HRC can be obtained.

The depth of current penetration depends upon workpiece permeability, resistivity and alternate current frequency. As frequency increases, the depth of current penetration decreases and hence for a shallow case depth, high frequency current is used and intermediate and low frequencies are used in applications requiring deeper case depth.

A shallow hardened case of 0.25–1.5 mm is planned for components subjected to light to moderate loading. In applications where parts are subjected to heavy or impact type of loading, a higher case depth of 1.5–6 mm is planned for adequate support and wear resistance.

Medium carbon steel with 0.4–0.5% carbon are ideal for induction hardening. Alloy steels, grey cast iron, ductile iron and pearlite malleable irons having a combined carbon of 0.4–0.8% can be induction hardened to 50–55 HRC.

**(c) Liquid Carburising** Liquid carburising is a process of case hardening low carbon and low carbon alloy steels by heating above the critical range in the molten cyanide salt, which induces both carbon and nitrogen into the case. Carburised parts are quenched to produce hard case and tough core. Low temperature cyanide bath operates at 850–900 °C with protective carbon cover. The cyanide concentration is maintained at 10–12% and carbon potential at 0.8–0.9%.

In liquid carburising bath, several reactions occur simultaneously, depending on the bath composition. Sodium cyanide melts at 560°C and decomposes at higher temperature. The following chemical reaction occurs in low temperature liquid carburising bath.

$$2NaCN \rightleftharpoons Na_2CN_2 + C$$
$$\text{(Sodium cyanamide)}$$

$$2NaCN + O_2 \longrightarrow 2NaCNO$$
$$\text{(sodium cyanate)}$$

$$3NaCNO \longrightarrow Na_2CO_3 + NaCN + 2N + C \downarrow \text{ (absorbs)}$$
$$\text{(Sodium carbonate)}$$

$$2NaCN + 2O_2 \longrightarrow Na_2CO_3 + CO + N_2 \downarrow \text{ (absorbs)}$$

Liquid carburising bath involves handling highly poisonous pure sodium cyanide and cyanide containing salt and hence safety precautions are mandatory. Liquid and solid effluents are to be neutralised as per pollution board specifications before disposal.

**(d) Nitriding** Nitriding is a process of case hardening a special alloy steel (nitralloy) in an atmosphere of ammonia or in contact with nitrogenous material at a temperature below the transformation range. The surface hardness is produced by the absorption of nitrogen without quenching. Single stage nitriding is being done at 500–510 °C with 30% dissociation of ammonia.

Ammonia dissociates according to the following reaction

$$2NH_3 \rightleftharpoons 3H_2 + N_2 \downarrow \text{ (absorbs)}$$

The nascent nitrogen combines with the elements of steel to form nitrides. Nitriding steels are hardened and tempered at 560–650°C to minimise distortion and to have a uniform high surface hardness. Hence the process is very ideal for components having complex shapes/critical section to control distortion. Medium carbon chrome steel (En19), medium carbon nickel chrome molybdenum steel (En24), medium carbon chrome-moly-vanadium steel (En40C) and medium carbon chrome moly aluminium steel (En41B) are very popular nitriding steels. Aluminium containing steel (En41B) can be nitrided to 1000–1100 HV (~ 71HRC). Hot work die steel (AISI H11/H12), high carbon high chrome steel (AISI $D_2/D_3$) and pearlitic spheroidal graphite iron (ASTM class 80-60-03) can also be nitrided.

*Application of gas nitriding* The components requiring very high wear resistance, antigalling, improved fatigue life, improved corrosion resistance can be nitrided to meet the service conditions.

*Tufftriding* Tufftriding is a salt bath nitriding process, operating at 510–570 °C. It is a chemical diffusion process of absorbing more nitrogen and very less carbon from nitrogen bearing fused salt containing sodium and potassium cyanides (NaCN and KCN) and sodium and potassium cyanates (NACNO and KCNO). The activity of the bath is controlled by cyanate content and the ratio of cyanide to cyanate is critical. Hence liquid nitriding baths are aged by maintaining the bath at 565–595 °C for 12 hours, to decrease cyanide content and to increase cyanate and carbonate content. Efficiency of the bath is maintained by retaining the cyanate content to 25% and carbonate to less than 2.5%. The ratio of cyanide to cyanate is maintained at 1.4–1.5.

At operating temperature 570°C, the process is controlled by oxidation reaction and catalytic reaction. In the oxidation process, cyanide is oxidised to cyanate while in catalytic reaction, in the presence of steel this cyanate disintegrates to carbonate and cyanide as per the following reaction.

$$4NaCN + 2O_2 \longrightarrow 4NaCNO \quad \text{Sodium cyanate}$$

$$2KCN + O_2 \longrightarrow 2KCNO \quad \text{Potassium cyanate}$$

$$8NaCNO \longrightarrow 2Na_2CO_3 + 4NaCN + CO_2 + (C)_{Fe} + 4(N)_{Fe}$$

$$8KCNO \longrightarrow 2K_2CO_3 + 4KCN + CO_2 + (C)_{Fe} + 4(N)_{Fe}$$

Steels are usually hardened and tempered before liquid nitriding to maintain the dimensional stability and core toughness by tempering at 600–650 °C. Liquid nitrided components are finally quenched in water or oil to achieve the desired surface hardness. Liquid nitriding salt contains highly poisonous and highly toxic sodium and potassium cyanides and great care should be exercised in handling this salt.

*Application of tufftriding* Liquid nitriding process is used to improve wear resistance and antigalling properties, fatigue resistance and corrosion resistance. This process is not suitable for many applications requiring deep case.

## 4.4 ELECTROPLATING

### 4.4.1 Principles

Electroplating or electrodeposition is a branch of electrometallurgy which deals with the art of depositing metals by means of electric current. The mechanism of electroplating is based on the theory of *ionic dissociation* and *electrolysis* process.

Ionic dissociation is a process of splitting the electrolyte into ions or radicals or group of atoms. Each ion carries a small electric charge. The acid radicals are negatively charged or they carry additional electrons whereas, the metal ions, hydrogen and basic radicals are positively charged or are minus an electron. Ionic dissociation has a very important bearing upon the electrodeposition.

In electrolytic dissociation (electrolysis), the electrolyte—the aqueous solution of a metal salt—undergoes electrochemical change or electrolysis when current is applied through the electrodes. The components to be plated are taken as cathode, whereas sheets or bars of pure metals to be deposited are taken as anode. During electrolysis the positively charged ions known as cations are attracted to cathode and negatively charged ions, anions migrate to anode. In this process, metal ions are plated on the cathode.

Electroplating is done for corrosion resistance, wear resistance and to salvage the undersized/worn components and also for decorative purpose.

The most commonly used plating is chrome plating with an under coat of copper and nickel.

### 4.4.2 Copper Plating

Copper can be electroplated on both ferrous and nonferrous metals by cyanide copper or acid copper plating solutions. Commercial electroplaters are using cyanide bath for copper plating, since the copper cyanide solution has much higher throwing power than acid type bath and is most suitable for complex-shaped components.

For heavy or high speed copper deposition, acid copper plating solution is generally preferred. Nonferrous metals and alloys such as bronze, brass, nickel and silver may be directly copper plated in acid solution. Acid copper plating solutions are more stable than cyanide solution with lesser cyanide effluent problem and simple to operate.

Copper plating is done as an under coat for chromium plating and also as an anticarburising media in carburising process. Generally, 50 µm thickness of copper plating is sufficient to prevent carbon diffusion.

### 4.4.3 Nickel Plating

The majority of nickel plating salts are a mixture of nickel sulphate and nickel chloride, boric acid and stabilisers to control the quality of nickel deposit. Boric acid is a weak acid and acts as a buffer to control the pH of the solution. The efficiency of nickel plating bath is controlled by maintaining the pH value. For a dull nickel plating solution, the pH is maintained between 5.6 and 5.8. Bright nickel plating solutions are operated at a lower pH value, usually about 4. If the pH value is more, the solution becomes more alkaline and nickel will be precipitated from the solution as nickel hydroxide and plated deposit is unsatisfactory. If the solution is more acidic (i.e., pH less than 4), than the efficiency of the solution and the possibility of corrosive attack on the base metal, i.e. the components being plated is reduced.

Bright nickel plating solutions are extensively used for decorative plating and are generally plated at higher current densities than necessary for dull plating. Efficient filtration and solution agitation are essential for bright nickel plating process. Great care must be exercised to reduce contamination of plating bath. Bright nickel plating is done as an under coat for chromium plating.

### 4.4.4 Chromium Plating

Chromium plating is done for decorative purposes or to improve corrosion resistance or to produce hard abrasion resistant surface or to salvage undersized and worn out parts.

Generally, dull chrome plating and flash chrome platings are done for decorative purposes, whereas hard chrome plating is done to increase wear resistance and to salvage undersized and worn out components and antiseize or antifriction applications. Chromium plating is not recommended for parts subjected to compression or severe shock loads.

In hard chrome plating, the deposits are very much harder and thicker and are directly applied to the base metal without any under coating of nickel/copper. It is found that hard chrome plating thickness of 0.3–0.38 mm is sufficient for any purpose. Normally, 100–120 μm of defect free hard chrome plating is recommended for salvaging undersized or worn out parts. A defect free, hard chrome plating will have a hardness of 1000 HV.

Hard chrome plated parts are subjected to de-embrittlement treatment by heat treating the plated components at 170–180°C for two hours to diffuse out the dissolved hydrogen gas.

Typical applications and thickness of plating of some of the commonly used platings are given in Table 4.6.

TABLE 4.6

*Application data of electroplating (commonly used)*

| Type of plating | Colour | Thickness | Reference | Applications |
|---|---|---|---|---|
| Tin plating | Mat grey | 4 μm min. | ISO:2093 | For surface protection and to improve solderability, e.g. inductors used in heat treatment |

*(Contd)*

Table 4.6  (Contd)

| | | | | |
|---|---|---|---|---|
| Chromium plating: | | | | |
| Dull/bright with under coat of copper and nickel | Dull/bright Dull : grey Bright : Blue-white | 0.3 μm min. Copper : 20 μm min. Nickel : 10 μm min. | ISO:1457 IS:1337 | For surface protection and to improved appearance, e.g. knobs, hand wheels, flexible conduits |
| Hard chromium for wear resistance | White | 50 μm min. | ISO:1457 IS:1337 | For wear resistance, e.g. piston rods |
| Hard chromium for dimensional build-up | White | 75 μm min. 600 μm max. | ISO:1457 IS:1337 | Dimensional build-up for under sized components, e.g. spindles, shafts |
| Cadmium plating | Silver-grey | 5 μm min. | ISO:2082 | For surface protection and improved electrical conductivity, e.g. electrical components such as screws, entry bush, tie bar nuts and valves. |

## 4.5  STANDARDS

### 4.5.1  Tolerances

The production of a part with the exact dimensions, repetitively is often difficult. Though this can be achieved, it requires care and skill and is not economical for regular production. Hence it is sufficient to produce parts with dimensions accurate to within two permissible limits of size. This difference in the limits of size is called the tolerance.

The enveloping surface of a cylindrical part is a hole; the enveloped surface is a shaft. For the system of limits and fits, all internal surfaces of other shapes are also referred to as holes and their mating external surfaces as shafts, e.g. a keyway and its key.

Figure 4.28 illustrates the basic size, upper deviation (*E'cart Superieur*, ES), lower deviation (*E'Cart Inferieur*, EI) tolerance, and maximum and minimum diameters of a hole (ES) and (EI) of a shaft,

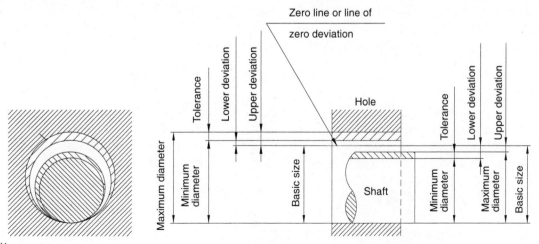

**Fig. 4.28**  Basic size, upper and lower deviation, tolerance and diameters of a hole and of a shaft

In Fig. 4.28 the two deviations of the shaft are negative, while those of the hole are positive.

The tolerance class is designated by letter(s) representing the fundamental deviation followed by the number representing the standard tolerance grade.

*Example*: H7 (holes)   h7 (shafts)

The position of the tolerance zone with respect to the zero line is a function of the basic size. It is designated by an upper case letter (in some cases two letters) for holes (A...ZC) or a lower case letter (in some cases two letters) for shaft (a...zc) as shown in Fig. 4.29.

### Toleranced Size

Toleranced size is designated by the basic size followed by the designation of the required tolerance class or the explicit deviations.

*Example*:   32H7         100g6         $100^{+0.012}_{-0.008}$

### Linear and Angular Tolerancing

Indication of linear and angular tolerancing on technical drawings is covered in IS: 11667-1991/ISO: 406-1987. The tolerances for various classes and grades are covered in IS: 919 (Part 2)-1993/ISO 286-2: 1998.

In some applications, tolerances generally achievable in a given type of industry are sufficient. Such dimensions, not specified with the tolerance are called general tolerances and are covered in IS: 2102 (Part 1)-1993/ISO: 2768-1: 1989.

### Standard Tolerance

The ISO system of tolerance provides for a total of 20 standard tolerance grades, of which international tolerance (IT) grades IT1 to IT18 are in general use. IT0 and IT01 are not commonly used.

A group of tolerances (e.g., IT7) are considered as corresponding to the same level of accuracy for all basic sizes.

IT grades for dimensions up to 500 mm are covered in IS: 919 (Part-1)-1993/ISO: 286/1-1988.

### Geometrical Tolerancing

In any engineering industry, the components manufactured should also satisfy the geometrical tolerances, in addition to the commonly used dimensional tolerances. This is on account of the fact that however accurately the two mating parts are manufactured with reference to dimensional tolerances, the two parts cannot be assembled if the geometrical tolerances are not met with. The various geometrical tolerances and their attributes for a component are listed below.

1. Form tolerances
    - straightness
    - flatness

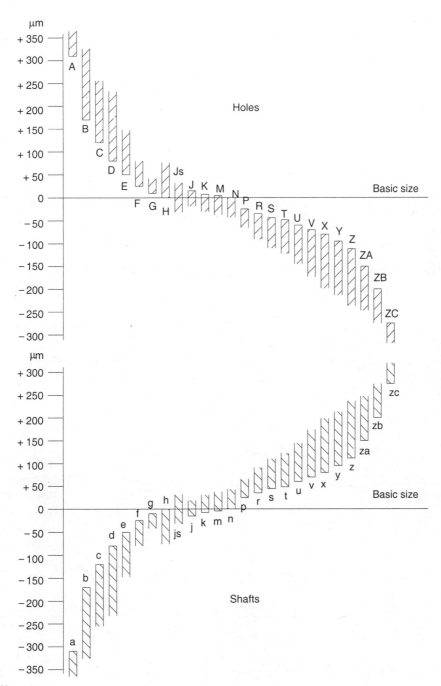

**Fig. 4.29** Letter symbols for tolerances (Examples for diameters 30-40 mm)

- circularity
- cylindricity
- profile of any line
- profile of any surface

2. Orientation tolerances
   - parallelism
   - perpendicularity
   - angularity

3. Location tolerances
   - position
   - concentricity
   - co-axiality
   - symmetry (2 × Incentre tolerance)

4. Run out tolerances
   - circular run out: radial axial total

Symbols of toleranced characteristics and examples of their usage in technical drawings are given in Table 4.7.

**TABLE 4.7**

| Toleranced characteristics | Symbol | Examples of indication and interpretation of the symbols | |
| --- | --- | --- | --- |
| | | Indication on the drawing | Interpretation |
| **SINGLE FEATURES** | | | |
| *Form Tolerances:* Straightness | — | φ 0.08 | The axis of the cylinder to which the tolerance frame is connected should be contained in a cylindrical zone of diameter ($\phi$) 0.08 mm. |
| Flatness | ▱ | 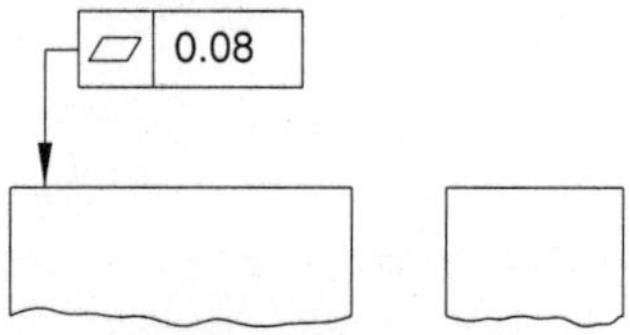 0.08 | The surface should be contained between two parallel planes 0.08 mm apart. |
| Circularity | ○ | 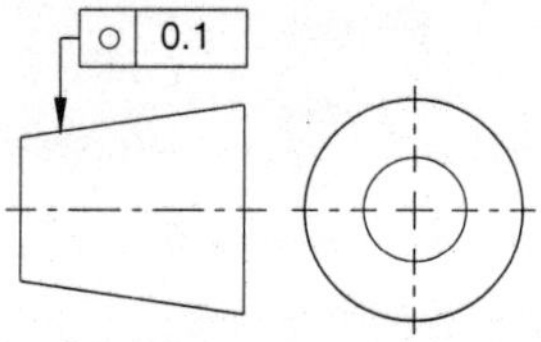 0.1 | The circumference of each cross-section should be contained between two coplanar concentric circles 0.1 mm apart |

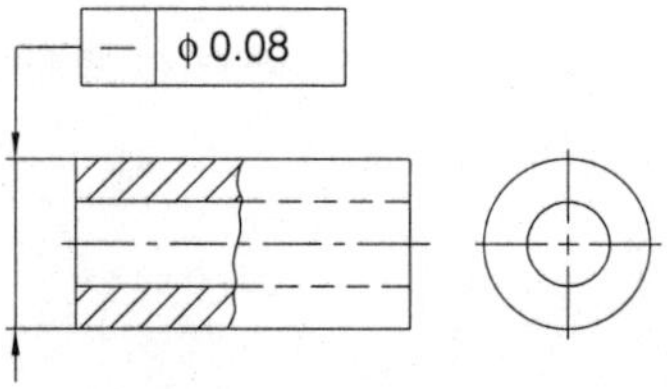

*(Contd)*

Table 4.7 (Contd)

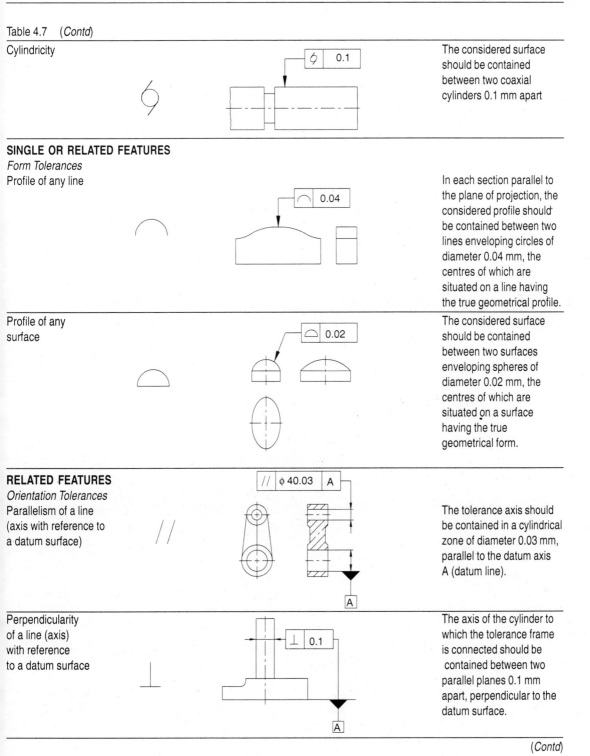

| Feature | | Description |
|---|---|---|
| Cylindricity | | The considered surface should be contained between two coaxial cylinders 0.1 mm apart |

**SINGLE OR RELATED FEATURES**
*Form Tolerances*

| Profile of any line | | In each section parallel to the plane of projection, the considered profile should be contained between two lines enveloping circles of diameter 0.04 mm, the centres of which are situated on a line having the true geometrical profile. |
|---|---|---|
| Profile of any surface | | The considered surface should be contained between two surfaces enveloping spheres of diameter 0.02 mm, the centres of which are situated on a surface having the true geometrical form. |

**RELATED FEATURES**
*Orientation Tolerances*

| Parallelism of a line (axis with reference to a datum surface) | | The tolerance axis should be contained in a cylindrical zone of diameter 0.03 mm, parallel to the datum axis A (datum line). |
|---|---|---|
| Perpendicularity of a line (axis) with reference to a datum surface | | The axis of the cylinder to which the tolerance frame is connected should be contained between two parallel planes 0.1 mm apart, perpendicular to the datum surface. |

(Contd)

Table 4.7  (Contd)

| | | |
|---|---|---|
| Angularity of a line (axis) with reference to a datum surface | 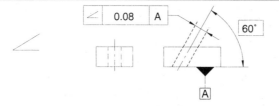 | The axis of the hole should be contained between two parallel planes 0.08 mm apart, which are inclined at 60° to the surface A (datum surface) |
| *Location Tolerances* Position of a line | 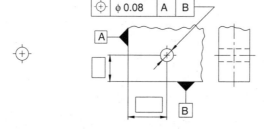 | The axis of the hole should be contained within a cylindrical zone of diameter 0.08 mm, the axis of which is in the theoretically exact position of the considered line with reference to the surfaces A and B (datum planes) |
| Coaxiality of an axis | 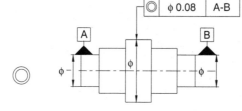 | The axis of the cylinder to which the tolerance frame is connected should be contained in a cylindrical zone of diameter 0.08 mm coaxial with the datum axis A-B. |
| Symmetry of a median plane | 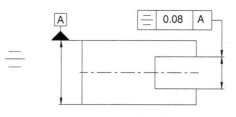 | The median plane of the slot should be contained between two parallel planes which are 0.08 mm apart and symmetrically disposed about the median plane with respect to the datum feature A. |
| *Run out Tolerances* Circular run out radial | 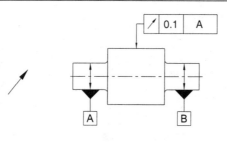 | The radial run out should not be greater than 0.1 mm in any plane of measurement during one revolution about the datum axis A-B. |

(Contd)

| Table 4.7 | (Contd) | |
|---|---|---|
| Total radial run out | 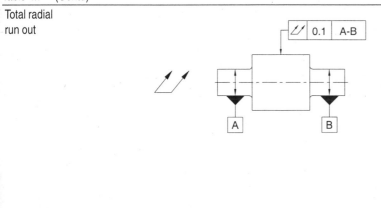 | The total radial run out should not be greater than 0.1 mm at any point on the specified surface during several revolutions about the datum axis A-B and during relative axial movement between a part and measuring instrument. The movement should be guided along a line having a theoretically perfect form of the contour and being in correct position to the datum axis. |

The definitions, symbols, methods of indication on drawings and interpretation of various geometrical tolerancing features are covered in IS: 8000 (Part 1)-1985/ISO: 1101-1983.

Geometrical tolerances should be specified only where they are essential, i.e. keeping in view functional requirements, interchangeability and probable manufacturing circumstances.

General tolerances—geometrical tolerances for features without individual tolerance indications, under three tolerance classes, viz. H, K and L are covered in IS: 2102 (Part-2)-1993/ISO: 2768-2: 1989.

### 4.5.2 Fits

The relationship resulting from the difference between the sizes of the two features (the hole and the shaft) is called a fit. The two mating parts of a fit have a common basic size.

### *Classification of Fits*

Fits are broadly classified as clearance fit, transition fit and interference fit.

***Clearance fit***  A clearance fit is one that always provides a clearance between the hole and the shaft when assembled, i.e. the minimum size of the hole is either greater than or, in the extreme case, equal to the maximum size of the shaft (Figs 4.30 and 4.31). Shafts *a* to *h* produce a clearance fit with the basic hole *H*.

***Transition fit***  A transition fit is one which may provide either a clearance or an interference between the hole and the shaft when assembled, depending on the actual sizes of the hole and the shaft, i.e. the tolerance zones of the hole and the shaft overlap completely or in part (Figs 4.32 and 4.33). Shafts *js* to *n* produce transition fits with the basic hole *H*.

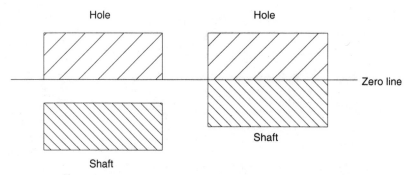

**Fig. 4.30** Schematic representation of clearance fits

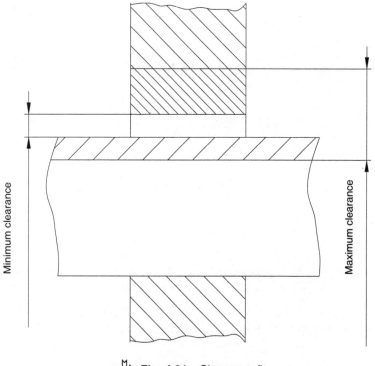

**Fig. 4.31** Clearance fit

***Interference fit*** An interference fit is one which provides an interference all along between the hole and the shaft when assembled, i.e. the maximum size of the hole is either smaller than or, in the extreme case, equal to the minimum size of the shaft (Figs 4.34 and 4.35). Shafts $p$ to $u$ produce an interference fit with the basic hole $H$.

### System of Fits

The hole basis system is a system of fits in which the design size of the hole is the basic size and the size of the shaft is allowed to vary according to the type of fit (Fig. 4.36). In

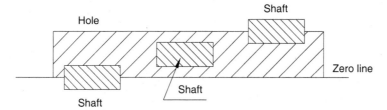

**Fig. 4.32** Schematic representation of transition fits

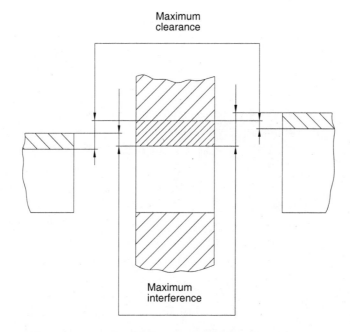

**Fig. 4.33** Transition fit

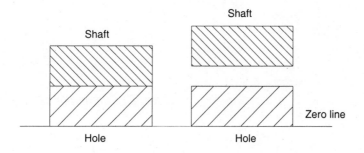

**Fig. 4.34** Schematic representation of interference fits

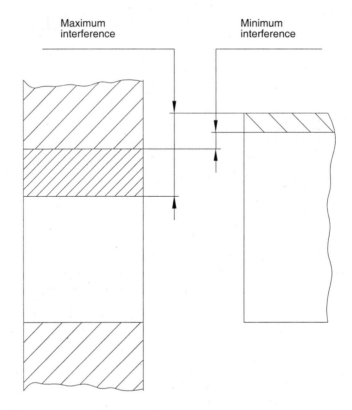

**Fig. 4.35** Interference fit

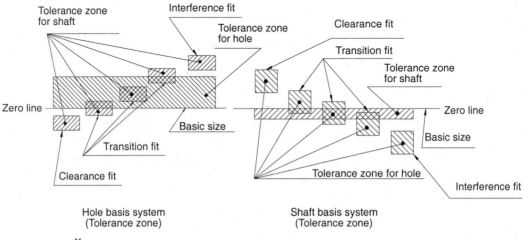

**Fig. 4.36** Example illustrating hole basis and shaft basis system

the shaft basis system, the design size of the shaft is the basic size and the size of hole is allowed to vary according to the type of fit.

The hole basis system is generally used in all the industries. This is because a shaft is easy to manufacture with a variety of tolerances to a given tolerance of a hole. The hole with a given tolerance can be produced by fixed diameter tools—twist drill, reamers. The shaft basis system should only be used where unquestionable economic advantages will accrue (for example, where it is necessary to be able to mount several parts with holes having different deviations on a single shaft of drawn steel bar without machining the latter.

*Example* Piston pin having different fits with the piston and the connecting rod. A guide for selection of fits is covered in IS: 2708-1982.

### 4.5.3 Surface Parameters

*Surface Texture*

Any surface, cylindrical, or flat has to be defined for the requirement of surface roughness (more precisely surface texture as a surface is fully defined not only with surface roughness value, but also, direction of lay, production method, etc.) proceeding from its functional application to guarantee the given quality of products.

The surface roughness should not be specified if it is not required. Whenever an ordinary manufacturing process ensures an acceptable surface roughness, the roughness of such surfaces should not be inspected.

Although a surface may appear smooth, any manufactured surface always departs to some extent from absolute perfection. The imperfections are characteristic of the machining process and take the form of a succession of hills and valleys, which may vary both in height and in spacing.

Surface roughness is specified in absolute values, though earlier machining symbols (roughness symbols) are used on a drawing to specify the type of finish required on a component. These are:

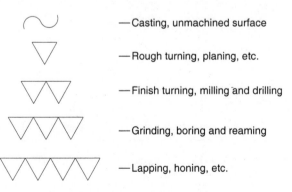

— Casting, unmachined surface
— Rough turning, planing, etc.
— Finish turning, milling and drilling
— Grinding, boring and reaming
— Lapping, honing, etc.

*Surface Roughness Parameters*

Surface roughness can be measured and specified with different parameters. However, $R_a$, the arithmetic mean deviation of a profile is the universally recognised, and most used international parameter of roughness. It is defined as the arithmetic mean of the absolute values of the profile departures $y$ within the sampling length $l$. This is also called the centre line average (CLA) value (Fig. 4.37).

$$R_a = \frac{1}{l}\int_0^l |y(x)|\,dx \quad \text{or approximately}$$

$$R_a = \frac{1}{n}\sum_{i=1}^n |y_i|$$

where $n$ is the number of discrete deviations $y$ of the profile.

The roughness symbol (machining symbols) and their equivalent roughness values $R_a$ (μm) and roughness grade numbers are given in Table 4.8.

TABLE 4.8

Equivalent roughness grades (based on IS: 3073-1967)

| Roughness symbol | Roughness value $R_a$ (μm) (max) | Roughness grade numbers |
|---|---|---|
| ∿ | 50 | N 12 |
| ▽ | 25 | N 11 |
|  | 12.5 | N 10 |
| ▽▽ | 6.3 | N 9 |
|  | 3.2 | N 8 |
|  | 1.6 | N 7 |
| ▽▽▽ | 0.8 | N 6 |
|  | 0.4 | N 5 |
| ▽▽▽▽ | 0.2 | N 4 |
|  | 0.1 | N 3 |
|  | 0.05 | N 2 |
|  | 0.025 | N 1 |

The method of indication of surface texture in technical drawings as per ISO standards is given below.

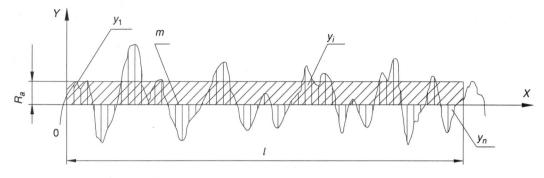

**Fig. 4.37** Surface roughness parameter

**Symbols with no inscription**

| Symbol | Meaning |
|---|---|
| ∨ | Basic symbol. It may be used alone only when its meaning is explained by a note or when its meaning is the surface under consideration. |
| ▽ | A machined surface with no indication of any other detail or when its meaning is a surface to be machined. |
| ⩗ | A surface from which the removal of material is prohibited. This symbol may also be used in drawings relating to a production process to indicate that a surface is to be left in the state resulting from a preceding manufacturing process, whether this state was achieved by removal of material or otherwise. |

**Symbols with indication of the principal criterion of roughness, $R_a$**

| Symbol — Removal of material by machining is | | | Meaning |
|---|---|---|---|
| Optional | Obligatory | Prohibited | |
| $R_a$ 3.2 ∨ | $R_a$ 3.2 ▽ | $R_a$ 3.2 ⩗ | A surface with a maximum surface roughness value $R_a$ of 3.2 μm. |
| $R_a$ 6.3 $R_a$ 1.6 ∨ | $R_a$ 6.3 $R_a$ 1.6 ▽ | $R_a$ 6.3 $R_a$ 1.6 ⩗ | A surface with a maximum surface roughness value $R_a$ of 6.3 μm and a minimum of 1.6 μm. |

## Additional symbols

| Symbol | Meaning |
|---|---|
| ∀ Milled | Production method: milling |
| ∀ 2.5 | Sampling length: 2.5 mm |
| ∀ ⊥ | Direction of lay: perpendicular to the plane of projection of the view |
| ∀ 2 | Machining allowance: 2 mm |
| ∀ ($R_t$ = 0.4) | Indication (in parantheses) of a criterion of roughness other than that used for $R_a$, e.g., $R_t$ = 0.4 μm. |

These may be used singly, in combination, or combined with an appropriate symbol used for indicating principle roughness. For more details refer to ISO: 1302-1992: Method of indicating surface texture on technical drawings.

Surface roughness achievable from different manufacturing processes is given in Table 4.9.

### Quality of Scraped Surface

The classification and application of scraped surface is given in Fig. 4.38.

### Method of Establishing the Quality of Scraped Surface

Each quality class is established by the number of contact points between the scraped surface and the inspection surface plate. The contact points may be almost uniformly distributed in an area of 25 mm × 25 mm and also over the entire area of the scraped surface and checked by applying blue paste. In cases of small normal surfaces, interrupted surfaces, etc., where determination of the number of contact points on an area of 25 mm × 25 mm is not reliable, they should be determined on a surface of different size and then recalculated to a surface of 25 mm$^2$ area.

### Method of Checking the Bearing Area

For normal applications, the method adopted is that blue is applied to the master and the workpiece is checked against it. For finer scraping, it is recommended to use blue for the master and a thin layer of red oxide mixed with linseed oil for the workpiece. The red oxide gives a contrasting background and also prevents the blue from spreading

## TABLE 4.9

### Surface roughness achievable from different manufacturing processes

| Manufacturing process | N1 (0.012) | N2 (0.025) | N3 (0.05) | N4 (0.1) | N5 (0.2) | N6 (0.4) | N7 (0.8) | N8 (1.6) | N9 (3.2) | N10 (6.3) | N11 (12.5) | N12 (25) | (50) | (100) | (200) |
|---|---|---|---|---|---|---|---|---|---|---|---|---|---|---|---|
| Sand casting | | | | | | | | | 5 ▨▨▨▨ | ▨▨▨ | ▨▨ | ▨ | 50 | | |
| Permanent mould casting | | | | | | | 0.8 ▨▨▨ | ▨▨▨ | ▨▨▨ | 6.3 | | | | | |
| Die casting | | | | | | | 0.8 ▨▨▨ | ▨▨▨ | 3.2 | | | | | | |
| High pressure casting | | | | | | 0.32 ▨▨▨ | ▨▨▨ | 2 | | | | | | | |
| Hot rolling | | | | | | | | | 2.5 ▨▨ | ▨▨▨ | ▨▨▨ | ▨▨ | 50 | | |
| Forging | | | | | | | | 1.6 ▨▨ | ▨▨▨ | ▨▨▨ | ▨ | 25 | | | |
| Extrusion | | | | | | 0.16 ▨▨▨ | ▨▨▨ | ▨▨▨ | 5 | | | | | | |
| Flame cutting, sawing and chipping | | | | | | | | | | 6.3 ▨▨ | ▨▨▨ | ▨▨▨ | ▨▨ | 100 | |
| Radial cut-off sawing | | | | | | | | 1 ▨▨ | ▨▨ | 6.3 | | | | | |
| Hand grinding | | | | | | | | | | 6.3 ▨ | ▨ | 25 | | | |
| Disc grinding | | | | | | | | 1.6 ▨▨ | ▨▨▨ | ▨▨ | ▨ | 25 | | | |
| Filing | | | | | | 0.25 ▨▨▨ | ▨▨▨ | ▨▨▨ | ▨▨ | ▨ | ▨ | 25 | | | |
| Planing | | | | | | | | 1.6 ▨▨ | ▨▨▨ | ▨▨▨ | ▨▨ | ▨ | 50 | | |
| Shaping | | | | | | | | 1.6 ▨▨ | ▨▨▨ | ▨▨ | ▨ | 25 | | | |
| Drilling | | | | | | | | 1.6 ▨▨ | ▨▨▨ | ▨▨ | 20 | | | | |
| Turning and milling | | | | | | 0.32 ▨▨▨ | ▨▨▨ | ▨▨▨ | ▨▨ | ▨ | ▨ | 25 | | | |
| Boring | | | | | | 0.4 ▨▨ | ▨▨▨ | ▨▨▨ | ▨▨ | 6.3 | | | | | |
| Reaming | | | | | | 0.4 ▨▨ | ▨▨▨ | ▨▨ | 3.2 | | | | | | |
| Broaching | | | | | | 0.4 ▨▨ | ▨▨▨ | ▨▨ | 3.2 | | | | | | |
| Hobbing | | | | | | 0.4 ▨▨ | ▨▨▨ | ▨▨ | 3.2 | | | | | | |
| Surface grinding | | | | 0.063 ▨▨ | ▨▨▨ | ▨▨▨ | ▨▨▨ | ▨▨ | 5 | | | | | | |
| Cylindrical grinding | | | | 0.063 ▨▨ | ▨▨▨ | ▨▨▨ | ▨▨▨ | ▨▨ | 5 | | | | | | |
| Honing | | 0.025 ▨▨ | ▨▨ | ▨▨ | ▨ | 0.4 | | | | | | | | | |
| Lapping | 0.012 ▨ | ▨▨ | ▨▨ | 0.16 | | | | | | | | | | | |
| Polishing | | | 0.04 ▨ | ▨ 0.16 | | | | | | | | | | | |
| Burnishing | | | 0.04 ▨ | ▨▨ | ▨▨ | ▨ | 0.8 | | | | | | | | |
| Superfinishing | 0.016 ▨ | ▨▨ | ▨▨ | ▨ 0.32 | | | | | | | | | | | |

when the master is rubbed against the workpiece. For checking and achieving still finer and larger number of bearing points, red oxide is applied on the workpiece and a clean master without any colour on it is rubbed against the workpiece. This will clearly show the shining points on the workpiece.

| Scraped quality grade | Approximate distribution of contact points | Number of contact points | Surface roughness $R_a$ (μm) | Application |
|---|---|---|---|---|
| 1 | 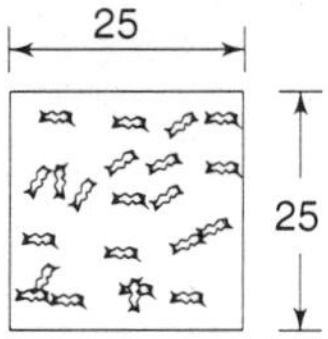 25 × 25 | 22–32 | 0.1 | Precision measuring machines, instruments, surface of oil tight joints |
| 2 | 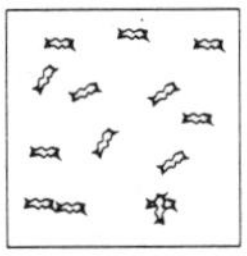 | 15–21 | 0.2 | Workshop inspection instrument and tools (surface plate, straight edge, etc.). Guideways for precision machines (machine tool slides of hydraulic equipment), bearing bushes |
| 3 |  | 10–14 | 0.4 | Workshop jigs and fixtures and guideways of machine tools |
| 4 |  | 6–9 | 0.8 | Surface of scraped tables, surface for clamping |
| 5 | 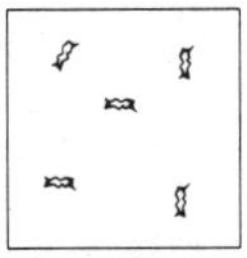 | 4–5 | 1.6 | Contact surfaces of rotary tables, covers of gearboxes, etc. Surface for clamping |

**Note:**
(i) The size of inspection surface used is recommended to be larger than the size of scraped surface.
(ii) It is necessary to have uniform distribution of contact point, as shown in the figure over the entire area.
(iii) In 4th and 5th the quality of scraped surfaces, contact points which protrude half or more than half of their own area beyond the periphery of measuring square is counted as half point and less than half as whole point.
(iv) The surface roughness of machined surface before scraping shall not be coarser than $R_a = 3.2$ μm.

**Fig. 4.38** Quality of scraped surfaces

## ANNEXURE 4.1

**Steels as per DIN standards and their approximate equivalent national standards**

| DIN (Germany) | IS (India) | BS (UK) | AFNOR (France) | UNI (Italy) | JIS (Japan) | GOST (Russia) | AISI (USA) |
|---|---|---|---|---|---|---|---|
| C45 | 45C8 | 080M46 | AF65C45 | C45 | S45C | 45 | 1045 |
| 45S20 | 40C10S18 | 212A42 | 45MF4 | — | — | — | 1146 |
| C60 | 60C4 | — | AF70C55 | C60 | — | 60 | 1060 |
| 15CrNi6 | 15Ni7CrMo2 | — | — | — | — | — | — |
| 36CrNiMo4 | 40Ni6Cr4Mo3 | 817M40 | 40NCD3 | 38NiCrMo4(KB) | — | 40ChN2MA | 9840 |
| 50CrV4 | 50Cr4V2 | 735A50 | 50CV4 | 50CrV4 | SUP10 | 50ChGFA | 6150 |
| 34CrAlMo5 | 40Cr7A110Mo2 | 905M39 | 30CAD6.12 | 34CrAlMo7 | — | — | A355 Cl.D |
| 105WCr6 | T110W6Cr4 | — | 105WC13 | 107WCr5KU | SKS31 | ChWG | — |
| 100Cr6 | T105Cr5Mn2 | BL3 | Y100C6 | — | SUJ2 | — | L3 |
| X210Cr12 | XT215Cr12 | BD3 | Z200C12 | X205Cr12KU | SKD1 | Ch12 | D3 |
| S6-5-2 | XT87W6Mo55Cr4V2 | BM2 | Z85WDCV06 05-04-02 | HS6-5-2 | SKH51 | R6AM5 | M2 |
| S6-5-2-5 | XT90W6CoMo5Cr4V2 | — | D Z90WKCV 06-05-05-04-02 | HS6-5-2-5 | SKH55 | R6M5K5 | — |
| S10-4-3-10 | XT125WCo10CrMo4V3 | BT42 | Z130WKCDV 10-10-04-04-03 | HS10-4-3-10 | SKH57 | — | — |

## ANNEXURE 4.2

**Cast iron as per DIN standards and their approximate equivalent national standards**

| DIN (Germany) | IS (India) | BS (UK) | AFNOR (France) | UNI (Italy) | JIS (Japan) | GOST (Russia) | AISI (USA) | MNC (Sweden) | UNE (Spain) | ISO |
|---|---|---|---|---|---|---|---|---|---|---|
| GG-15 | FG150 | Grade 150 | Ft15D | G15 | FC15 | Sc15 | 25B | 0115-00 | FG15 | Grade 150 |
| GG-20 | FG200 | Grade 220 | Ft20D | G20 | FC20 | Sc20 | 30B | 0120-00 | FG20 | Grade 200 |
| GG-25 | FG260 | Grade 260 | Ft25D | G25 | FC25 | Sc25 | 40B | 0125-00 | FG25 | Grade 250 |
| GG-30 | FG300 | Grade 300 | Ft30D | G30 | FC30 | Sc30 | 45B | 0130-00 | FG30 | Grade 300 |
| GGG-40 | SG400/12 | 420/12 | FGS400-12 | GS400-12 | FCD40 | VC42-12 | 60-4-18 | 0717-02 | — | 400-12 |
| GGG-50 | SG500/7 | 500/7 | FGS500-7 | GS500-7 | FCD50 | VC50-2 | 65-45-12 | 0727-02 | — | 500-7 |
| GGG-60 | SG600/3 | 600/3 | FGS600-3 | GS600-3 | FCD60 | VC60-2 | 80-55-06 | 0732-03 | — | 600-3 |

CHAPTER 5

# Design of Modern CNC Machines and Mechatronic Elements

## 5.1 INTRODUCTION

The design and construction of computer numerically controlled (CNC) machines differs greatly from that of conventional machine tools. This difference arises from the requirement of higher performance levels. The CNC machines often employ the various mechatronic elements that have been developed over the years. However, the quality and reliability of these machines depends on the various machine elements and subsystems of the machine. The following are some of the important constituent parts, and aspects of CNC machines to be considered in their designing.

(a) Machine structure
(b) Guideways
(c) Feed drives
(d) Spindle and spindle bearings
(e) Measuring systems
(f) Controls, software and operator interface
(g) Gauging
(h) Tool monitoring

## 5.2 MACHINE STRUCTURE

The machine structure is the load carrying and supporting member of the machine tool. All the motors, drive mechanisms and other functional assemblies of machine tools are aligned to each other and rigidly fixed to the machine structure. The machine structure is subjected to static and dynamic forces and it is, therefore, essential that the structure does not deform or vibrate beyond the permissible limits under the action of these forces. All components of the machine must remain in correct relative positions to maintain the geometric accuracy, regardless of the magnitude and direction of these forces. The machine structure configuration is also influenced by the considerations of manufacture, assembly and operation.

The basic design factors involved in the design of a machine structure are discussed below.

### 5.2.1 Static Load

The static load of a machine tool results from the weights of slides and the workpiece, and the forces due to cutting. To keep the deformation of the structure due to static loading within permissible limits, the structure should have adequate stiffness and a proper structural configuration. Generally there are two basic configurations of machine tools as depicted in Fig. 5.1.

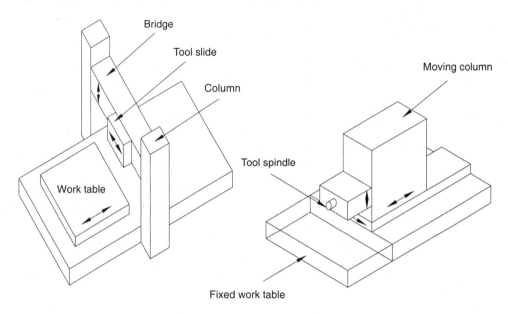

**Fig. 5.1** Commonly used configurations of machine tools

### 5.2.2 Dynamic Load

Dynamic load is a term used for the constantly changing forces acting on the structure while movement is taking place. These forces cause the whole machine system to vibrate. The origin of such vibrations is:

(a) Unbalanced rotating parts
(b) Improper meshing of gears
(c) Bearing irregularities
(d) Interrupted cuts while machining (like in milling)

The effect of these vibrations on the machine performance is reduced by:

(a) Reducing the mass of the structure
(b) Increasing the stiffness of the structure
(c) Improving the damping properties

## 5.2.3 Thermal Load

In a machine tool there are a number of local heat sources which set up thermal gradients within the machine. Some of these sources are:

(a) Electric motor
(b) Friction in mechanical drives and gear boxes
(c) Friction in bearings and guideways
(d) Machining process
(e) Temperature of surrounding objects

These heat sources cause localised deformation, resulting in considerable inaccuracies in machine performance. The following steps are generally followed to reduce thermal deformation.

(a) External mounting of drives, i.e. motors and gear boxes
(b) Removing frictional heat from bearings and guideways by a proper lubrication system
(c) Efficient coolant and swarf removal system for the dissipation of heat generated from the machining process
(d) Thermo-symmetric designing of the structure

## 5.3 GUIDEWAYS

Guideways are used in machine tools to:

(a) Control the direction or line of action of the carriage or the table on which a tool or a workpiece is held
(b) To absorb all the static and dynamic forces

The shape and size of the work produced depends on the accuracy of the movement and on the geometric and kinematic accuracy of the guideway. The geometric relationship of the slide (the moving part) and the guideway (stationary part) to the machine base determines the geometric accuracy of the machine. Kinematic accuracy depends on the straightness, flatness and parallelism errors in the guideway. These errors further result in a variety of tracking errors like pitch, yaw and roll that are difficult to measure and correct. Further, over a period of usage, any kind of wear in the guideway reduces the accuracy of the guide motion, resulting in errors in movement and positioning.

When machining is taking place, the rate of translational movement (feed rate) can be as low as 20 mm/min. During non-machining operations such as positioning, the feed rate can be as high as 50 m/min. For fine machined surfaces and for accurate positioning during machining, the movement must be smooth and continuous, and free from any jerky movements.

The following points must be considered while designing guideways.

- Rigidity
- Damping capability

- Geometric and kinematic accuracy
- Velocity of slide
- Friction characteristics
- Wear resistance
- Provision for adjustment of play
- Position in relation to work area
- Protection against swarf and damage

These criteria vary in importance depending upon a particular application and hence the selection of guideways and their geometry can be quite critical in some cases. The position of the drive mechanism relative to the guiding faces of the slideway is very important. Ideally, the drive mechanism should be placed in such a manner that the reaction and hence the frictional forces are uniform in the guiding system. This will ensure uniform wear on guideways.

Guideways are primarily of two types.

(a) Friction guideways
(b) Antifriction linear motion (LM) guideways

### 5.3.1 Friction Guideways

Friction guideways are most widely applied in conventional machine tools due to their low manufacturing cost and good damping properties. These guideways operate under conditions of sliding friction and do not have a constant coefficient of friction. The coefficient of friction varies with the sliding velocity as shown in Fig. 5.2.

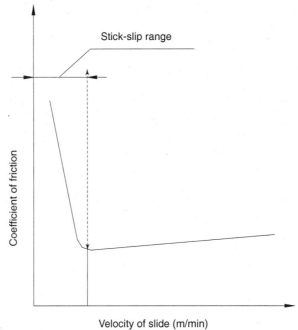

**Fig. 5.2** Relation of coefficient of friction and slide velocity for friction guideways

The coefficient of friction is very high when the movement commences and as the speed of the slide increases, it rapidly falls and beyond a certain critical velocity it remains almost constant. Therefore to start the movement, the force to overcome friction has to be correspondingly high. This force results in the drive mechanism, such as a screw, being elastically deformed. The energy thus stored in the screw, together with the applied force, causes the carriage to slip and move at a faster rate. As the speed increases the friction decreases and a greater amount of movement than that intended for the slide takes place. There is a possibility of this cycle of events repeating itself and resulting in errors in positioning and consequently in a jerky motion. This phenomenon is known as the *stick-slip* phenomenon.

To reduce the possibility of stick-slip, there should be a minimum but constant friction between the surfaces in contact. This is achieved in friction guideways by using strips of material such as poly tetra fluoro ethylene (PTFE) or turcite lining at the guideway interface. Turcite is a special type of plastic with particles of graphite embedded on its surface. These materials have a low and constant coefficient of friction (of the order of 0.1).

In addition to this, when the strips wear to such an extent that the alignment error goes beyond the permissible limits, then they can be replaced and the accuracy restored. The guideways commonly used on machine tools have a number of different forms, such as cylindrical, vee, flat and dovetail. Figure 5.3 shows cross-sections of coated guideways. Coatings can be carried out on vee, flat as well as on dovetail guideways.

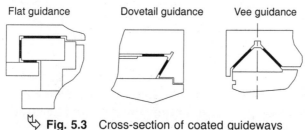

**Fig. 5.3** Cross-section of coated guideways

## 5.3.2 Vee Guideways

The vee or inverted vee is widely used on machine tools, especially on lathe beds. One of the advantages of the vee or inverted vee is that the parallel alignment of the guideway with the spindle axis is not affected by wear. There is a closing action as the upper member settles on the lower member, and this automatically maintains the alignment. Jibs are, therefore, not required with the vee guideway to take up the clearance caused by wear. On some machines, the angles of the vee are different so as to reduce the possibility of uneven wearing of the vee sides. The majority of lathes have a combination of vee and flat guideways to prevent the twisting of the slide, as shown in Fig. 5.4. Provision has also to be made to prevent the carriage from lifting off the guideway.

### 5.3.3 Flat and Dovetail Guideways

Although the vee type has certain advantages, it is the flat or dovetail forms (shown in Figs 5.5 and 5.6) which are used on CNC machine tools. The flat guideways have better load-bearing capabilities than the other guideways.

After a period of use, wearing may occur owing to the sliding of the surfaces over each other. Jibs are used to ensure accurate fitting of the slide to both the flat and dovetail guideways. The jibs are tapered and can be adjusted to reduce excessive clearance caused by wear.

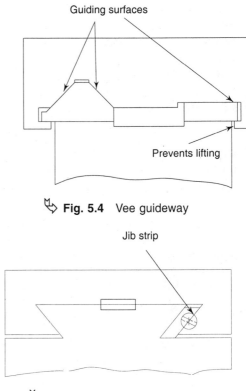

**Fig. 5.4** Vee guideway

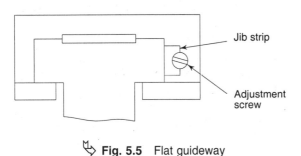

**Fig. 5.5** Flat guideway

**Fig. 5.6** Dovetail guideway

The metal-to-metal contact on the vee, flat and dovetail types of guideway is normally cast iron to cast iron. The cast iron may be heat treated (flame hardened) to increase its hardness, and the surfaces ground to obtain the required accuracy.

In the past, after machining the guideways on the castings of a machine bed or column, it was a common practice to hand scrape them to ensure efficient bearing of the slide on the guideways. But now, though a limited amount of hand scraping is still carried out, machining the guideways on precision grinding machines has reduced the amount of scraping required.

When the guideways are an integral part of the castings and get worn out after a period of time, it is necessary to dismantle the machine to remachine the guideways so that their accuracy is restored. To overcome this difficulty, pre-machined hardened steel guideways are fastened to the main castings which can be replaced if they are worn out or damaged.

### 5.3.4 Cylindrical Guideways

In cylindrical guideways, the bore in the carriage housing provides support all around the guideway as shown in Fig. 5.7. For relatively short traverses and light loads, cylindrical guideways are very efficient. A limitation on the use of these guideways for long traverses is that if the guide bar is supported only at each end, it may sag or bend in the centre of the span under a load.

### 5.3.5 Antifriction Linear Motion (LM) Guideways

Antifriction linear motion guideways are used on CNC machine tools to:

(a) Reduce the amount of wear
(b) Improve the smoothness of the movement
(c) Reduce friction
(d) Reduce heat generation

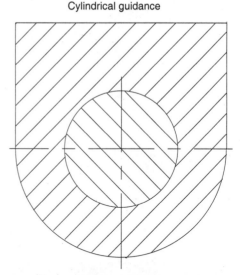

Fig. 5.7 Cylindrical guideway

Antifriction guideways are used to overcome the relatively high coefficient of friction in metal-to-metal contacts and the resulting limitations addressed in the above list.

They use rolling elements in between the moving and the stationary elements of the machine. They provide the following advantages when compared with friction guides.

(a) Low frictional resistance
(b) No stick-slip
(c) Ease of assembly
(d) Commercially available in ready-to-fit condition
(e) High load carrying capacity
(f) Heavier preloading possibility
(g) High traverse speeds

The main disadvantage of these guideways as compared to friction guideways is their lower damping capacity.

The manufacturers of machine tools use several options for antifriction linear motion guideways like recirculating ball bushings, linear bearings with balls and rollers such as recirculating LM guides, recirculating roller bearings and cross roller bearings. Although the rolling element bearings have less damping characteristics than friction guideways, linear motion guideways have become very common in machine tools on account of their higher rapid traverse rates.

### *Recirculating Ball Bushings*

Figure 5.8 illustrates the inner construction of a recirculating ball bush. Various kinds of ball bushes are available like closed type (Fig. 5.9), open type (Fig. 5.10), etc. They are available with or without seals. Different methods of mounting the shafts and ball bushes are explained in Figs 5.11–5.15. The shafts can be ground to a tolerance as fine as 0.005 mm depending on the application.

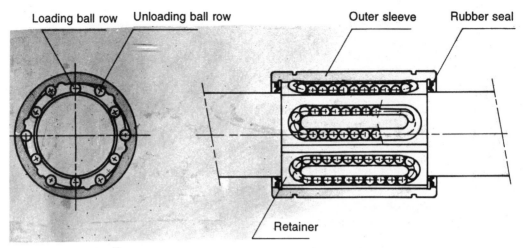

**Fig. 5.8** Details of inner construction of ball bushing

**Fig. 5.9** Closed type ball bushing

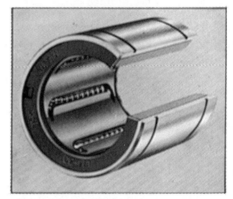

**Fig. 5.10** Open type ball bushing

### Linear Bearings with Balls and Rollers

A number of machines use rollers to provide for a rolling motion rather than a sliding motion. The rollers are in contact with the guideway machined on a casting of the machine. These are very effective in providing smooth and easy movement, but still require an accurate form to be machined on castings. The surfaces in contact with the rollers have to be hardened and should have a smooth texture.

To reduce the problem of machining, an accurate form on the bed of a machine, hardened steel rails with special guide forms (Fig. 5.16) may be fastened to the castings of the machine. Special blocks, a pair along each guide rail, with recirculating balls can move along the rails. The balls provide the rolling motion and since the contact form on the rail is a mating form of the balls, there is a line contact between the balls and the rails, a pair along each guide rail. The coefficient of friction is reduced and there is no stick-slip in this arrangement. These guideway sets are precision elements. The tolerance on overall height $H$ (Fig. 5.16 ) is as fine as $\pm$ 10 µm and the height difference for

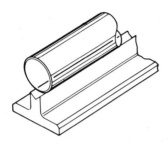
Fig. 5.11  Shaft mounting using rail

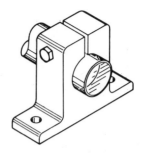
Fig. 5.12  Shaft mounting using block

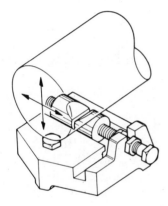
Fig. 5.13  Shaft mounting on adjustable block

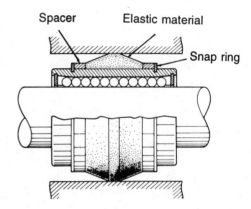
Fig. 5.14  Ball bushing mounting on elastic body

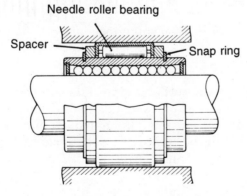
Fig. 5.15  Bush mounting for linear as well as rotary movements

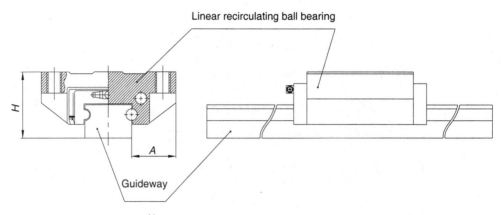

**Fig. 5.16** Linear bearing with balls

a given set is within 5 μm (maximum). The tolerance on dimension $A$ (Fig. 5.16) is as fine as ± 15 μm with a variation of 10 μm (maximum) for a given set of rails and bearing blocks.

Various forms of linear guides are illustrated in Fig. 5.17. Their applications and mounting methods are explained in Figs 5.18 and 5.19.

Figure 5.20 shows another form of a guide rail with a rectangular cross-section. This is made of hardened steel which can be fastened on the bed. On the guiding surface, a rectangular block with recirculating rollers/needles runs along the length of the guide providing a rolling contact. This rectangular block is fastened to the slide. A pair of rolling blocks (a distance apart) is fixed on each guiding surface. A minimum of 12 rolling blocks are required to make one slide/guide system. These guideways are used for relatively high loads. Figures 5.21 and 5.22 illustrate some applications of these guides. A pictorial representation of these bearings is given in Fig. 5.23.

### 5.3.6 Other Guideways

Other types of guideways used in machine tools are

(a) Hydrostatic guideways
(b) Aerostatic guideways

In hydrostatic guideways, the surface of the slide is separated from the guideway by a very thin film of fluid supplied at pressures as high as 300 bar (1 bar = 0.1 MPa). Frictional wear and stick slip are entirely eliminated. A high degree of dynamic stiffness and damping is obtained with these guideways, both characteristics contributing to good machining capabilities. Their application is limited due to high cost and difficulty in assembly.

In aerostatic guideways, the slide is raised on a cushion of compressed air which entirely separates the slide and the guideway surfaces. The major limitation of these type of guideways is their low stiffness, which limits their use to positioning applications only, e.g. a Coordinate Measuring Machine (CMM).

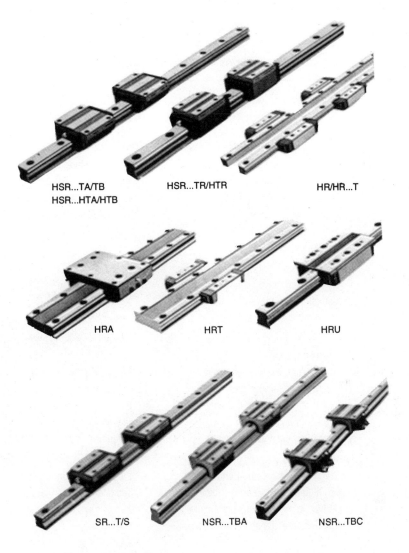

**Fig. 5.17** Types of linear bearings with balls

The selection of guideways for a particular application basically depends upon the requirements of the load carrying capacity, damping and the traverse speed. In order to get the maximum benefit, most of the machine tool manufacturers make use of a combination of antifriction and friction guideways with turcite/PTFE lining. Such a combination improves the load carrying capacity through the use of antifriction guideways and the damping property through the use of friction guideways.

The features and main applications of linear bearings illustrated in Fig. 5.17 are given in Table 5.1.

## TABLE 5.1

**Features and main applications of linear bearings illustrated in Fig. 5.17**

| Type | Features | Main applications |
|---|---|---|
| HSR... TA<br>HSR... TB<br>HSR...TR<br><br>(self adjusting type<br>—uniform loading capacity in four directions | Heavy load,. Super rigid type with high strength design.<br>Long life and widely increased rated load because of large ball diameter and increased number of effective balls.<br>Wide application range because of uniform loading capacity in four directions. Sufficient strength in reverse radial direction. | Machining centres<br>CNC lathes; X, Y, Z axes of machine tools for heavy cutting<br>Grinding wheel infeed axis for grinding machines<br>Places where high accuracy is required with large moments. |
| HR<br>(self adjusting type—separable) | Extremely thin type with high rigidity. Most suitable for places with restrictions on space availability. Preload adjustment is possible. | X, Y, Z axes of electric discharge machines<br>Precision tables<br>X and Z axes of CNC lathes<br>Assembly robots; Transport devices. |
| HRA<br>HR7<br>(self adjusting type—separable)<br>HRU<br>(self adjusting type—separable) | Wide enough to form a guideway with a single rail.<br>Preload adjustment is possible.<br>Automatic assembly machines;<br>Extremely heavy load capacity.<br>High mounting rigidity.<br>Preload adjustment is possible. | Assembly devices for printed circuit boards<br>Measuring instruments<br>Tool changers Automatic assembly machines<br>Grinding wheel feed axis for grinding machines<br>Machines where high accuracy is required with heavy moments |
| SR...T<br>(self adjusting-radial type) | Self adjustment, smooth movement with high accuracy can be obtained.<br>LM blocks are compact and highly rigid because of their construction. | Industrial robots<br>Table feed for grinders<br>X and Y axes of electric discharge machines<br>Transport equipment in flexible machining system<br>Precision tables<br>Drilling machines for printed circuit boards<br>X and Z axes of jig grinders<br>Punching presses |
| NSR...TBC<br>(self aligning type) | For coarse application<br>Preload adjustment is possible | Welding machines<br>Transport equipment<br>Automatic painting machines<br>Tool changers<br>Robot transfer equipments |

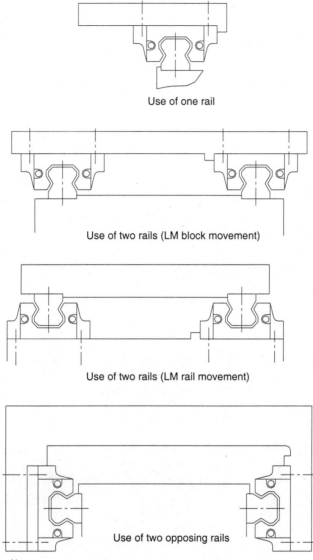

Fig. 5.18 Applications of linear bearings with balls

## 5.4 FEED DRIVES

On a CNC machine the function of feed drive is to provide motion to the slide as per the motion commands. Since the degree of accuracy requirements are high, the feed drive should have high efficiency and response. The feed drive consists of:

(a) Servomotor
(b) Mechanical transmission system

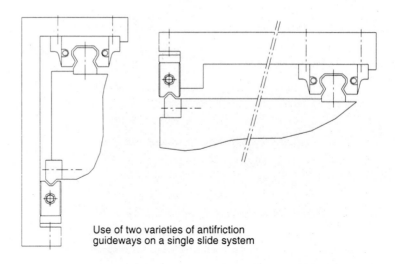

Use of two varieties of antifriction guideways on a single slide system

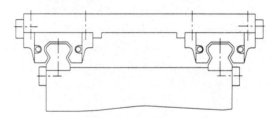

Total surface fixed installation

**Fig. 5.19** Applications of linear bearings

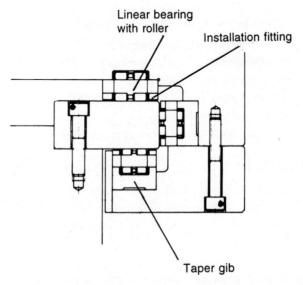

**Fig. 5.20** Linear bearing with rollers

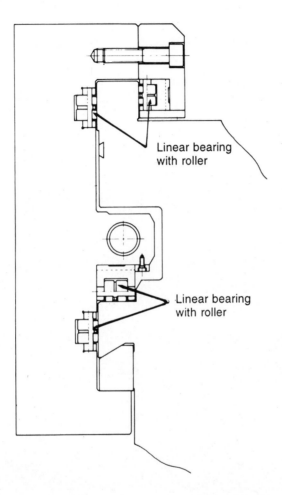

**Fig. 5.21** Applications of linear bearings with rollers

## 5.4.1 Servomotor

Commonly used feed drive motors for CNC machines are direct current (dc) servomotors and alternating current (ac) servomotors.

Initially, dc servomotors and drives were used most commonly on all CNC machines. These servomotors provide excellent speed regulation, high torque and efficiency. With the development of ac servos, at a cost comparable with dc servos, the former became more popular for machine tool applications. This is because ac servomotors provide a constant torque over their entire speed range, require less maintenance due to brushless operation, have a better response and dynamic stiffness, and a higher reliability compared to dc servomotors.

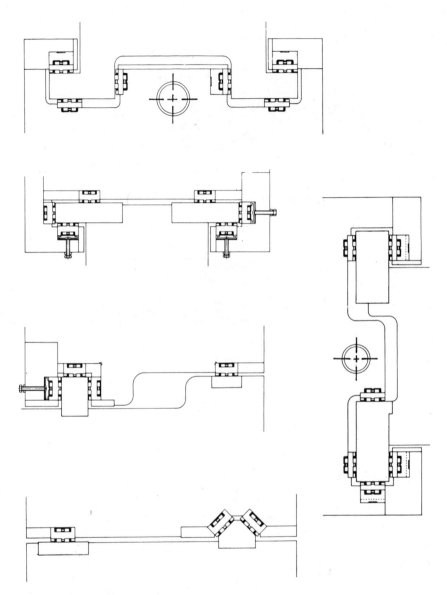

**Fig. 5.22** Applications of linear bearings with rollers

### 5.4.2 Mechanical Transmission System

The mechanical transmission system of a feed drive comprises all the components which are in the force and motion transmission paths from drive motor to the slide. They are:

(a) Elements to convert the rotary motion to a linear motion (recirculating ballscrew-nut or rack-and-pinion system)

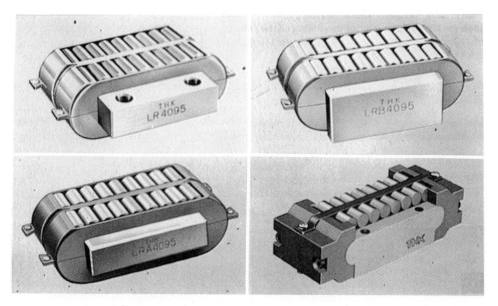

**Fig. 5.23** Types of linear bearings with rollers

(b) Torque transmitting elements (gear box or timing belt and couplings)

The main criterion to be considered in the design of a mechanical transmission system is to keep the transmission errors to a minimum. The essential requirements for this are:

(a) High natural frequency
(b) High stiffness
(c) Sufficient damping
(d) Low friction
(e) Backlash free operation

## Elements Used to Convert the Rotary Motion to a Linear Motion

Several actuating mechanisms are used in CNC machines to convert the rotational movement to a translational movement. The efficiency and responsiveness of the actuating mechanism have the greatest influence on the accuracy of the work produced.

The actuating mechanisms used for the slides of CNC machines are screw and nut, and rack and pinion.

**Screw and nut** The screw and nut system is effective for medium traverses. With longer traverses the screw sags under its own weight. Longer the screw length, lower is the upper limit of traverse rates due to reduction in the critical speed.

Conventional vee, trapezoidal or square thread forms are not suitable for CNC machines, because the sliding action of the contacting surfaces of the thread on the screw and nut results in rapid wear and the friction is high. The efficiency of these screws is only of the order of 40%.

There are two types of screw and nut systems used on the CNC machine tools which provide low wear, accuracy over a long life, reduced friction, higher efficiency and better reliability. These are recirculating ballscrews and roller screws.

*Recirculating Ballscrews* In ballscrews, the sliding friction encountered in conventional screws and nuts is replaced by rolling friction in a manner analogous to the replacement of simple journal bearings by ball bearings.

The recirculating ballscrews are widely used on CNC machine tools because of the following advantages:

- Low frictional resistance
- Low drive power requirement
- Little temperature rise
- Less wear and hence longer life
- No stick-slip effect
- High traverse speed
- High efficiency

Two types of thread forms are used on these screws. They are *gothic arc* and *circular arc*. These forms are illustrated in Figs 5.24 and 5.25 respectively.

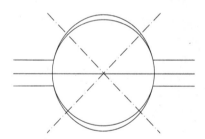

**Fig. 5.24** Thread form—gothic arc

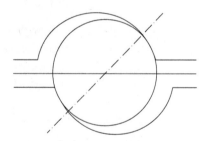

**Fig. 5.25** Thread form—circular arc

The balls rotate between the screw and the nut and at some point are returned to the start of the thread in the nut. Two types of arrangements for returning the balls, i.e. the recirculating arrangements are shown in Figs 5.26 and 5.27.

In the arrangement shown in Fig. 5.26 the balls are returned through an external tube, whereas, in the arrangement shown in Fig. 5.27 the balls are returned to the start through a channel inside the nut. To enable the movement of the slide to be bidirectional without any significant errors in positioning, there must be a minimum of backlash in the screw and nut. One method of achieving a virtually zero backlash with ball screws is by fitting two nuts as shown in Fig. 5.28. The nuts are forced apart, or alternatively squeezed together, so that the balls in one nut contact one side of the threads in the nut and the balls in the other nut contact the opposite side of the threads.

The efficiency of a recirculating ballscrew is of the order of 90% and is obtained by the balls providing a rolling motion between the screw and the nut. The mounting

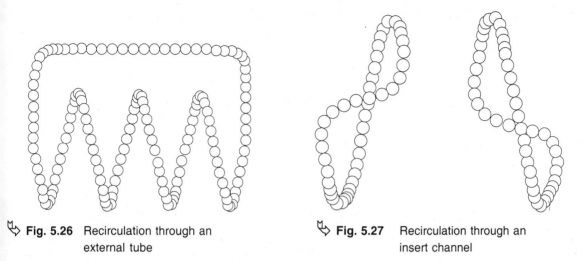

**Fig. 5.26** Recirculation through an external tube

**Fig. 5.27** Recirculation through an insert channel

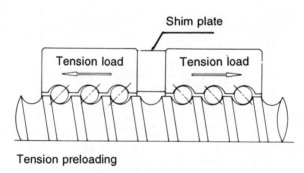

Tension preloading

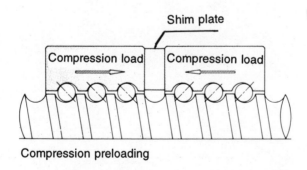

Compression preloading

**Fig. 5.28** Types of nut preloading

arrangement of a ballscrew depends on its required speed, length and size. Figures 5.29–5.31 show various methods of mounting a ballscrew in machine tools. The position of the ballscrew should be near the line of the resultant force arising from cutting, frictional and inertial forces.

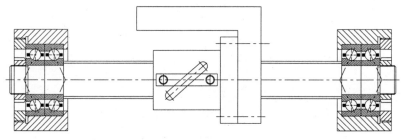

Both ends fixed

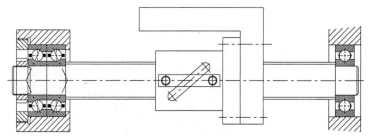

One end fixed other end supported

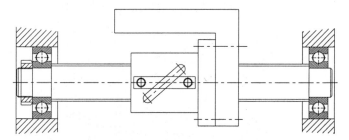

Both ends supported

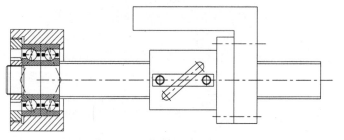

One end fixed other end free

**Fig. 5.29** Recommended mounting methods of ballscrews

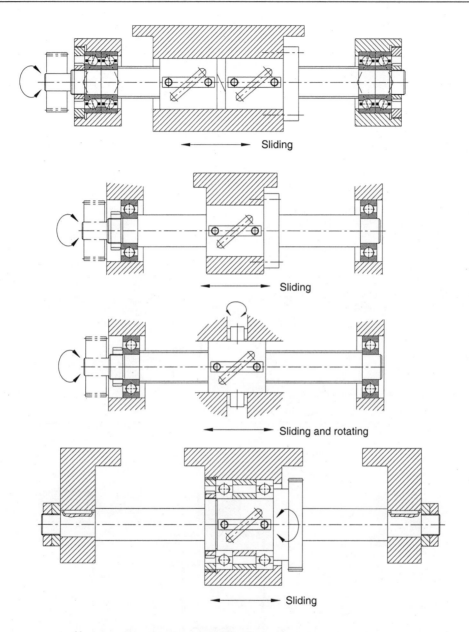

**Fig. 5.30** Ballscrew shaft and nut supporting methods

In a ballscrew system, attention should be paid to the selection of end bearings to minimise the positioning inaccuracies. The function of the bearings in a ballscrew is to locate the screw radially and resist the axial thrust force. These bearings should have high load capacity, high axial stiffness and low axial run-outs (of the order of 2 μm). Commonly used ballscrew end bearings are (Fig. 5.32):

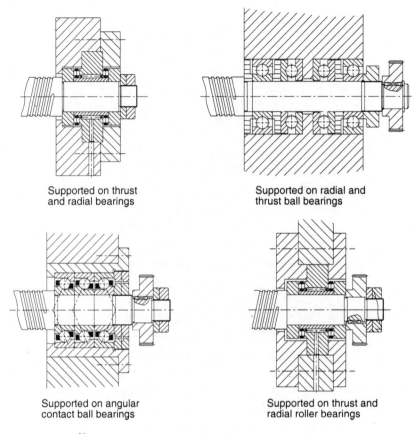

Fig. 5.31  Ballscrew shaft supporting methods

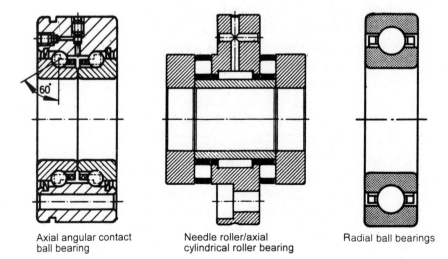

Fig. 5.32  Ballscrew for ballscrew support Ballscrew shaft supporting methods—deep groove bell bearing

(a) Set of angular contact ball bearings
(b) Set of thrust and radial roller bearings
(c) Precision deep groove ball bearings

Due to friction within a ballscrew and nut, the movement of the slide produces a temperature rise in the ballscrew leading to its expansion. This results in compressive loading on the ballscrew in case of fixed-fixed mounting. For this reason, in such mounting arrangements, ballscrews are stretched to the extent of the expected thermal expansion. Figure 5.33 illustrates some examples of the stretching or pretensioning.

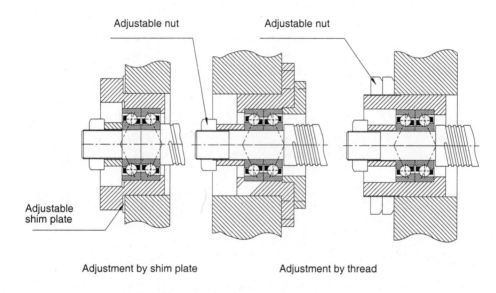

**Fig. 5.33** Methods of pretensioning

The ball nut should never be removed from the ballscrew by the user as the balls will fall out of the ball nut. There is a very special method to dismantle the ball nut from the ballscrew. A tube whose outside diameter is equal to the root diameter of ballscrew is brought close to the end of ballscrew threads and the nut is driven onto this tube so that the ball in the nut is supported by the outside surface of the tube. Various types of ball nuts are illustrated in Figs. 5.34 and 5.35.

Ballscrews are manufactured in a wide range of lead accuracies to meet the demands of general engineering industries. The required movement accuracy, smooth operation and interaction of various elements depend on the manufacturing accuracy of elements like the ballscrew, nut and balls. Errors in manufacturing cause unequal loading of the balls and seriously affect the load carrying capacity and rigidity of transmission.

Depending upon the accuracy, ballscrews are generally classified as of commercial or of precision grades. In commercial grades, the threads are invariably rolled while in the precision classes, threads are cut and ground to obtain the required accuracy. The ballscrews used on CNC machines are usually of the precision grade.

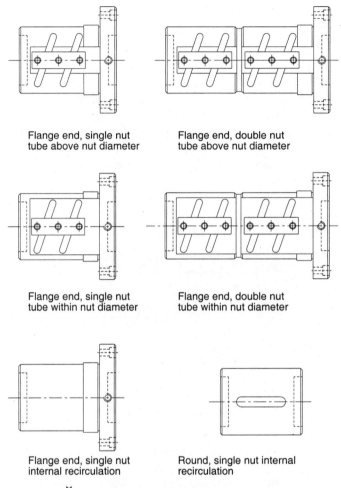

**Fig. 5.34** Types of ballscrew nuts

The accuracy of a ballscrew can be specified as:
(a) Cumulative lead accuracy over a specified length
(b) Total cumulative lead accuracy
(c) Fluctuations of cumulative lead accuracy over one revolution

Depending on the above accuracies, the ballscrews are classified into seven grades C0, C1, C2, C3, C4, C5 and C7. Table 5.2 gives the recommended accuracy grades for various applications.

*Roller Screws* Generally, two types of roller screws are used—planetary and recirculating. Both types provide backlash-free movement and their efficiency is of the same order as of ballscrews. An advantage of roller screws is that because the pitch of the screw is smaller than the minimum pitch of the ballscrew, the less complex

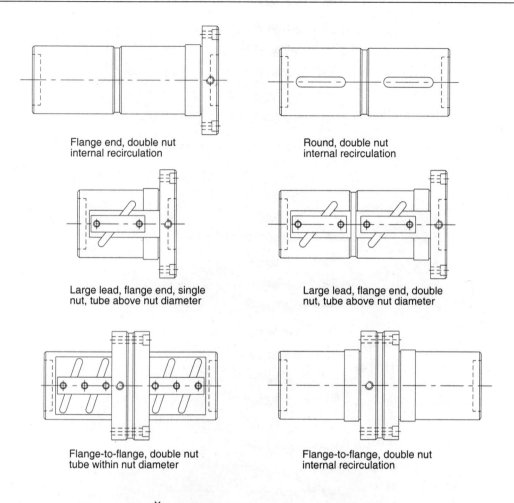

Fig. 5.35  Types of ballscrew nuts

electronic circuitry will provide more accurate positional control. Roller screws are, however, much costlier than the ballscrews. The rollers in both types of screw are positioned between the nut and the screw, and engage with the thread from inside the nut and on the outside of the screw.

The planetary roller screw is shown in Fig. 5.36. The rollers (3) are threaded as shown. At each end of the rollers, gear teeth (7) are cut. The gear teeth mesh with an internally toothed ring (8) on the nut (2) which drives the rollers to provide a rolling motion between the nut and the screw (1). The rollers are equally spaced around the shaft and are retained in their circumferential positions by spigots (4) which engage themselves in the locating rings(5) at each end of the nut. There is no axial movement of the rollers relative to the nut. The planetary roller screws are capable of transmitting high loads at fast speeds.

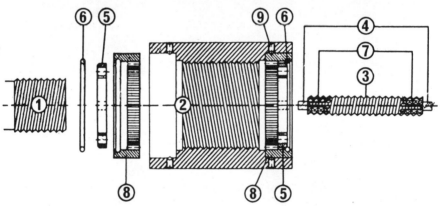

**Fig. 5.36** Elements of a planetary roller screw

The recirculating roller screw is shown in Fig. 5.37. Here, the rollers (3) are not threaded but have circular grooves of thread form along their length as shown. The rollers are equally spaced around the shaft (1) and are kept in their circumferential position by a cage (4). In operation, the rollers move axially relative to the nut (2) at a distance equal to the pitch of the screw for each rotation of screw or nut. There is an axial recess (5) cut along the inside of the nut. After one rotation of the drive screw, the rollers pass into this recess and disengage from the thread on the screw and the nut. While they are in the recess, an edge cam (6) on a ring inside the nut causes them to move back to their starting positions. While one roller is disengaged, other rollers provide the driving power. The recirculating roller screws are slower in operation than the planetary type, but are capable of taking high loads with greater accuracy.

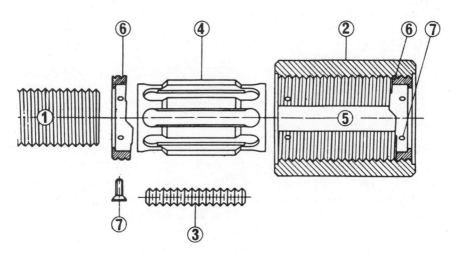

**Fig. 5.37** Elements of a recirculating roller screw

TABLE 5.2

*Example of recommended lead accuracy for aplications*

|  | C0 | C1 | C2 | C3 | C4 | C5 | C7 |
|---|---|---|---|---|---|---|---|
| Lathes | • | • | • | • | • | • |  |
| Machining centre |  | • | • | • | • | • |  |
| Grinding machines | • | • | • | • |  |  |  |
| Electric Discharge Machines |  | • | • | • | • |  |  |
| Milling machines |  | • | • | • | • | • |  |
| Boring machines |  | • | • | • | • | • |  |
| Drilling machines |  |  |  | • | • | • | • |
| Punching machines |  |  |  | • | • | • |  |
| Semiconductor manufacturing equipment | • | • | • |  |  |  |  |
| General use machine |  |  |  |  | • | • | • |
| Special use machine |  |  |  |  |  |  |  |
| Machines for steel equipment |  |  |  | • | • | • | • |
| Three dimensional measuring equipment | • | • | • |  |  |  |  |
| Aircraft |  |  |  | • | • | • | • |
| Control axis of nuclear power |  |  |  | • | • | • | • |
| Right angle coordinate type of robot |  | • | • | • | • | • |  |

**Rack-and-pinion** For longer strokes, a ballscrew needs to be supported at intermediate points to minimise deflection due to its own weight over the length and a large diameter has to be used to reduce torsional deflection. In addition, there is a limitation on the operating speed of a ballscrew due to its lower critical speed. These factors restrict the use of ballscrews for machines with longer strokes. Rack-and-pinion drives are particularly suitable for longer strokes. A slide operated by a rack-and-pinion drive has the advantage that the stiffness of the drive is independent of the length of the stroke. The rack-and-pinion system is also cheaper as compared to the ballscrew system. There are special pinions which provide for a minimum backlash. These pinions are in two sections across the width: teeth on one side of the pinion mesh with one side of the rack teeth, and teeth on the other side of the pinion mesh with the other side of the rack teeth.

*Torque Transmission Elements* The torque is transmitted from a prime-mover shaft to an output shaft through torque transmission elements. The output shaft may be a pinion or a ballscrew.

Various elements are used on CNC machines to transmit the torque, viz., gears, timing belts, flexible couplings, etc.

*Gear Box* Depending on the requirement, the drive to the ballscrew may be through a gear box or a timing belt. A gear box is required to reduce the high motor speed to a speed suitable for the feed drive, to reduce the load inertia referred to on the motor shaft and to reduce the torque on the motor shaft. They are more frequently used where the reduction is required between the shafts which are not coaxial or parallel.

*Timing Belts* These are endless toothed belts. The teeth engage with a timing pulley having teeth on its periphery. The teeth profiles on the belt and the pulley are compatible with each other. They are becoming more popular due to their inherent advantages of low cost, less noise, elimination of lubrication, less maintenance, and higher efficiency. Various manufacturers have their own teeth profiles. One of the profiles is shown in Fig. 5.38. Figures 5.39 and 5.40 show a set of timing belt and pulley. It is a positive transmission system. The timing belt has a steel wire reinforcement (as shown in Fig. 5.41). For efficient working, it is necessary to set the initial tension in the belt. The extent of the tension and the method of verifying it are recommended in the supplier's catalogue (Fig. 5.42).

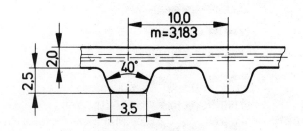

**Fig. 5.38** Timing belt profile

*Selection of Timing Belt* Speed and power to be transmitted are the two factors which influence the selection of a timing belt.

The timing belts are manufactured in various standard tooth pitches. The catalogue of belts deals with the guidelines regarding selection of pitch of belts for various applications. The width and the length of the belt depends on specific applications.

An illustration of calculation for a timing belt is given below.

## Known Parameters

(i) *Power source*: Electric motor, power $P = 3$ kW, speed $n = 3000$ rpm.
Moment of inertia–low

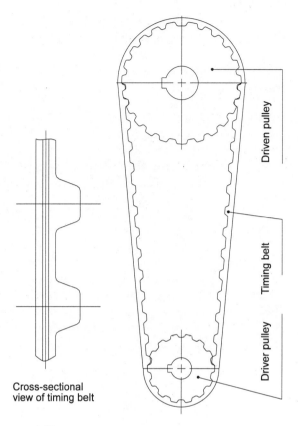

**Fig. 5.39** Timing belt drive system

(ii) *Operating conditions*: Daily running 12 hours; no idler, approximate centre distance $a$ = 290 mm, pulley diameter (max) = 200 mm.

(iii) *Application*: Output speed $n$ = 1000 rpm, Machine tool–Drilling machine

(a) As per the guidelines provided by the catalogue, the timing belt chosen is type T10, i.e., pitch of the teeth ($t$) is 10 mm.

$$m = \frac{t}{\pi} = \frac{10}{\pi} = 3.183$$

(b) The speed ratio $i = \dfrac{n_{\text{driver}}}{n_{\text{driven}}} = \dfrac{3000}{1000} = 3$

The net factor of safety is arrived at on the basis of manufacturer's guidelines.

(c) $S = S_1 + S_2 + S_3 = 1.4 + 0 + 0 = 1.4$. $S_1$ is the safety factor which depends upon both type of motor and type of machine; $S_2$ is the factor which is only for speed up drives and is zero for speed down drives; $S_3$ is the fatigue factor which is only for drives with contraflexure, i.e. drives using idler pulleys. It is zero otherwise.

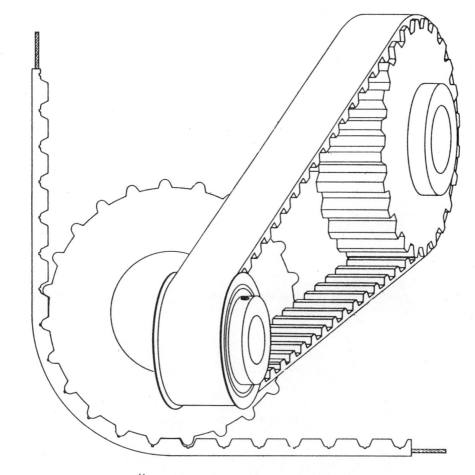

⇨ **Fig. 5.40** Timing belt drive system

⇨ **Fig. 5.41** Timing belt—Constructional details

(d) Number of teeth for $i = 3$. The chosen number of teeth are $Z_1 = 20$ and $Z_2 = 60$.

Hence $\quad i = \dfrac{60}{20} = 3$

(e) Pitch circle diameters $\quad d_0 = \dfrac{z \cdot t}{\pi}$

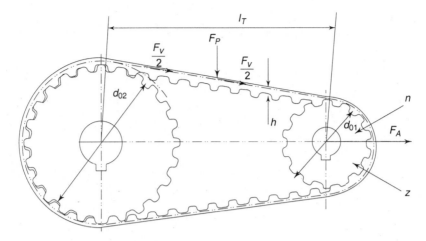

**Fig. 5.42** Method of verifying timing belt tension
$d_0$: Pitch circle diameter of pulley; $l_T$: Centre distance; $z$: Number of teeth on smaller pulley; $n$: Speed of smaller pulley; $F_A$: Force required to adjust the tension; $F_V$: Static tension in the belt; $F_p$: Testing force for verification of tension; $h$: Deflection of belt under force $F_p$

$$d_{01} = \frac{20 \times 10}{\pi} = 63.66 \text{ mm}$$

$$d_{02} = \frac{60 \times 10}{\pi} = 190.99 \text{ mm (limiting diameter of pulley being 200 mm)}$$

(f) Belt length is determined with the help of centre distance tables given in catalogue.

Factor $\quad \frac{a}{t} = \frac{290}{10} = 29$ pitches

Factor $\quad Z_2 - Z_1 = 60 - 20 = 40$

From the table a value of $\frac{a}{t}$ which is nearest to 29 is chosen against $Z_2 - Z_1 = 40$. Such a value is 28.794 pitches under factor $Z_B - Z_1 = 79$.

Hence $\quad Z_B = 79 + Z_1 = 79 + 20 = 99$

(g) Choose the nearest standard belt length from the catalogue, i.e.,

$\quad Z_B = 98 \quad$ or $\quad L_B = 980$ mm (belt length)

(h) Calculate the accurate centre distance using chosen belt length. The factors are $Z_2 - Z_1 = 60 - 20 = 40$ and $Z_B - Z_1 = 98 - 20 = 78$. From the table used at step (f), the intersection of the above two factors 40 and 78 results in $a/t = 28.281$. Hence, $a = 28.281 \times 10 = 282.81$ mm (approximate centre distance being 290 mm).

(i) Number of teeth in mesh on smaller pulley

$$Z_e = \frac{Z_1 \left[ 180 - 2 \arcsin \frac{(Z_2 - Z_1) \times m}{2 \times a} \right]}{360}$$

i.e. 
$$Z_e = \frac{20 \left[ 180 - 2 \arcsin \frac{(60 - 20) \times 3.183}{2 \times 282.81} \right]}{360} = 8.55, \text{ i.e. } Z_e = 8 \text{ teeth}$$

(j) Calculation of the belt width $b$ from the power table (provided in the catalogue) using $n_1$ and $Z_1$.

Intersection of $n_1 = 3000$ rpm and $Z_1 = 20$ teeth on power table gives specific power,

$$P_{spez} = 0.226 \text{ kW/cm}$$

The required belt width for given power with factor of safety 1.4 (as per step (c)) is

$$b = \frac{P \times 10}{P_{spez} \times Z_e} \times S = \frac{3 \times 10}{0.226 \times 8} \times 1.4 = 23.23 \text{ mm}$$

Choosing the nearest standard belt width from catalogue,

$$b = 25 \text{ mm}$$

*Result*

One synchroflex timing belt T10/980, width 25 mm;
One timing pulley (driver) T10, 20 teeth;
One timing pulley (driven) T10, 60 teeth.

**Flexible Couplings** These couplings are used when the driver and driven shafts are coaxial, since it is difficult to align the driver and driven shafts perfectly on the same axis. Further, heat and elastic deformation cause additional misalignments between the two coaxial shafts. A certain amount of flexibility is built into the couplings to compensate for these errors. The supplier of the couplings provides data on the permissible misalignments between two shafts for a particular coupling.

The servomotor and ballscrew are coupled directly using this coupling. The following three kinds of errors (Fig. 5.43) can be compensated by using the flexible couplings.

(a) Radial misalignment ($\lambda$)
(b) Angular misalignment ($\alpha$)
(c) Axial shift ($\delta$)

The various couplings are illustrated in Fig. 5.44. Figure 5.45 shows the use of flexible couplings for connecting a ballscrew and a servomotor. These couplings behave like

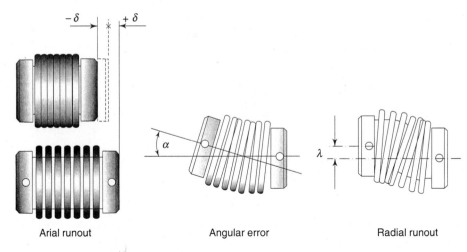

**Fig. 5.43** Types of shaft misalignment

a rigid element in the direction of rotation. In the axial and angular directions, however, they have elastic properties. Misalignments caused by fitting errors or other influences can be compensated by these elastic properties.

Similar couplings are used for connecting the ballscrew and encoder which are illustrated in Fig. 5.46.

*Taper Lock Bushes* These elements are used to couple a shaft and a hub of a gear or timing pulley, etc. They are built in the form of taper rings with self-releasing tapers. Both male and female tapers are supplied by the supplier of these elements. For the purpose of assembly, the user has to machine only cylindrical shafts and bores. Figure 5.47 shows a cross-section of the element and Fig. 5.48 illustrates its applications.

The taper lock bushes are capable of transmitting a torque from the shaft to the hub (or vice versa) without any backlash. For assembling the bushes, the male and female tapers are forced onto each other by tightening the screws axially. This expands the bushes and generates an enormous radial force due to small taper angle which locks the hub and the shaft, and the torque is transmitted through friction.

## Fluid Pressurised Hub—Shaft Connections

This bush consists of a double-walled hardened steel sleeve filled with a pressurised liquid medium, a sealing ring, a piston, a pressure flange, and clamping screws.

When tightening the screws, the bush expands uniformly against the shaft and the hub and creates a rigid joint. On loosening the screws, the bush returns to its original position and can easily be dismantled. A fluid pressurised bush is illustrated in Fig. 5.49.

The working principle of this type of bush is illustrated in Fig. 5.50, and its appplication is similar to that of a taper lock bush.

### Design of Modern CNC Machines and Mechatronic Elements 169

(a)

(b)

(c)  (d)

**Fig. 5.44** Types of flexible couplings (a) Axial locking type (b) Flange type (c) and (d) Tangential locking type

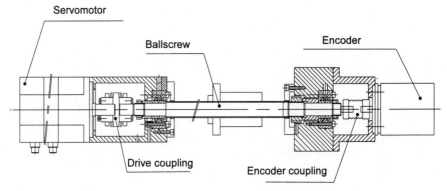

**Fig. 5.45** Application of flexible couplings

Approximately 1:1

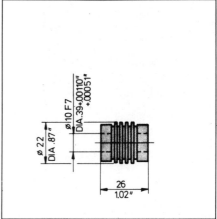

Approximately 1:1

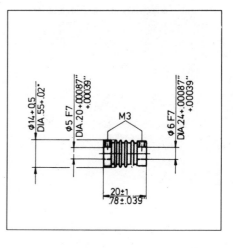

**Fig. 5.46** Types of flexible couplings for connecting encoder

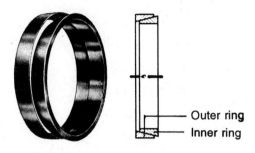

 Fig. 5.47  Taper lockbush

## 5.5  SPINDLE/SPINDLE BEARINGS

Material removal using single point or multipoint workpiece requires rotational speeds of the order of 30–6000 rpm and even higher. All work or tool carrying spindles rotating at these speeds are subjected to torsional and radial deflections. They are also subjected to thrust forces depending on the nature of the metal cutting operation being performed. To increase the stiffness and minimise torsional strain on the spindles they are designed to be as stiff as possible with a minimum overhang. Also, the final drive to the spindle should be located as near as possible to the bearings.

When a workpiece holder (such as a chuck) is mounted on the spindle, the accuracy of rotation is extremely important as it affects the roundness of the components produced. The rotational accuracy of the spindle is dependent on the quality and design of the bear-

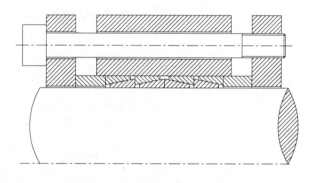
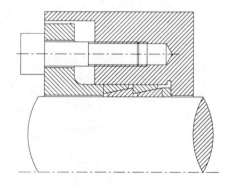
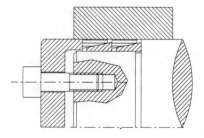

 Fig. 5.48  Application of taper lock bushes

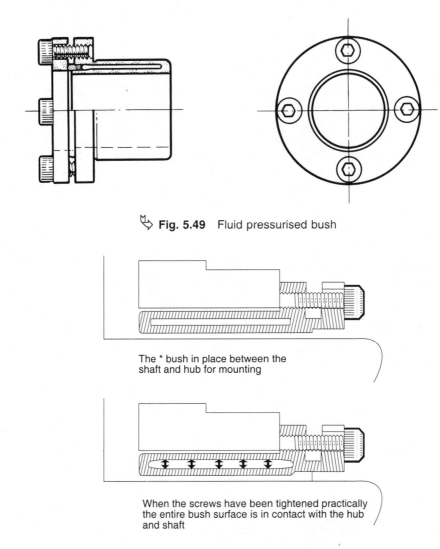

Fig. 5.49 Fluid pressurised bush

Fig. 5.50 Inner details of fluid pressurised bush

ings used and the preloading. The bearings should support the spindle radially and axially.

The accuracy and the quality of the work produced depends directly on the geometrical accuracy, running accuracy and the stiffness of the spindle assembly.

The various types of spindle bearings used in the design of a spindle for machine tools are:

(a) Hydrodynamic bearings
(b) Hydrostatic bearings
(c) Antifriction bearings

## 5.5.1 Hydrodynamic Bearings

Hydrodynamic bearings are journal bearings with a thin film of oil between the spindle and the journal. These are used where the load carrying capacities are low and frequent, starting and stopping of the spindle is not required; for example in grinding machines. The essential features of these bearings include simplicity, good damping properties, and a good running accuracy.

The principle of the hydrodynamic bearing is illustrated in Fig. 5.51. The pressure of the oil is created within the bearing by the rotation of the spindle. As the spindle rotates, the oil in contact with the spindle is carried into wedge-shaped cavities between the spindle and the bearing. The oil pressure is increased as the oil is forced through the small clearances between the bearing and the spindle.

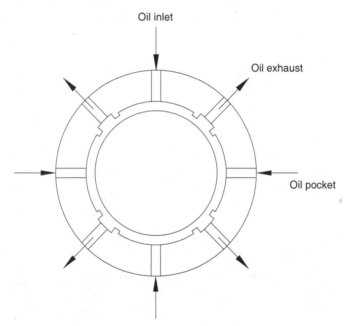

**Fig. 5.51** Principle of hydrodynamic bearings

The main limitation in a hydrodynamic bearing is that a definite clearance must be provided for the oil film to be maintained between the spindle and the bearing. This clearance may result in the centre of a spindle in the bearing to change its position owing to variation in the applied force. Clearances normally provided between the spindle and the bore of the bearing for the oil film vary from 50 μm up to 200 μm depending upon the diameter of the journal.

## 5.5.2 Hydrostatic Bearings

For a hydrostatic bearing, the spindle is supported by a relatively thick film of oil supplied under pressure, similar to that used in the bearings for linear movements. The oil is pressurised by a pump external to the bearing. The principle of the hydrostatic bearing is illustrated in Fig. 5.52. The load carrying capacity of this type of bearing is independent of the rotational speed. They have high damping properties, high running accuracy, high wear resistance, but are very expensive. These are used only where temperature effects cause problems in the part accuracy as in the case of grinding machines and fine boring machines.

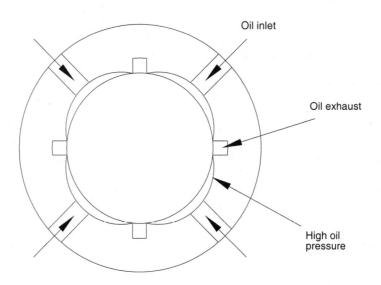

**Fig. 5.52** Principle of hydrostatic bearings

### 5.5.3 Antifriction Bearings

These are suitable for high speeds and high loads. They are often used in preference to hydrodynamic bearings because of their low friction, moderate dimensions, lesser liability to suffer from wear or incorrect adjustment, ease of replacement, and high reliability.

For efficient service, it is essential that all the components of the ball and roller bearings, particularly the rolling elements, and the inner and outer bearing tracks are of the highest accuracy. An error in one component can affect the quality of the work produced.

Several kinds of ball and roller bearings are used for spindles on CNC machines.
1. Ball bearings
    (i) deep groove ball bearings (Fig. 5.53)
    (ii) angular contact ball bearings (Fig. 5.54)
2. Roller bearings
    (i) cylindrical roller bearings (Fig. 5.55)
    (ii) cylindrical roller bearings (double row) with tapered bore (Fig. 5.56)
    (iii) tapered roller bearings (Fig. 5.57)

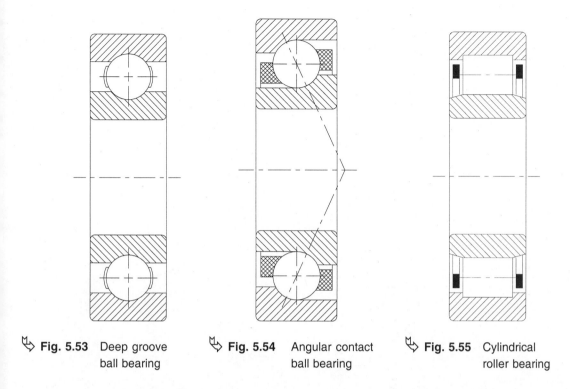

**Fig. 5.53** Deep groove ball bearing

**Fig. 5.54** Angular contact ball bearing

**Fig. 5.55** Cylindrical roller bearing

The selection of a particular type of bearing for the spindle depends on the requirements of the particular machine, like speeds of operation, accuracy of the spindle and stiffness of the spindle.

## Preloading

The ball and roller bearings have some amount of radial and axial clearances. When a main spindle is mounted on bearings, there should not be any axial or radial play in the main spindle assembly. To achieve this, the clearances in the bearings have to be taken up by preloading them. In case of tapered roller bearings and angular contact ball bearings, the axial and radial clearances can be taken up simultaneously by preloading. Cylindrical roller bearings (double row) with tapered bores are radially preloaded by pushing the inner race against the taper on the spindle.

## Accuracies

On main spindles the following two accuracies of bearings are defined.
(a) Radial run out
(b) Axial run out

The accuracy of a spindle during running depends on the thermal stability of the spindle. This is the most important aspect of spindle design, especially for high speed and high accuracy spindles. Considering the above, appropriate bearing lubrication and

**Fig. 5.56** Cylindrical roller bearing with taper bore

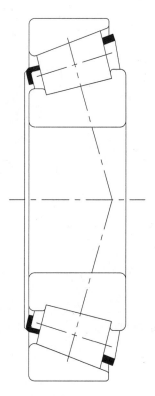

**Fig. 5.57** Tapered roller bearing

spindle cooling systems should be adopted in the design of spindles for machine tools. A typical main spindle bearing arrangement is illustrated in Fig. 5.58.

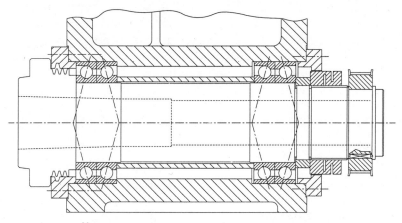

**Fig. 5.58** Typical spindle bearing arrangement

One of the recent developments in bearings is the introduction of ceramic bearings for higher spindle speeds of the order of 10,000–20,000 rpm for machining of light alloys. Ceramic bearings offer advantages such as low coefficient of friction, greater thermal stability and higher hardness. Lower density of the ceramic balls reduces the centrifugal forces and hence are suitable for higher operating speeds.

## 5.6 MEASURING SYSTEMS

On all CNC machines, to an electronic measuring system is employed on each controlled axis to monitor the movement and to compare the position of the slide and the spindle with the desired position. Measuring systems are used on CNC machines for:

(a) Monitoring the positioning of a slide on a slideway
(b) Orienting the spindle/table and measuring the speed of the spindle.

There are two terms that are used with measuring systems: accuracy and resolution. The measuring accuracy of a measuring system is the smallest unit of movement that it can consistently and repeatedly discriminate. The resolution is the smallest unit of a dimension that can be recognised by the measuring system.

Measuring devices used in measuring system are classified as rotary and linear measuring devices. One of the rotary measuring devices is an incremental rotary encoder which is widely used on CNC machines. A linear scale is a linear measuring device which is used very often.

The ideal method for monitoring the slide position would be to continuously measure the distance of the tool or cutting edge from the workpiece datum, thus eliminating tool wear errors as well as the deflection due to misalignment. Since the ideal monitoring system is not yet available due to many problems—interference caused by the presence of swarf and cutting fluid, the following methods are generally used.

### 5.6.1 Direct Measuring System

In this, the linear displacement is measured directly at the slide. Since backlash errors in axis drive elements do not affect the measuring process, a high degree of accuracy is achieved. The measuring device is fixed onto the moving machine element which detects the actual distance travelled by the machine slide. Examples include linear scales and inductosyn.

### 5.6.2 Indirect Measuring System

Here, the slide position is determined by the rotation of the ball screw/pinion or the drive motor. This system is more convenient and less costly as compared to a direct measurement system. However, in such systems additional sources of error like backlash and torsional deformation on the drive system may creep in. These errors can be reduced using various error compensation features available on CNC systems. Encoders and resolvers are some of the feedback devices used in indirect measuring systems.

**Mounting of measuring system**  Orientation of spindle (C-axis) as in the case of turning centres is required for off-axis drilling and milling operations as well as for machining different profiles. Similarly, positioning of a rotary table in a machining centre is required for machining the component in different orientations. The measurement of the speed of rotation of a spindle is required when a synchronisation of spindle speed with slide movement is called for as in the case of threading on CNC turning machines or tapping on CNC machining centres. Usually encoders are used as measuring elements in such cases.

**Mounting system for spindle monitoring**  The mounting position of a feedback device is important as it may lead to additional errors. Linear measuring devices should be positioned near the guideway and ballscrew and in an accessible position for maintenance purposes. Rotary measuring devices can be located at the free end or at the driving end of the screw or on the motor shaft.

### 5.6.3 Incremental Rotary Encoders

Incremental measurement means measurement by counting, i.e. the output signals of incremental rotary encoders are fed to an electronic counter from which the measured value is obtained by counting the individual increments.

The details of an incremental rotary encoder are depicted in Fig. 5.59. The visible exterior of the unit includes the shaft, the flange, the rotor housing and the output cable. The shaft rotates in the flange on preloaded ball bearings and carries a graduated glass disc with radial gratings. The disc has two types of markings — one for measurement and another for referencing.

**Principle of the Working of an Encoder**

Figure 5.59 shows an incremental rotary encoder on which a graduated disc is scanned using the transmitted light technique. Graduated disc has gratings marked radially ranging from 200 parts per revolution (ppr) to 18,000 ppr. Measurements are obtained on a disc scale through a scanning plate and photoelectric cells (solar cells). The active surface of the photoelectric cells spans several gratings on the disc and a scanning plate of corresponding width is placed between the light sources and the cells. The movement of the disc in relation to the scanning unit produces an output of sinusoidal signals from the cells which are converted into rectangular wave form in the read-out unit. The number of pulses is a measure of the length of displacement. Reference markers are used to indicate a definite starting point.

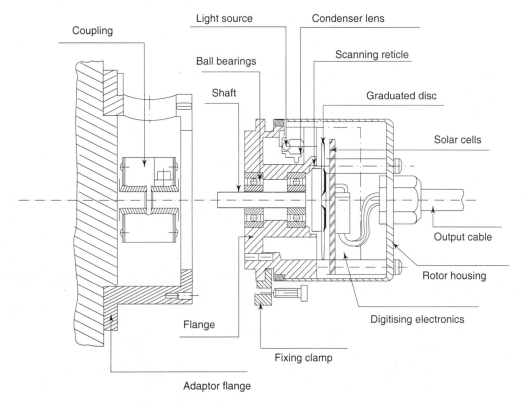

**Fig. 5.59** Details of incremental rotary encoder

Figure 5.60 gives the schematic details of the disc. The encoder is connected mechanically to the ballscrew or any rotating shaft through a flexible coupling. Most commonly used flexible couplings for connecting the encoders are illustrated in Fig. 5.46.

It is very important to ensure that the axis of an encoder and that of a connecting shaft are aligned both angularly and radially within the permissible limits imposed by the flexible couplings. If the above alignments are beyond the permissible limits there will be an undue mechanical load on the bearings of the encoder. These are illustrated in Fig. 5.43. A typical value for the allowable misalignment is ± 0.2 mm. The clamps used for mounting are shown in Fig. 5.61. In some cases the encoders are continuously purged by using clean and dry air to keep the interior dust-free.

## 5.6.4 Linear Scale

Linear scales measure the actual movement of the slide because they move with the slide. For this reason linear scales provide more accurate results than rotating encoders. But, because they are longer, they tend to be more expensive. The material of the scale should be selected so that its thermal expansion is equal to that of the machine tool. Inaccuracies could arise on long measurement if the thermal expansion rates are different. Another problem with linear scale is protecting them from oil, coolant, swarf, etc.

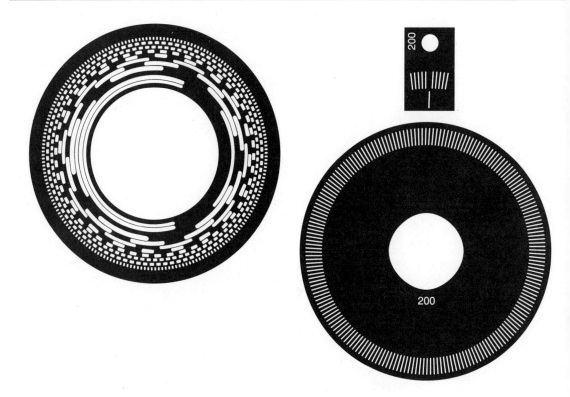

**Fig. 5.60** Schematic details of encoder disc

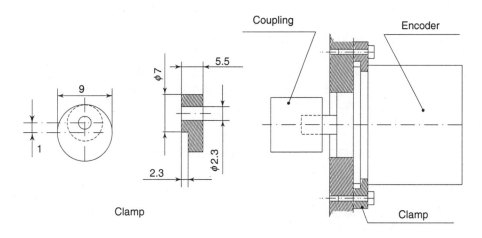

**Fig. 5.61** Clamps for mounting encoder

The linear scale consists of a glass scale with gratings and a reading head. One of the above two elements is mounted on a fixed member and another on the moving slide. The glass scale has gratings. The reading head contains the light source, a condenser, lens for collimating the light beam, the scanning reticle with index gratings, and cells. Figure 5.62 illustrates schematically the details of a linear scale.

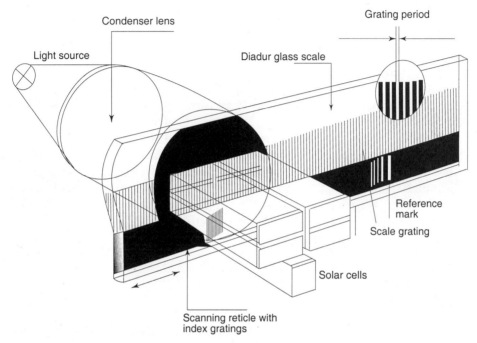

**Fig. 5.62** Details of linear scale

When the scale is moved relative to the scanning unit, the lines and spaces of the scale alternately coincide with those of the index grating. The corresponding fluctuations of light are sensed by the cells which generate the signals. These signals are further processed for measurements in the same manner as in the case of rotary encoders.

## 5.7 CONTROLS, SOFTWARE AND USER INTERFACE

CNC controls are the heart of the CNC machines. The early CNC controls were developed for simple applications in turning, machining centres and grinding. But with increased capabilities on modern machine tools such as higher spindle speeds, higher rapid traverses and more number of axes, CNC systems have been developed to meet these needs.

The new generation computer numerical controls allow simultaneous control of more axes, interpolate positions faster, and use more data points for precise control. These processors perform multitasks — run one program while programming and simulating a second — which maximises the machine use. Most CNC machine tools in the market permit data input manually on the shopfloor as well as from a distributed controller via a communication link. The new controllers offer advanced graphic interfaces, program simulation and some cutter selecting capabilities.

## 5.8 GAUGING

Better workpiece quality is one of the most important advantages in using a hi-tech CNC machine. To maintain quality the effect of parameters like tool wear and thermal growth can be eliminated by automatic gauging system.

Gauging on a machine tool is basically used for workpiece inspection, for defining tool offsets and for tool breakage detection. It is further extended for other functions like workpiece presence detection, automatic alignment of workpiece and detection of stock variation. A touch trigger probe is normally employed for measurement and inspection. These probes are basically very sensitive switches with a long spring loaded stylus. They are capable of detecting small deflections of the stylus from the home position once the contact is made. The signal is transmitted to the machine system which processes the data as per the gauging requirements. Depending on the machine and function, these probes are mounted on the turret, table or headstock housing or can be transferred from the tool magazine to the spindle for use.

## 5.9 TOOL MONITORING SYSTEM

During machining the tools wear out or even break. Continued use of worn out tools result in poor part quality. Tool breakage, unless recognised at the correct time, can lead to irrepairable damage to the workpiece or the machine. Hence, tool wear and breakage, if not properly monitored, affect the productivity of the machine and the quality of the component produced.

Presently, established tool monitoring sensors and systems are available commercially for integrating with CNC machines. Tool monitoring systems enable the introduction of adaptive controls on machines for optimising the cutting parameters.

A tool monitoring system monitors the tool wear and tool breakage. There are two ways of monitoring tool wear and breakage.

### 5.9.1 Direct Monitoring

Monitoring of the tool condition is done directly using a touch probe by checking the tool edge position and checking for the existence of a tool edge.

### 5.9.2 Indirect Monitoring

The tool condition is checked indirectly by monitoring the change in certain parameter whose value when affected reflects the condition of the tool. Some of the parameters used to monitor tool condition are:

(a) Life of the tool
(b) Power of the spindle or a feed drive or a driven tool
(c) Forces due to cutting
(d) Noise emission during cutting
(e) Dimensions of the workpiece

CHAPTER **6**

# Assembly Techniques

## 6.1 GUIDEWAYS

The guideways are normally employed on machine tools to obtain a rotating motion or a linear motion. For the linear motion, a precise guiding system is used. The main types of linear guides employed in machine tools are:
- Metallic guideways
- Metallic guides with low friction pads (turcite pads)
- Antifriction guideways
  - linear motion (LM) guides
  - tychoways
  - needle roller and flat cage assembly (M and V guides, M and L guides, and flat guides)

The following are the basic prerequisites of a guiding system for linear motion as applicable to computer numerically controlled (CNC) machines.
- Stiffness in all directions
- Movement with a constant low friction and freedom from stick-slip
- Better accuracy in movement, restricting rolling, pitching and yawing
- Retention of accuracy over a longer period
- Lower cost and less lead time for manufacture and assembly
- Easy to assemble and install
- Simple and easy to prepare mounting surfaces and to machine
- Easy maintenance and noiseless performance
- Economy in energy consumption

### 6.1.1 LM Guides

The basic requirements of most machines are met by using commercially available antifriction guideways known as linear motion guides.

The LM guides are available in the following configurations
- Ball type—four-row, six-row
- Roller type

Thye are available in different lengths which make them more suitable for the following machines.

- Machining centres
- Turning centres
- Grinding machines
- Wire electric discharge machines (WDM)
- Robots

## Assembly of LM Guides

The desired accuracy, rigidity and load carrying capacity directly influence the quality of the mounting surfaces. Also, the location of the guides, the lubrication and the sealing of the bearings are important.

### Mounting and Locating Surface Accuracy

To obtain a higher positioning accuracy and a smooth running of the guiding system, accurate mounting and locating surfaces are required. Typical values of mounting accuracies are indicated in Fig. 6.1. These can be obtained by either precision milling or grinding.

### Parallelism of Mounted Guideways

The parallelism of mounted guideways are checked using standard fixture and dial indicators. The allowable values depend on the type of the guideway and its clearance. The typical value varies from 6–20 µm.

### Location

In order to achieve high rigidity and high load carrying capacity, the guiding elements should be supported on both the sides against the locating surfaces. The locating heights ($h_1$, $h_2$) and corner radii ($r_1$, $r_2$) of the locating surfaces are given in Fig. 6.1.

### Lubrication

Normally, LM guides are provided with a lubrication nipple mounted on the end face of the carriage. Guides are available with grease lubricated for life or with oil lubrication. The guidelines for lubrication are given in Annexure 6.1.

### Assembly Precautions

(a) Foreign materials and flashes on the mounting surface of the guide should be removed with an oilstone.
(b) The straightness of the mounting surface in both the vertical and the horizontal direction should be measured by using an autocollimator at an interval of 120 mm.
(c) Rust preventive coatings applied on LM guides should be thoroughly wiped. A thin coat of oil is applied on the resting surfaces of LM guides and the mounting surfaces.

# Mechatronics

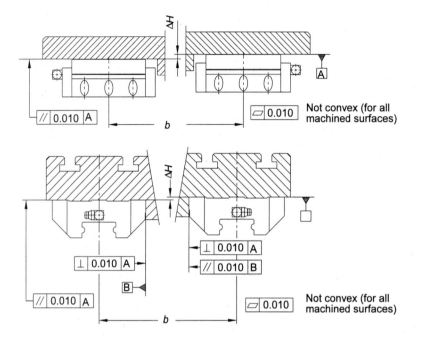

Straightness of mounting faces, 10-15 micron/depending on the nature of the machine

$\Delta H = 0, 2b$, for $v_1$ preload class
       $= 0, 1b$, for $v_2$ preload class
If height difference exceeds this limit if will affect the operation life

Tolerances for adjacent surfaces

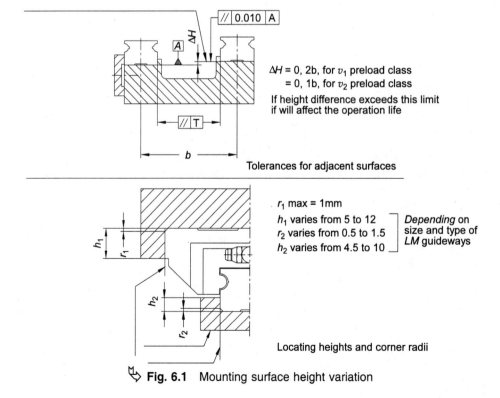

$r_1$ max = 1mm
$h_1$ varies from 5 to 12
$r_2$ varies from 0.5 to 1.5
$h_2$ varies from 4.5 to 10

Depending on size and type of *LM* guideways

Locating heights and corner radii

**Fig. 6.1** Mounting surface height variation

(d) Clean and correct threaded length of screws (for fixing LM rails) should be used. Forcible tightening of a screw having incorrect thread length will result in inaccuracy. Sequential tightening from the centre to either end has to be ensured. LM rail push screws can be used to ensure correct butting of the sides. Tightening should be done using a torque wrench with the specified torque.

(e) Presence of dust or foreign matter inside the LM guide drastically increases the wear resulting in reduced life. In order to prevent this, use of telescopic covers or bellows is mandatory. Also, the assembly room and the surrounding area must be clean.

(f) Care should be taken to prevent damage to the lip seal during assembly. A thin sheet of spring steel should be used between the guideway surface and the bearing block. The closing plugs should be pressed into the fixing holes to avoid any damage to the lip seals.

## Effect of Mounting Misalignments on the Performance of the LM Guides

(a) *Parallelism Error* Parallelism error causes the rolling resistance and decreases the life of the bearing. In addition there is a loss in straightness of the motion.

(b) *Mounting Surface Height Variation (Fig. 6.1)* As the difference in height $H$ increases, the rolling resistance also increases due to the moment load imposed on the bearings, resulting in a reduction in life of the guides.

(c) *Preload* Preload increases the dynamic stiffness of the guides, natural frequency and damping. Increase in preload beyond a certain value will result in higher rolling resistance causing a reduction in life. The preload value should be approximately one-third the operating load.

(d) *Track Base Joint Deviation* When assembling multiple rails, track joint deviation should be kept to a minimum. Increase in track joint deviation causes resistance on the slides.

A checklist for mounting of the linear guideways is given in Annexure 6.1.

### 6.1.2 Tychoways

The tychoway linear bearings were introduced in the 1950s with the inception of the numerically controlled (NC) machines. These bearings have the capacity to carry very large static and dynamic loads. With the development of LM guides, the use of tychoways has decreased because of the advantages of LM guides over tychoways. However, for heavy duty applications tychoways are still being used.

Typical mounting of tychoways for horizontal guiding application is shown in Fig. 5.20. On all guiding surfaces, precision hardened and ground guide strips are required to run the tychoways. The accuracy requirements of tychoway mounting surfaces and guide strips as assembled on the machine are given in Fig. 6.2.

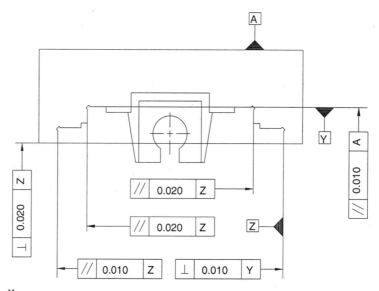

**Fig. 6.2** Accuracy requirements of tychoway mounting surfaces

### Assembly Precautions

(a) As far as possible, select equal height tychoways for each plane of guiding from a selective assembly, in order to ensure equal loading of the bearings.

(b) Parallelism of the guide strips is the most important parameter where the preload is controlled within the limits to achieve higher acccuracy of the guiding system. A graph of the recommended tolerance of guideways with respect to their length is given in Fig. 6.3.

(c) Hardness of the guide strips should be well controlled and should be in the range 58-62 HRC (650-750 HV30) (through hardened). The hardness has to be checked by non-destructive method before assembly. For a hardness value less than the above, there is a corresponding reduction in static and dynamic capacities. The hardness factors are shown in Fig. 6.4.

(d) The squareness of a guide strip and the associated mounting surface of the tychoway should be within limits. Any deviation on this count will result in uneven loading of the rollers, which can be seen by the track impression on the strips.

(e) Preloading of tychoways is very difficult in the sense that this cannot be measured directly and preloading has to be done only during assembly. The normal preload value for most of the machine tool applications is 20% of the dynamic load rating. The preload is applied from one side only and is achieved through the adjustment of the tychoway mounted surface or shim plates or spring pads on which the tychoway is mounted. The preload can be checked by measuring the force required to move the preloaded assembly. A coefficient of friction of approximately 0.005 may be assumed. This includes friction of the wipers. Premounting the bearings and forcing the assembly onto the machine ways should be avoided.

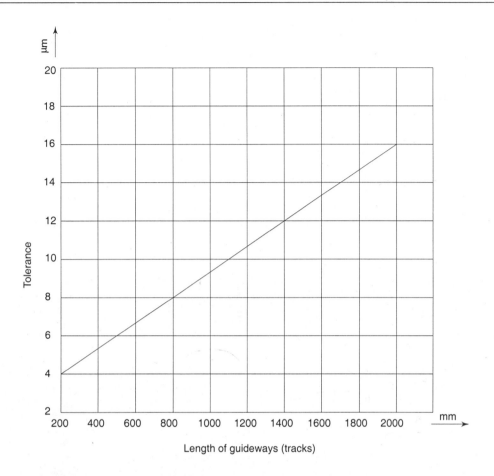

**Fig. 6.3** Recommended tolerance for the guideways

(f) Precaution has to be taken to align the tychoways in the direction of traverse. For most applications this alignment should be 0.002 mm/100 mm. The parallelism between mounting faces of the tychoway and the guideway should be within 0.008/100 mm (Fig. 6.5).
(g) The assembly environment should be clean and free from dirt, swarf, or abrasive material, which may find its way onto the surface of the guideway and cause serious damage to the bearings. Way covers or bellows are mandatory for protection.
(h) Tychoway bearings generally require a minimum amount of lubrication. Oil lubrication is normally preferred. The lubricant must be clean and any dirt in the oil can seriously damage the bearings or strips. A water absorbent oil such as cheviot c.x. machine tool lubricant with a viscosity of 68 cSt is recommended. Guidelines for lubrication are given in Annexure 6.1.

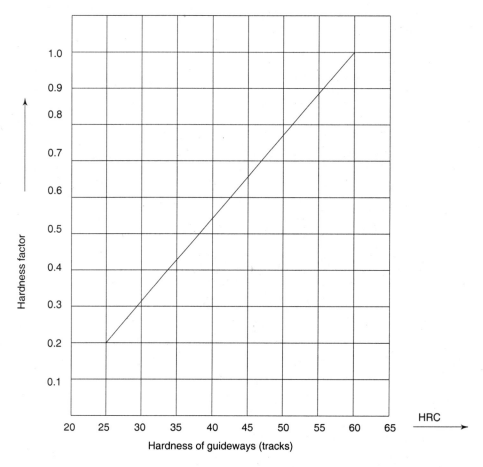

**Fig. 6.4** Hardness factor with respect to hardness of guideways

*Factors Which Could Affect Performance*

(a) Skidding due to high acceleration, deceleration or very high velocity and due to incorrect assembly
(b) Skewing due to incorrect shape of track caused by incorrect assembly and poor alignment
(c) Incorrect preloading: too high, too low, uneven
(d) Overloading or overstressing
(e) Offset load
(f) Shock caused by dropping of masses during assembly
(g) Vibration
(h) Fluctuating load
(i) Uneven load distribution

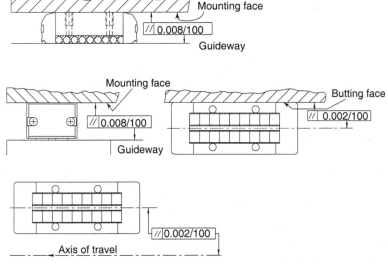

**Fig. 6.5** Alignment of centreline of the tychoway units to the axis of travel

(j) Incorrect lubrication
(k) Presence of dirt, swarf and moisture
(l) High and low temperature
(m) Corrosive or hostile environment

### 6.1.3 Needle Roller and Flat Cage Assemblies

Most commonly these guideways come in M and V shapes. The most common arrangement in M and V type is the closed arrangement (Fig. 6.3). M and V guides are combined with angled flat cage assemblies to produce linear locating bearing arrangements. For light load applications needle flat cages are used and for heavy loads cylindrical roller flat cages are used. The typical mounting accuracy for needle flat cage M and V guide assembly is shown in Fig. 6.7.

*Accuracy of Guideways*

The M and V guideways are available in three quality grades, viz., Q2, Q6 and Q10. The normally used quality for the machine tools is Q6. Figures 6.8 and 6.9 show the tolerance values of these three qualities in relation to the guideway length and profile tolerances.

*Matching of Guideways*

M and V guides are always used in pairs. Each pair of guideways with the same profile size has the same support distances $h_m$ and $h_v$ (Fig. 6.10). The matching accuracy is with respect to the quality grade (e.g. for quality Q6 it is 6 μm).

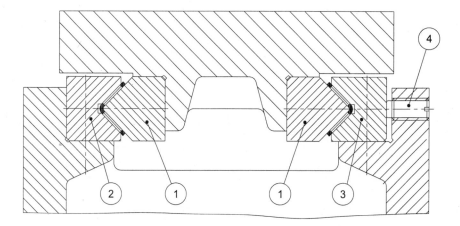

**Fig. 6.6** Assembly of a closed arrangement

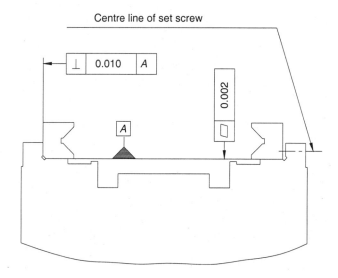

**Fig. 6.7** Mounting accuracy for needle flat cage M and V guide assembly

### Assembly Precautions

Fixing holes can either be drilled in the machine bed during the manufacturing stage if the tolerance permits or can be transferred from guideways during assembly. These holes must be carefully deburred to ensure proper seating of the guideways. The seating surfaces of the guideways are preferably ground.

It is absolutely essential that the guideways marked and packed as matched pairs are fitted in the same guidance assembly. The mounting face of a guideway can be identified by the larger chamferred edge (unmarked face). For proper butting of the vertical face, clamps may be used against the vertical seating face before tightening the screws.

When multipiece guideways are assembled it should be noted that the pieces which make up two matched guideways and form one guideway set are packed together. All guideways which belong to the same set have the same set number. In addition all the ends of the guideway pieces which abut are identified by marking with letters in alphabetical order (Fig. 6.11).

The assembly of guideways must follow a certain sequence as mentioned below (Fig. 6.6).

- Mount the fixed guideway pair (1) onto the movable machine member.

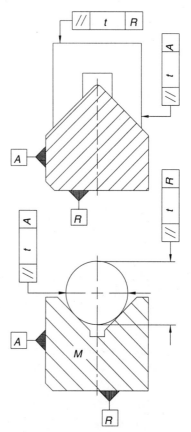

**Fig. 6.8** Tolerance values

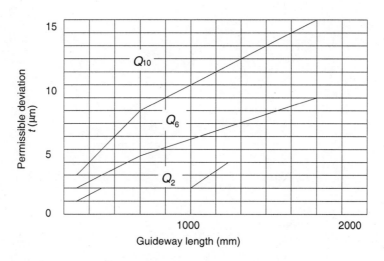

**Fig. 6.9** Tolerance values

- Check the guideway pair (1) for parallelism (Fig. 6.12).
- Mount the opposite fixed guideway (2).
- Mount the adjustable guideway (3) but do not fully tighten the screws so that the guideway can be moved.
- Insert the movable machine member in the longitudinal direction.
- Slide the flat cage assemblies in between the guideways and position them accurately.
- Set the adjustable guideway (3) with the adjusting screws (4) so that it is either clearance free or preloaded, then tighten the fixing screws.
- Fix the end pieces or end piece and the wiper units.

The parallelism error of the seating surface should be close to that of the corresponding guideways as given in Fig. 6.10. The squareness of the seating surface should be within ± 0.0003 radian.

To ensure a uniform load distribution over the rolling element length, the difference in height of the seating surface should be $\Delta h(\mu m) < 0.1b$ for a needle cage and $\Delta h(\mu m) < 0.3\,b$ for roller cage (Fig. 6.13).

B
$h_v$
H
$h_m$
h
Width, $B\ ^{0}_{0.02}$ mm
Assembly height, $H\ ^{0}_{0.2}$ mm
Support distance, $h = h_m + h_v\ ^{0}_{-0.2}$ mm

**Fig. 6.10** Matching of guideways

1A
1A
1B
1B
Pair
1C
1C
1D
1D

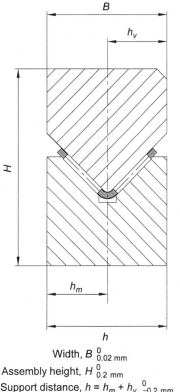

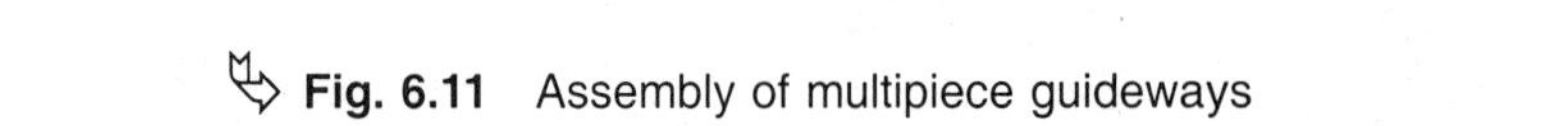

**Fig. 6.11** Assembly of multipiece guideways

If the preload value is low, the preload is applied by adjusting the set screws on the back face of the guideway and these screws are provided along the length of the guideway

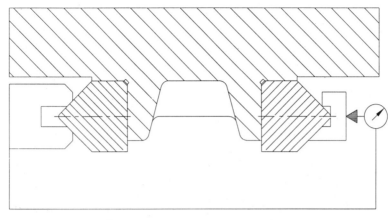

**Fig. 6.12** Checking parallelism

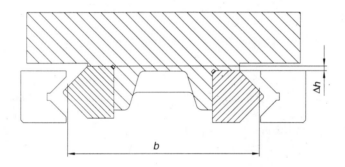

**Fig. 6.13** Permissible difference in heights of the seating surfaces

at a regular pitch. The preloading through the screws needs expertise and skill. For higher preloads, ML guideways with taper jib are used.

Lubricants are generally introduced through the lateral gap between the M and V guideways except in the vertical configuration where oil is fed through lubrication holes in the guideways. The lubrication path has to be thoroughly cleaned and should be free from any metallic particles or swarfs that may damage the bearings. Guidelines for lubrication are given in Annexure 6.1.

One of the most important conditions for trouble-free operation of these guideways is effective sealing or covering of the arrangement. For this purpose, telescopic covers or bellows should be provided.

## 6.2 BALLSCREW AND NUT

### 6.2.1 Basic Requirements of Ballscrew Assembly

(a) For a backlash-free movement all mounting brackets such as end bearing assembly, nut mounting, etc. should be rigid.

(b) The ballscrew and nut assembly should have good axial rigidity and torsional stiffness.
(c) The preload on ballscrew and nut must be optimum, so that it will not affect the movement characteristics and does not induce excessive heating of the ballscrew during rapid traverse.
(d) The end mounting bearings should have true axial running accuracy. The bearings are to be preloaded optimally for rigidity and backlash-free operation.
(e) Movement must be with a constant drag torque, free from stick-slip and must aid precise positioning.
(f) Maintenance of accuracy over a long period.
(g) Easy to assemble and dismantle.
(h) Easy maintenance and noiseless performance.
(i) Good service life.

### 6.2.2 Assembly Techniques of Ballscrew and Nut Mounting

#### Mounting Parallelism between a Ballscrew and the Guideways

When a ballscrew is used with a linear motion guide such as a ball way, ball spline or turcite, it is necessary to pay attention to the mounting parallelism of ballscrew with respect to the guideways.

If the mounting parallelism is incorrect it will result in a twisting load and hence temperature rise and loss of accuracy.

#### Influence of Mounting Error

Eccentric load (moment load and radial load) on the ballscrew adversely affects not only operability, but also fatigue life. Figure 6.14 shows a typical calculation of fatigue life when the moment load is applied to the ballscrew. The figure presents the calculated value by assuming the support stiffness of the mounting section (screw shaft, support bearing and guide) as infinitely large.

#### Eccentric Load (Moment and Radial Loads)

The ballscrew demonstrates its fullest performance when the load distribution on the steel balls rolling between the screw shaft and the nut is as even as possible. Moment load on the nut may cause concentrated loading on a small portion of the steel balls, thereby affecting the operating characteristics and shortening the service life. It is also essential to avoid a radial load to the extent possible.

Generally, an inclination error of $\frac{1}{2000}$ (Fig. 6.15) $\left(\text{target } \frac{1}{5000}\right)$ or less and a misalignment of the centre by about 20 μm or less are recommended as the control values for the precision class of ballscrews.

Overturning or twisting loads should be avoided wherever possible and the screw and the nut must be fixed squarely so as to load coaxially.

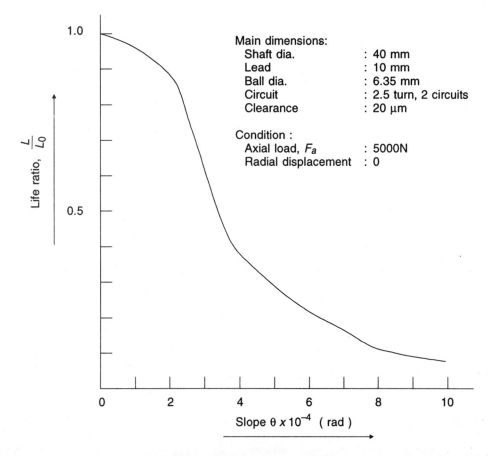

**Fig. 6.14** Calculation of fatigue life when the moment load is applied to the ballscrew

## 6.2.3 Ballscrew Alignment Assembly Precautions

(a) The ballscrew support requires alignment with respect to the guideways both vertically and horizontally. In practice, an error of 0.02 mm/m span may be allowed.

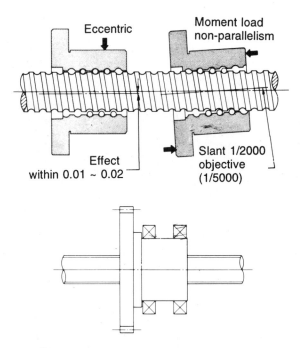

**Fig. 6.15** Eccentric load in ballscrews

(b) For alignment, a precision mandrel must be used and not the ballscrew. The mandrel support may be designed to suit the ballscrew support. Preloading of the support bearing must be ensured before alignment. Any error in the alignment may be adjusted by scraping the bearing support bracket seating or else it has to be adjusted within the screw clearance hole on the bearing mounting or bracket.

(c) The ball nut seating surface has to be prepared in assembly, taking the seating surface as the master reference. The allowable tilt or the slope in mounting is $\frac{1}{5000}$. If the nut outside the diameter is guided, it should freely enter the bore. Any shift in the axis will result in an interference and a radial misalignment force on the ballscrew and nut assembly.

(d) The fixing holes must match with the clearance holes on the ball nut. Any forceful assembly of mismatched holes induces a moment on the ball nut which will affect the accuracy and life of the ballscrew.

## *Effects of Ballscrew Misalignment (Fig. 6.16)*

If the ballscrew is not aligned with respect to the guideways, a moment load is induced on the ballscrew assembly resulting in:
  (a) Loss of accuracy
  (b) Temperature rise
  (c) Deterioration of fatigue life

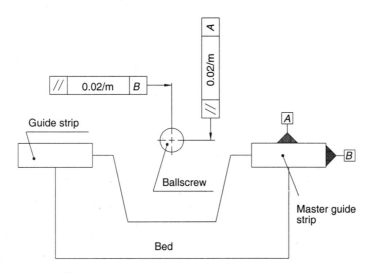

**Fig. 6.16** Allowable ballscrew misalignment

(d) Sticky and unsteady motion near the support ends
(e) Pitching in straightness of motion if guideways are not rigid enough or if guideways have a clearance

### 6.2.4 Assembly Technique of Ballscrew Bearing Mounting

To explain the assembly procedure of a ballscrew end bearing mounting, the following are included, drawing a reference as specified by the popular bearing manufacturer viz., INA (Germany). For a bearing assembly of ballscrews the tolerances and surface quality requirements of the housing and the shafts are important and critical. These are indicated in Figs 6.17; and 6.18. The values of these depend on the bearing size and the application, both of which are usually specified by the manufacturers in their product catalogues.

*Fitting and Dismantling*

***Axial angular contact ball bearings*** When mounting bearings, it should be ensured that the mounting forces are not directed through the rolling elements or sealing rings.

Fixing screws for flange mounted bearings should be tightened in a crosswise sequence. The fixing screws may be tensioned to a value up to 70% of their elastic limit.

For quick and easy dismantling of INA bearings of series ZKLF...2RS an extraction groove is provided on the outside surface of the outer ring.

***Needle roller and axial cylindrical roller bearings*** Before mounting the bearings, the closing plugs supplied with the bearings must be pressed into the redundant lubrication holes (Fig. 6.19).

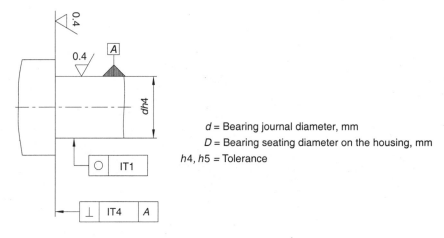

Fig. 6.17  Tolerance and surface quality requirements of the shafts

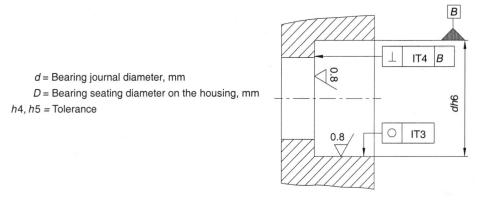

Fig. 6.18  Tolerance and surface equality requirements of the housing

Care should be taken not to mix individual bearing items from different assemblies during installation.

Fixing screws for flange mounted bearings should be tightened in a crosswise sequence. The fixing screws may be tensioned to a value up to 70% of their elastic limit.

**Locknuts**  The bearings are preloaded by tightening the locknut. The locknut tightening torque values should be as per the recommendations of the bearing supplier.

After tightening the locknut, both the socket set screws must be tightened with a socket wrench. It is recommended that these screws be tightened alternately. In order to counteract the setting phenomena, it is recommended to first tighten the locknut to a torque value, three times the recommended value and then to loosen it again. The locknut should then be re-tightened in accordance with the recommended torque value.

While dismantling, the order is reversed and both the socket set screws are loosened first, followed by the locknuts. INA locknuts can be re-used several times if proper fit-

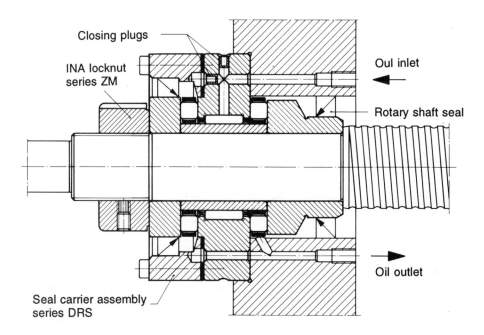

**Fig. 6.19** Mounting of needle roller and axial cylindrical roller bearings

ting and dismantling procedures are followed. The dimensions of the inner rings of INA bearings are matched so that on tightening the locknut a certain preload is produced in the bearings which is adequate for most applications.

If the frictional torque of the bearing is measured, the value obtained should be compared with the specified value for the bearing. When checking the frictional torque of needle roller and axial cylindrical roller bearings, care should be taken to ensure that the raceways are properly lubricated.

*Lubrication* Bearings for screw drives can be operated safely with grease lubrication or with oil lubrication (lubrication method: recirculating oil, oil bath or oil pulse). It should be noted that with grease or oil pulse lubrication no heat can be dissipated from the bearing and the bearing operating temperature in machine tools should not exceed 50 °C.

Axial angular contact ball bearings of series ZKLN...2RS, ZKLF..2RS and VKLF...2RS are supplied, lubricated for life with a grease KTCPE 2K to DIN 51 826. In this way all the disadvantages which can occur with relubrication, due to overgreasing or choosing an unsuitable type of grease are avoided. However, in unfavourable operating conditions the existing lubrication holes in bearings ZKLN...2RS and ZKLF...2RS make relubrication possible with the above mentioned grease.

If oil lubrication is provided, the limiting speed can be 1.5 times the limiting speed specified for grease lubrication.

For INA bearings of series ZARN and ZARF, oil lubrication should be preferred. Choice of the lubrication method (heat dissipation) and the oil viscosity must be made in accordance with the operating conditions. Proven media are lubrication oils HLP (DIN 51 525) and CLP (DIN 51 517) of ISO VG 32 to 100.

With grease lubrication, the following aspects must be considered with regard to the particular operating conditions: the type of grease and method of delivery, the quantity of grease and the relubrication interval. Proven lubricants are greases K2K or KTCPE 2 K to DIN 51 826. Grease replacement and distribution is best achieved if relubrication is carried out with the bearing rotating and in warm condition. Guidelines for lubrication are also given in Annexure 6.2.

### 6.2.5 Ballscrew Failures due to Improper Assembly

*Ballscrew Problems*

***Too much play*** The play is because of no preload or not enough preload. The ball nut will rotate and move downwards by its own weight when a non-preloaded ballscrew assembly is held vertically. A significant backlash may exist in a non-preloaded ballscrew unit. Therefore non-preloaded ballscrews are used only in those machines where there is a low operating resistance, and the positioning accuracy is not important.

***Inappropriate bearing installation*** The following are some examples:
  (a) The bearing is not attached to the screw shaft properly causing axial play under load. This problem may be caused if the bearing journal of the screw shaft is too long or the threaded part of the screw shaft is too short.
  (b) The perpendicularity between the seating face and the thread axis of the bearing locking nut or the parallelism between the opposite faces of the locking washer is out of the tolerance limits, causing the bearing to tilt. The thread and seating face of a bearing locking nut should be machined in one setting to ensure the perpendicularity. It is even better if they can be ground.
  (c) Two locking nuts and a spring washer should be used in the bearing installation to prevent them from getting loose in operation.

*Improper Mounting of Ball Nut Housing and Bearing Housing*
  (a) Components may loosen due to vibration. Solid pins should be used for locating purposes.
  (b) Ball nut fixing screws are not seated firmly because of the use of a longer screw or due to a shorter tapped hole depth in the housing.
  (c) Ball nut fixing screws loosen due to vibration, since spring washers are not used.

***Parallelism or flatness of the guideway—out of tolerance*** In a machine assembly, a mandrel is frequently used for alignment purposes. The clearance of table movement may vary at different locations if the parallelism or flatness of any of those matching components is out of tolerance, no matter whether they are ground or scraped.

### Misalignment of the Motor and the Ballscrew Shaft Coupling

(a) There will be a relative rotation between the motor shaft and the ballscrew shaft if they are not coupled firmly by the coupling or the coupling itself is not rigid enough.

(b) Driving gears are not engaged properly or driving mechanism is not compatible. A timing belt should be used to prevent slipping if the ballscrew is to be driven by a belt.

(c) Locating key is too loose in key seat. Any mismatch among the hub, key, and key seat may cause those components to get loosened and generate a backlash.

### Foreign Objects Trapped inside the Ball path

(a) *Packing material is trapped in the ball path*  Various materials and anti-rust paper are normally used to pack the ballscrew units for shipment. It is possible that some of these foreign materials or other objects are trapped inside the ball path, if proper procedures are not followed while installing or aligning a ballscrew unit. This may cause the bearing balls to slide instead of rolling, or even cause the ball nut to jam up completely.

(b) *Machined chips are trapped in the ball path*  The chips or dust generated during machining processes may be trapped inside the bearing ball path if wiper kits are not used to keep them away from the rubbing surface of a ballscrew unit. This may cause unsmooth operation, deteriorated accuracy and reduced service life.

**Over-travel**  Over-travel may occur during the set-up procedure in the machine or as a result of a limit switch failure or a machine collision. To prevent further damage, an over-travelled ballscrew unit should be checked or repaired by the manufacturer before reinstallation.

An over-travel can damage the returning tube and cause it to collapse or even break. When this happens, the bearing balls will not circulate smoothly. They may break and damage the groove on the ballnut or the ballscrew shaft under severe circumstances.

**Misalignment**  A radial load will develop if the axis of the ballnut housing and the screw shaft are not aligned properly. The ballscrew unit may bend at either end of travel if this misalignment exceeds limits. An abnormal wear may result even if the misalignment is not significant enough to cause bending. The accuracy of a ballscrew unit will deteriorate rapidly if it remains misaligned. Higher the preload, more is the precision required in the alignment of the ballscrew assembly.

### Misalignment of Ball Nut Assembly

A ball nut will exert abnormal load if it is tilted or misaligned. When this problem arises, the motor will get overloaded in both the directions of rotation.

## Causes and Precautions against the Failure of Ballscrew and Nut

**Broken bearing ball** The temperature of an under-lubricated or non-lubricated ballscrew unit rises substantially during operation. This temperature elevation could make the bearing balls brittle or even cause the balls to break, resulting in damage to the grooves of the ball nut or the ballscrew shaft.

Therefore, lubricant replenishment should be considered during the design process. If an automatic lubricating system is not available, a periodical replenishment of grease should be done as part of the maintenance programme.

**Collapsed or broken return tube** An over-travel of the ball nut or an impact on the return tube could cause the return tube to collapse or break. This may block the pathway of the bearing balls, causing them to slide rather than to roll. The balls will then eventually break.

**Ballscrew shaft-end breaks** *Inappropriate design* Sharp corners and undercuts on the ballscrew shaft should be avoided to reduce local stress concentration. Figure 6.20 shows some of the appropriate screw end designs.

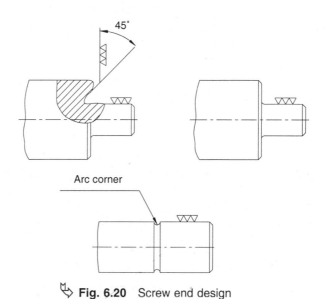

**Fig. 6.20** Screw end design

*Bent screw shaft-end* It is possible that the seating surface and the thread axis of the bearing locking nut are not perpendicular to each other or the opposite surfaces of the locking washer are not parallel to each other. This will cause the end of the screw shaft to bend and eventually break. The amount of deflection at the end of the ballscrew shaft should not exceed 0.01 mm after tightening the locking nut (Fig. 6.21).

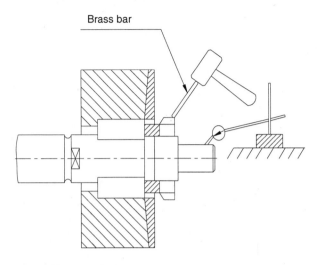

**Fig. 6.21** Bent screw shaft-end

*Radial force or fluctuating stress (fatigue)* Misalignment of a ballscrew installation creates abnormal shear or a fluctuating stress.

The following measurement procedures can be adopted to locate the cause of abnormal backlash in a ballscrew installation.

- Glue a gauge ball in the centre hole at one end of the screw shaft. Use the flat plate of a dial indicator to check the axial movement of this gauge ball in the axial direction, while rotating the screw shaft [Fig. 6.22(a)]. The movement should not exceed 0.003 mm, if the bearing hub, ball nut, and ball nut mounting housing are all installed properly.
- Use a dial indicator to check the relative movement between the bearing housing and the bearing hub while rotating the ballscrew [Fig. 6.22(b)]. Any dial indicator reading other than zero indicates that the bearing housing is either not rigid enough or is not installed properly.
- Check the relative movement between the machine table and the ball nut mounting housing [Fig. 6.22(c)].
- Check the relative movement between the ball nut mounting housing and the ball nut flange [Fig. 6.22(d)].
- Contact the ballscrew manufacturer if an unsatisfactory backlash still exists while all the above checks are correct. The preload or the rigidity design of the ball screw unit may have to be increased.

## 6.2.6 Thermal Displacement of Ballscrew

When a ballscrew is operated at high speed during rapid traverse, in the presence of preload, the temperature of the ballscrew rises.

The temperature rise causes a thermal displacement, calculated by the following formula:

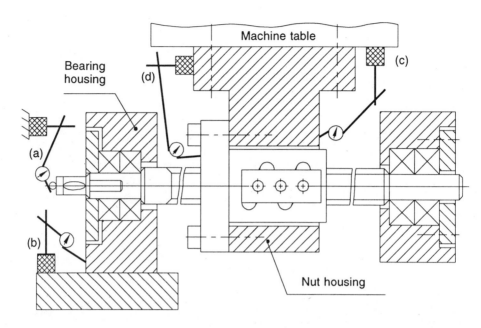

**Fig. 6.22** Locating the cause of abnormal backlash in a ballscrew installation

$$\Delta L_0 = POL$$

where, $\Delta L_0$ = thermal displacement
$P$ = coefficient of thermal expansion for steel 11.7 µm/m/°C
$O$ = temperature rise of screw shaft
$L$ = screw length

The thermal displacement for a 1 m long ballscrew and 1 °C temperature rise will be 11.7 µm. This rise will result in a positioning inaccuracy.

The counter measures are:

(a) Optimisation of preload of the ballscrew and the support bearing
(b) Correct selection of lubricant grade and quality
(c) Use of high lead ballscrew so as to reduce the rotational speed.
(d) Use of hollow screw and forced cooling
(e) Cooling of screw outside by air blast
(f) High speed warming up for use in a temperature stabilised state
(g) Use of minus target value of cumulative lead
(h) Thermal compensation of pitch error
(i) Stretching the ballscrew

Figure 6.23 shows the effect of forced cooling of hollow ballscrew and temperature rise in normal application.

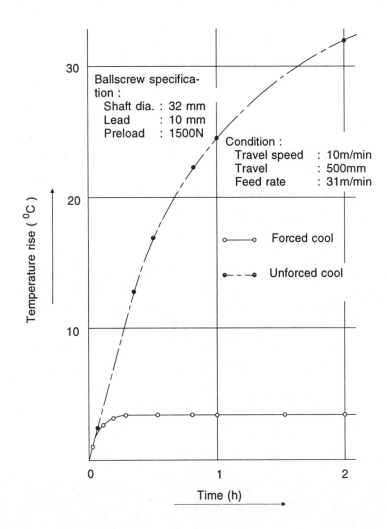

**Fig. 6.23** Effect of forced cooling of hollow ballscrew and temperature rise

## 6.3 FEEDBACK ELEMENTS

### 6.3.1 Mounting Instructions for Linear Scale

*Recommended Mounting Positions of Linear Scale Assembly (Fig. 6.24)*

The positions 'a', 'b' and 'c' offer protection against dust, chips, etc. Protection against fluid penetration into the measuring system is given in position 'a'. Position 'd' offers insufficient protection and should therefore be avoided.

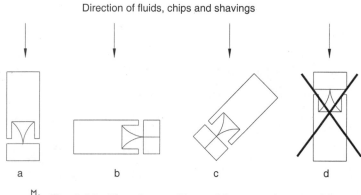

**Fig. 6.24** Mounting positions of linear scale assembly

### Precautions for Mounting the Scale Unit (Fig. 6.25)

Prepare mounting surfaces for scale unit. They should be free from paint. The mounting surfaces must be parallel to the machine guide within a tolerance of 0.25 mm and be in level with each other within 0.25 mm. For longer length of scales—1 m and above intermediate fixing brackets and support brackets may be used for alignment. After alignment, screws of these brackets should be tightened.

### Mounting the Scanning Head (Fig. 6.26)

Prepare mounting surfaces for the scanning head. The surfaces should be free from paint and must be parallel to the scale unit, otherwise the scanning head may either be caught between the cover and the housing or may get severed through.

### 6.3.2 Mounting of Incremental Angle Encoder

#### Mechanical Installation

Connection of the encoder shaft can be carried out via the metal bellows coupling or the precision diaphragm coupling (Figs 6.27 and 6.28).

If the encoder is to be driven with a gear, care must be taken to ensure that the gear wheel is precisely mounted onto the shaft. No backlash should be present while changing the direction of rotation. Under no circumstances should the permissible axial or radial shaft load be exceeded.

### 6.3.3 Assembly Care of Mounting of Proximity Switch

Before mounting a proximity switch, it is important to know whether the switch is of shielded construction or of unshielded construction. Also, switching distance, working distance, repeatability, switching hysteresis for proper and reliable signal generation over a long period are the important parameters to be known.

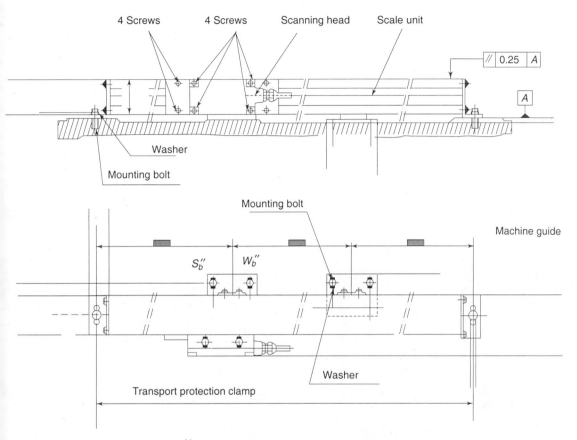

**Fig. 6.25** Mounting the scale unit

*Shielded construction* Shielded proximity switch can be mounted on a metallic material with no loss in sensor performance as long as the active surface of the sensor either protrudes, or is mounted flush with the metal surface. For standard models, each sensor must have at least one diameter of clearance between it and the other sensing surfaces (Fig. 6.29).

*Unshielded construction* Proximity switch models of this type require a metal free zone of three times its diameter, around its active sensing surface. The other alternative is to provide a nonmetallic insert with the required dimensions of the metal free zone (Fig. 6.30).

*Switching distance* The switching distance is that distance where an electrical state change is induced in the proximity switch when a target has come close enough to be recognised by the proximity sensor (Fig. 6.31).

*Normal switching distance ($S_n$)* The normal switching distance is defined as that distance which does not include variations due to temperature and voltage fluctuations, mounting media, and manufacturing tolerances.

# Mechatronics

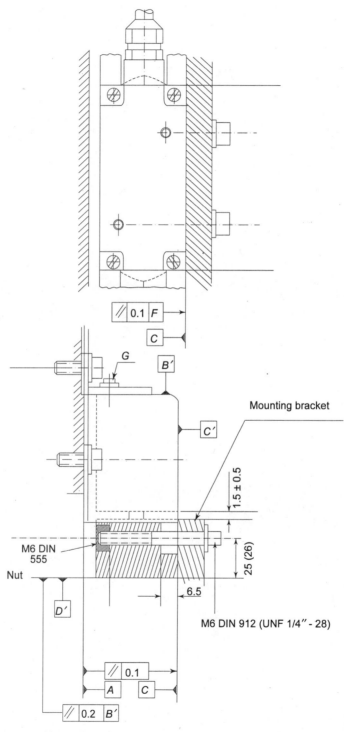

**Fig. 6.26** Mounting the scanning head

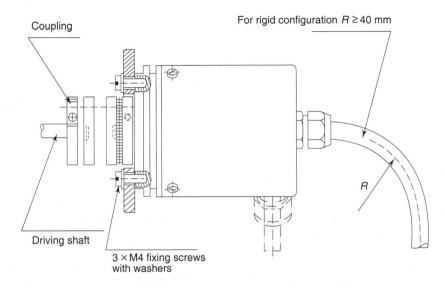

**Fig. 6.27** Connection of the encoder shaft via diaphragm coupling (by means of fixing holes)

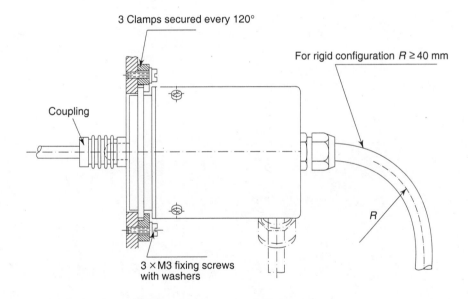

**Fig. 6.28** Connection of the encoder shaft via the metal bellows coupling (by means of clamps)

*Effective switching distance* $(S_r)$  The effective switching distance for an inductive proximity switch is that distance which accounts for permissible production tolerance variations at defined voltage and temperature conditions.

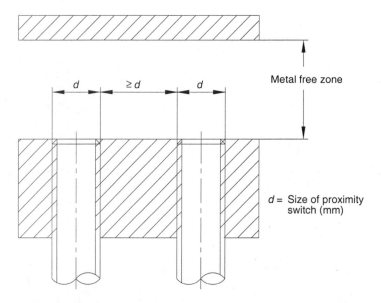

**Fig. 6.29** Mounting of proximity switch-shielded construction

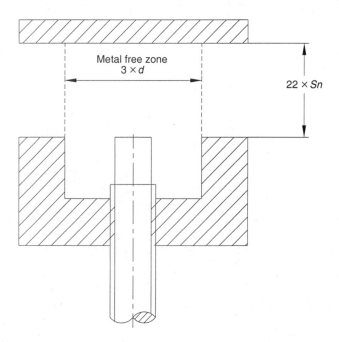

**Fig. 6.30** Mounting of proximity switch—unshielded construction

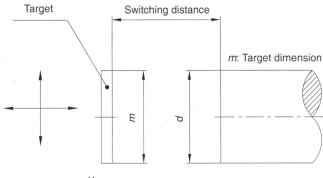

**Fig. 6.31** Switching distance

*Usable sensing distance*   The usable sensing distance is that distance of a proximity switch which is being measured over permissible temperature and voltage conditions.

*Working distance ($S_a$)*   The working distance is the distance which guarantees safe operation of the proximity switch under established temperature and voltage conditions. It is defined as being between 0 and 81% of the normal sensing distance.

## Target Considerations

For all published switching distances, a standard square target is made from 1 mm thick steel material with a side length $m$ equal to the diameter $d$ of the sensor's active surface.

Spherical or smaller targets will result in a smaller working distance and hence the normal sensing distances will be reduced. On the other hand, thin foils or large targets will slightly increase the sensing distance.

The use of other metals will reduce the switching distances. The following is a list of some common materials with their calculable correction factors (Fig. 6.32).

| | |
|---|---|
| Stainless Steel | 0.8 $S_n$ |
| Aluminium | 0.4 $S_n$ |
| Brass | 0.5 $S_n$ |
| Copper | 0.3 $S_n$ |

## Repeatability (R)

The repeatability of the usable switching distance is measured by performing two consecutive measurements within eight hours at a housing temperature of 15 to 30°C. During this test period, the supply voltage variation must not exceed +5%.

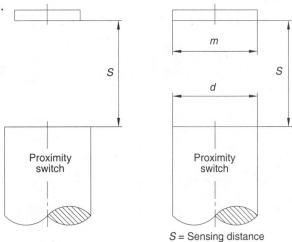

**Fig. 6.32** Target considerations in switching distance

## Switching Hysteresis (H)

The switching hysteresis is the difference between the switching activation distance, with the target approaching the active surface of the proximity switch and the deactivation distance, when the target is moved away from the sensing surface (Fig. 6.33).

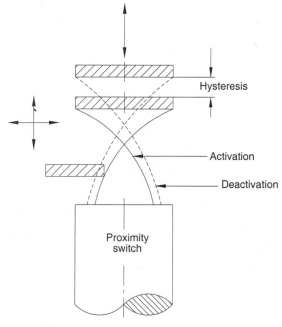

**Fig. 6.33** Switching hysteresis

## Switching Frequencies

The switching frequency of a proximity switch becomes a limitation when used to count the teeth on the spindle (Fig. 6.34). The product of the number of teeth and spindle speed (rotation/sec) gives the switching frequency to which the proximity switch should cater. Table 6.1 gives the switching frequency of proximity switches for dc and ac models. Figure 6.34 also gives the proportion of tooth and gash for such applications.

## 6.4 SPINDLE BEARINGS

The overall performance of a machine in terms of accuracy and productivity depends on the machine design including the spindle bearing system of the tool or the work spindle. The work or tool spindle and the bearing arrangements links the flow of force through the machine tool. As such the machine design and the spindle bearing system must be carefully adapted to one another.

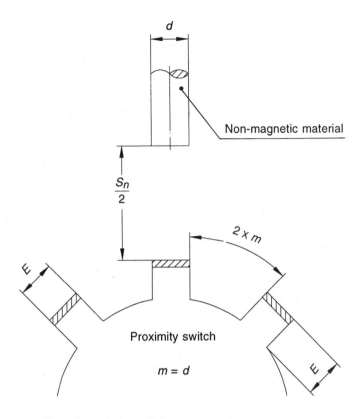

d = Size of proximity switch
E = Width of tooth
m = Gash width
$S_n$ = Normal sensing distance

**Fig. 6.34** Switching frequency

In addition to the cutting capacity, the standard of working accuracy is a major criterion of the machine tool. The working accuracy of a spindle depends on its radial and axial true running accuracy, static and dynamic rigidity and thermal behaviour. The basic requirements of a CNC machine tool spindle assembly are as follows:

(a) True running accuracy of spindle assembly—radial and axial
(b) Rigid, both statically and dynamically
(c) Good damping properties and dynamic behaviour to reduce chatter vibrations, even at high speeds.
(d) Good attainable speeds.
(e) Good thermal behaviour and optimum operating temperature.
(f) Good service life of spindle bearings.

## TABLE 6.1
### Switching frequency of proximity switches

| Normal switching distance | | Hz |
|---|---|---|
| Shielded (mm) | Unshielded (mm) | |
| Switching frequencies for dc models | | |
| 1 | — | 1000 |
| 2 | — | 800 |
| — | 4 | 400 |
| 5 | — | 500 |
| — | 8 | 200 |
| 10 | — | 300 |
| 15 | 15 | 100 |
| 20 | 20 | 100 |
| — | 25 | 100 |
| 25 | — | 50 |
| Switching frequencies for a models | | |
| 2 to 50 | — | 50 |

The factors influencing the working accuracy of a spindle assembly are given in Fig. 6.35. The true running accuracy of a spindle depends on the bearing class and on

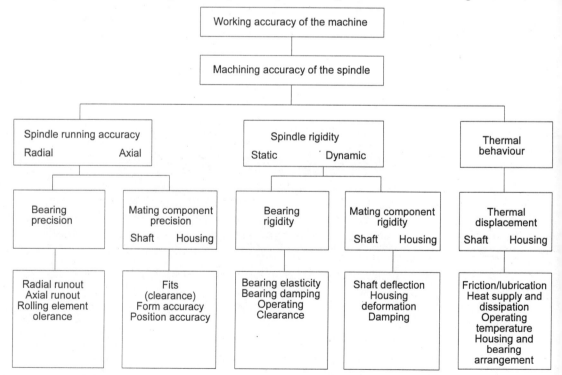

**Fig. 6.35** Factors influencing the working accuracy of a spindle assembly

the precision of the mating component. Figure 6.36 gives the bearings which are most frequently used for precision and high precision spindle bearing arrangements and the corresponding precision class of the mating components.

| | | | Precision bearing arrangement | High-precision bearing arrangement |
|---|---|---|---|---|
| Rolling bearing precision | | | | |
| | Tapered roller bearing | | P5, SP | |
| | Cylindrical roller bearing, double row | | SP | UP |
| | Angular contact thrust ball bearing double direction | | SP | UP |
| | Spindle bearing | | P4 | HG, P2 (T9) |
| Mating component accuracy | | | | |
| Spindle | Roundness | ○ | IT0 | IT01 |
| | Parallelism | // | IT1 | IT0 |
| | Run-out | | IT2 | IT1 |
| | Coaxiality | ◎ | IT3 | IT2 |
| Housing | Roundness | ○ | IT! | IT0 |
| | Parallelism | // | IT2 | IT1 |
| | Run-out | | IT3 | IT2 |
| | Coaxiality | ◎ | IT4 | IT3 |
| Attainable running accuracy of the spindle | | | | |
| | d max. 120 mm | | 2–5 µm | 1–3 µm |

**Fig. 6.36** Attainable running precision of a spindle as a function of the precision of the rolling bearings and the mating components

The following bearing arrangements for a spindle are frequently used on machine tools:
(a) Machine tool spindles supported by two tapered roller bearings in O arrangement. These bearings can take both the axial and radial loads and are most commonly used on turning centres (Fig. 6.37).
(b) Spindle bearing arrangement with two double-row cylindrical roller bearings and one double direction angular contact thrust ball bearing, which provides independent support for radial and axial loads. These bearings are suitable for high speed, available in ultra precision grade and are suitable for precise applications such as fine boring machine, tool room milling machine and machining centres (Fig. 6.38).

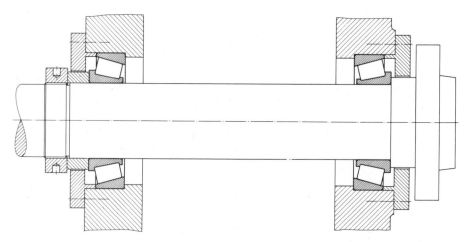

**Fig. 6.37** Machine tool spindle supported by two tapered roller bearings in O arrangement; the bearings accommodate radial and axial loads

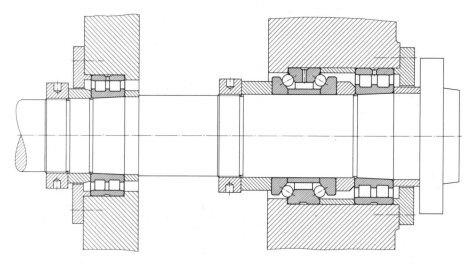

**Fig. 6.38** Spindle bearing arrangement with two double row cylindrical roller bearings and one double direction angular contact thrust ball bearings separate accommodation of radial and axial loads

(c) Spindle bearing arrangement with spindle bearings for combined loads incorporating preloaded angular contact bearings at the nose-end and cylindrical roller bearings at the drive-end. This is the most common arrangement for machining centres, CNC lathes, milling machines and wheel head of grinding machines (Fig. 6.39)

The comparative characteristics of these bearing arrangements are given in Figs 6.40–6.43.

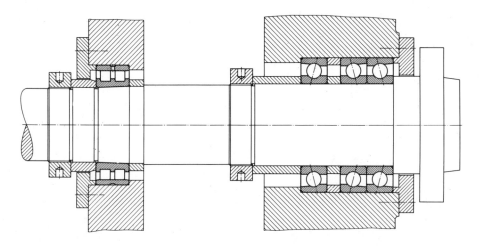

**Fig. 6.39** Spindle bearing arrangement with spindle bearings for combined loads at the work end; the double row cylindrical roller bearing at the tail end accommodates radial loads only

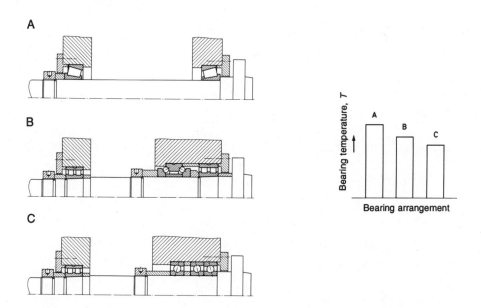

**Fig. 6.40** Comparison of the anticipated bearing temperatures of variants A (tapered roller bearings). B ball bearing, and C (spindle bearings/cylindirical roller bearing)

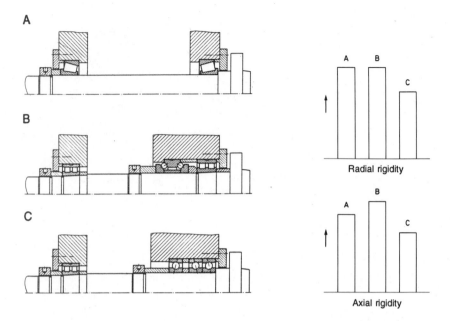

**Fig. 6.41** Overall radial and axial rigidity of three different spindle-bearing systems

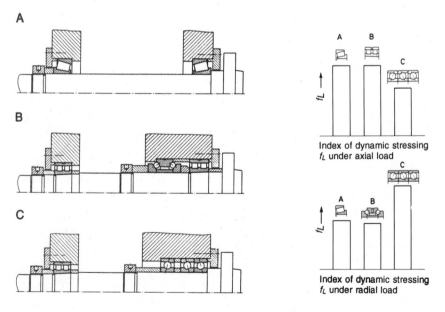

**Fig. 6.42** Rating life as a function of the spindle bearing assembly and the type of loading

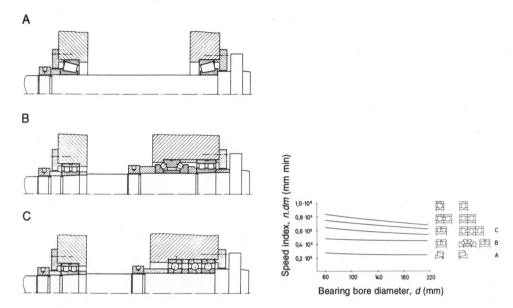

**Fig. 6.43** Attainable spindle speeds with grease lubrication as a function of bearing size, type, and arrangement

## 6.4.1 Mounting Accuracy of Spindle and Housing

Figure 6.44 shows the accuracy of form for spindles and Fig. 6.45 shows the accuracy of form for housing.

## 6.4.2 General Assembly Precautions for Assembling Spindle Bearings

*Storage*

Spindle bearings should be stored in a dry, clean room with a constant temperature if possible, and with a relative humidity not higher than 65%. Bearings are supplied with a preservative (mineral oil based anti-corrosive agent or dry vapour phase corrosive inhibitor (VCI paper preservative) and should not be removed from the original package until immediately before assembly.

Once bearings have been removed from a package, the packaging should be carefully resealed immediately, since the protective vapour phase of the VCI paper can only be maintained inside the sealed package. The bearings which were removed must be greased or oiled immediately.

The storage period for greased and sealed bearings should not exceed three years, even under ideal storage conditions.

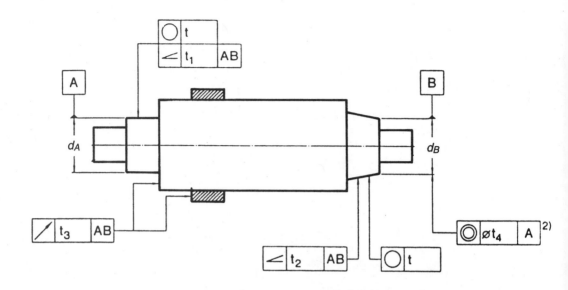

| Characteristic | Symbol for characteristic | Tolerance zone[3] | Permissible deviations Tolerance grade Bearings of tolerance class | | | Hydrostatic bearings | |
|---|---|---|---|---|---|---|---|
| | | | P5 | P4, SP | PA97, | PA9, UP | |
| Circularity | ○ | $t$ | $\dfrac{IT3}{2}$ | $\dfrac{IT2}{2}$ | $\dfrac{IT1}{2}$ | $\dfrac{IT0^{(4)}}{2}$ | $\dfrac{IT1}{2}$ |
| Angularity | ∠ | $t_1$ | $\dfrac{IT3}{2}$ | $\dfrac{IT2}{2}$ | $\dfrac{IT1}{2}$ | $\dfrac{IT0^{(4)}}{2}$ | $\dfrac{IT1}{2}$ |
| | ∠ | $t_2$ | — | $\dfrac{IT3}{2}$ | $\dfrac{IT2}{2}$ | — | — |
| Run-out | ╱ | $t_3$ | IT3 | IT3 | IT2 | — | IT1 |
| Coaxiality | ○ | $t_4$ | IT5 | IT4 | IT3 | — | IT3 |

(1) The spindle form tolerances, the symbols and the reference surfaces are in accordance with ISO/R 1101.
(2) The bearing seatings are generally too narrow to provide a sufficient basis for checking the coaxiality between the seatings. This tolerance only applies, therefore, if the seating has suffcient width to qualify as a reference surface.
(3) When determining the tolerances for the permissible form deviation shaft diameters $d_A$ and $d_B$ are reference dimensions: e.g., for a bearing of tolerance class P5 and a shaft diameter of 50 mm, the tolerance zone for the circularity $t = IT3/2 = 5/2 = 2.5$ μm.
(4) IT0 is preferred when the dimensional tolerance for the bearing seating is of grade IT3.

**Fig. 6.44** Accuracy of form for spindles

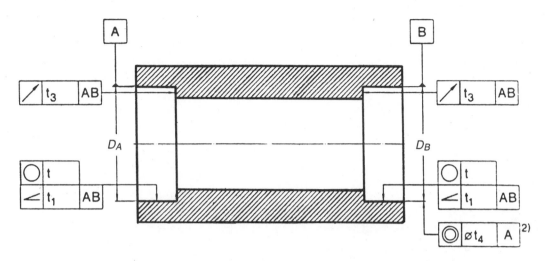

| Characteristic | Symbol for characteristic | Tolerance zone[3] | Permissible deviations Tolerance grade Bearings of tolerance class P5 | P4, SP | PA97, PA9, UP | Hydrostatic bearings |
|---|---|---|---|---|---|---|
| Circularity | ○ | $t$ | $\frac{IT3}{2}$ | $\frac{IT2}{2}$ | $\frac{IT1}{2}$ | $\frac{IT1}{2}$ |
| Angularity | ∠ | $t_1$ | $\frac{IT3}{2}$ | $\frac{IT2}{2}$ | $\frac{IT1}{2}$ | $\frac{IT1}{2}$ |
| Run-out | ↗ | $t_3$ | IT3 | IT3 | IT2 | IT1 |
| Coaxiality | ◎ | $t_4$ | IT5 | IT4 | IT3 | IT3 |

(1) The housing form tolerances, the symbols and the reference sufaces are in accordance with ISO/R 1101.
(2) The bearing seatings are generally too narrow to provide a sufficient basis for checking the coaxiality between the seatings. This tolerance only applies, therefore, if the seating has sufficient width to qualify as a reference surface.
(3) When determining the tolerances for the permissible form deviation housing bore diameters $D_A$ and $D_B$ are reference dimensions: e.g., for a bearing of tolerance class P5 and a housing bore of 80 mm, the tolerance zone for the circularity $t = IT3/2 = 5/2 = 2.5$ μm.

**Fig. 6.45** Accuracy of form for housings

Bearings should be regreased if:

- The colour of grease has changed
- Oil has separated from grease (oil stains on the packaging material)
- The lubricating grease has hardened (drying cracks) or the bearing is hard to turn.

## Cleaning

The anti-corrosive preservative used for oil-preserved bearings is compatible and miscible with lubricating greases and mineral based oils; therefore, as a rule, the bearings will not require special treatment before mounting. However, if any dirt or foreign material is found on the bearing, it may be washed using organic cleaning media such as acid and water free kerosene or naphtha. Chlorinated hydrocarbons are not recommended for washing since they increase the risk of corrosion and damage the seal.

## Assembly Area

The assembly area must be free from dirt and dust. Contaminants can damage rolling bearings. Automatic dust concentration measuring instrument FH 62 (FAG, Germany make) may be used for controlling the dust and dirt in the assembly area. Moisture and aggressive environments lead to corrosion. Abrasive tools such as files and sand paper should be kept away from the assembly areas; the same applies to linty cleaning rags. It is best to use light-coloured, clean and non-fibrous surfaces (e.g., plastic) on which the rolling bearings and tools can be neatly laid out. Good lighting will make assembly easier.

## Preparing for Assembly

The assembly sequence for the individual parts depends on the design. A drawing of the design often indicates equipment and additional tools which will facilitate assembly. All accessories must be cleaned.

Another prerequisite for proper bearing operation is dimensional accuracy of the matching parts; their surface quality and geometrical accuracy should be checked.

Radial running accuracy checking rig may be used to check the accuracy of bearings. Figure 6.46 shows the rig set-up and a talyrond plotter diagram for FAG spindle bearing B7017C TPAP4.

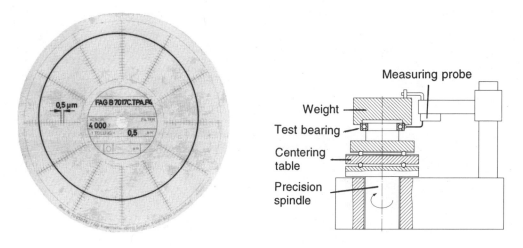

**Fig. 6.46** Radial running accuracy of a bearing

The contact surfaces on the shaft step or on the shoulder of the housing bore must be perpendicular to the cylindrical mounting surfaces and free from dirt. Excessive side run-outs in these contact surfaces or in the snap rings can cause the bearings to jam.

Hands should be clean and dry for assembly; sweat can cause corrosion. This is particularly important for bearings with a dry preservative treatment. In certain circumstances, cotton gloves must be worn. Oily preservation agents can neutralise sweat from hands.

## Assembly—General Information

(a) Assembly sleeves made of copper or brass, with a uniform contact surface over the entire end circumference, are well suited for assembly. A press will speed up the assembly of the bearings. If a press is not available, bearings can also be mounted with light, carefully centred taps on the mounting sleeve. In no case should mounting forces be directed through the rolling elements and direct impacts should never be applied to the bearing rings.

(b) The mounting of inner and outer rings is made easier if the seating surfaces are lightly oiled or rubbed with a solid lubricant, and designed with a 15–30° lead chamfer. The easiest way of mounting inner rings or bearings on a shaft, particularly with a light fit, is to first heat them with an induction heating unit. If such a heating unit is not available, the bearing can also be heated to approximately +80 °C in an oil bath or heating oven.

(c) Freezing of the bearing for easier mounting into a housing is not recommended, since condensed water can cause corrosion in the bearings and bearing seats.

(d) Greased bearings should not be heated.

(e) Flame heating is not permissible, since it produces local heating of the material, which results in a loss of hardness. In addition, deformations occur within the bearings.

(f) If the bearing rings are tilted during assembly, they must be carefully disassembled and refitted. A shaft should be inserted into the assembled bearing with a slight rotary motion; this reduces the risk of damage to the raceways, the rolling elements and the seals.

(g) When rolling bearings with a lubrication hole are installed, it should be ensured that the lubrication hole is located in the unloaded zone.

To prevent excessive re-lubrication pressures, lubrication holes in the housing or shaft should be aligned with those in the bearing.

After the bearing is mounted and lubricated, a functional test of the assembly is recommended.

## Functional Test

(a) When assembly is complete and the roller bearing has been lubricated, the shaft should initially be rotated by hand. It should run smoothly if no contact-type seals are present. The rotation speed can then be increased, while the operating characteristics are observed. For high-speed bearings, the unit should be run at

full speed and full output only after run-in is complete (once the initial higher temperature has dropped back to a constant level).
(b) Adapter sleeves, end flanges, and locknuts must be re-tightened after the test running period.
(c) A simple metal or wooden rod can be used to listen to the bearings. The same person should listen to the bearings, so that subtle abnormal noises are correctly recognised. The different types of noises can be interpreted as follows:
- A steady hum indicates that the rolling bearing is running properly.
- A scraping noise indicates the presence of contaminants in the bearing.
- A rumbling noise is attributable to damaged rolling elements or raceways.
- A metallic ringing or whistling sound indicates inadequate lubrication or insufficient operating clearance.
(d) Operating clearance should always be checked once the roller bearing is installed and rechecked after the bearing has operated for a short time.

## Dismantling

Provision for dismantling a rolling bearing should be made during the design stage of the bearing arrangement. If the rings are mounted with a tight interference fit, grooves should be provided shaft or housing to facilitate easy removal of rings with an extraction device. If the bearing is to be reused, direct impact on the bearing rings should be avoided and disassembly forces should not be applied through the rolling elements. For adapter sleeves, the sleeve nuts should be loosened several turns after the sleeve is removed, and the inner ring should be pounded loose by hammering a special impact adapter against the opposite side of the cone.

Before any roller bearing is reused, it should be thoroughly cleaned, preferably while disassembled. Never use hard or pointed objects to remove solidified lubricants or other impurities: these residues should be softened by placing the bearing in a solvent bath. Ultrasonic units, rotation, spraying, and wiping will all speed up the cleaning process. To prevent damage, dirty bearings should never be turned, even by hand. After every cleaning, the bearing should be blown out with dry, clean compressed air.

### 6.4.3 Misalignment in Bearing Assembly

Misalignment in a bearing arrangement is a result of the following:
(a) The bearing seating on the spindle is out of concentricity and squareness beyond permissible value.
(b) The bearing seating on the housing is out of concentricity and squareness beyond permissible value.
(c) Shaft or spindle is bent as a result of improper handling during assembly.
(d) Inaccuracy in associated components of bearing assembly like preloading spacer, locknuts, etc.
(e) Overhanging load exceeding the spindle capacity.

## 6.4.4 Effects of Misalignment

(a) The perfectly aligned bearing arrangement will have an equal load gradient over the entire bearing width. If misalignment exist, there will be local concentration of load on the roller end (Fig. 6.47), resulting in a decrease in life against the nominal value.

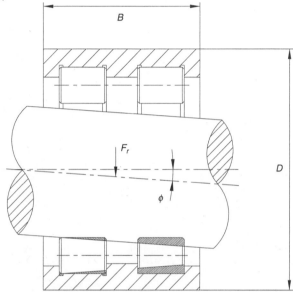

**Fig. 6.47** Effect of misalignment

(b) Misalignment increases the operating temperature of the bearing assembly due to an unequal load gradient. This causes undue heating of the spindle, which results in an unsteady temperature, thereby hampering the working accuracy.
(c) The misalignment causes undue run out on the spindle nose, affecting the accuracy and also causing noise and vibration.

The allowable misalignment for different type of bearing arrangement is given in Table 6.2.

## 6.4.5 Noise and Vibration—Causes, Effects and Reduction Measures

The causes of bearing noise and vibration can be divided into two categories—(i) Direct causes (ii) Indirect causes.

### Direct Causes
- Elastic deformation
- Surface defects
- Influence of lubrication
- Friction effects of cages and seals

**TABLE 6.2**

| Type of bearing | Angular misalignment | | Slope |
|---|---|---|---|
| | (degree/min) | (Radians) | |
| Deep groove ball bearings (62/63/64) | 5'–10' | 0.00145–0.0029 | 1.5–3 µ/mm |
| Deep groove ball bearing (618/160/60) | 2'–6' | 0.006–0.0017 | 0.6–1.7 µ/mm |
| Self-aligning ball bearing (open) | 4° | 0.07–0.026 | 70 µ/mm |
| Self-aligning ball bearing (sealed) | 1.5° | 26 µ/mm | |
| Cylindrical roller bearings | 2'–3' | 0.006–0.0009 | 0.6–0.9 µ/mm |
| Taper roller bearings | 1'–2' | 0.0029–0.006 | 0.3–0.6 µ/mm |
| Barrel roller bearing | 4° | 0.07 | 70 µ/mm |
| Needle roller bearings | 2'–3' | 0.006–0.009 | 1 µ/mm |
| Angular contact bearings fitted in pairs, double row needle bearings or needle or cylindrical roller thrust bearings | Misalignment not allowable | | — |
| Spherical roller bearings | 0.5–2° | 0.043 | 35–8 µ/mm |
| Deep groove ball bearings | | | |
|     C3 | 12' | | |
|     C4 | 16' | | |

*Elastic Deformation* The forces affecting the bearing can result in elastic deformation in the contact areas. The force to deformation ratio is not linear. This causes a variable shaft displacement due to various positions of the rolling element. As shown in Fig. 6.48, the deflection $X$, eventhough very small, changes due to external loading condition and causes vibrations. This effect can be reduced, to a little extent, by preloading of bearings.

*Surface Defects* The surface quality of rolling elements has the strongest influence on bearing noise. Three surfaces influence noise quality:

- Rolling element surface
- Outer ring surface
- Inner ring surface

The extent to which each of these surfaces affects noise varies. In addition, the surface roughness, in terms of regular waveforms on the surface, also has a very strong influence on bearing noise. The relationship between the number of waves and amplitudes at a constant vibration level is shown in Fig. 6.49. It can be seen from the graph that, as the number of waves reduces, the amplitude increases exponentially.

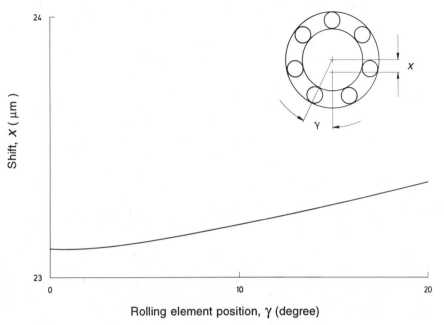

**Fig. 6.48** Shift of shaft centre due to various rolling element positions

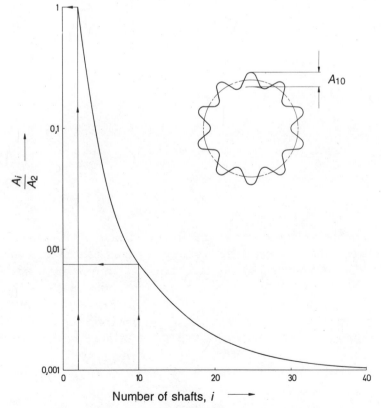

**Fig. 6.49** Relationship between shaft height and number of shafts at constant vibration level

*Influence of Lubrication* A lubricating film is formed between the parts in rolling contacts. The nature of this film depends on the geometry of the rolling contact, the load, the speed and the viscosity of the lubricant. The stronger lubricating film lessens the metal-to-metal contact of the rolling elements on account of the elasto-hydrodynamic property of the film. With speed and geometry remaining constant, there is a relation between the viscosity $V$ and the lubrication film thickness $h$ given by $h = V^{0.7}$.

In addition to separating the parts in rolling contact, the lubricating film has damping properties also. Figure 6.50 shows the dependence of lubricating film thickness on noise damping. It can be seen that beyond a certain film thickness there is no more reduction in noise. In practice, grease lubricated bearings are a little more noisy. For low bearing noise, use of special lithium soap base grease with a mineral oil base is recommended.

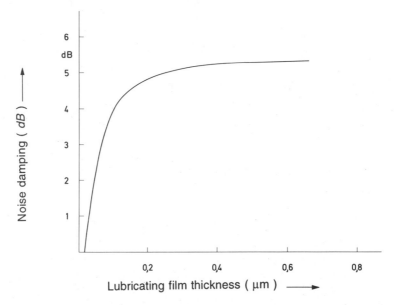

**Fig. 6.50** Influence of lubricating film thickness on noise damping in rolling bearings

## Friction Effects of Cages and Seals

The sliding friction of the cage which primarily occurs between the cage and the guiding surface and between the seals and the mating surfaces contributes to the creation of noise. The noise does not depend on the cage material. In practice, there is a slippage of the cage from the theoretical running speed due to the loading pattern and an insufficient preload. Depending upon the speed, cage guidance, viscosity and quantity of lubricant, the cage slippage can result in vibration and a loud irritating noise. To avoid this, there should be a minimum preload and a load of 1–2% of load rating in the case of deep groove ball bearings with an adequate supply of the lubricant.

## Indirect Causes

The causes of noise described so far refer directly to the rolling bearings. The bearing, however is also indirectly involved in generation of machine noise in the following two ways:

(a) Due to their low damping characteristics, the bearings transmit vibrations resulting from the chatter vibrations of the cutting process.
(b) The true running properties of a bearing influence noise generation in other areas. Gear noise depends on precise gear mesh, and in electric motors continuous change of gap geometry leads to excited vibrations.

The other cause of noise and vibration is as a result of false brinelling. This occurs when applying pressure or hitting the wrong ring during assembly. The result is indentation on the track which causes vibration and noise.

## Noise Reduction Measures

Noise and vibration reduction is achieved by:
- Proper assembly and maintenance
- Improvement in bearing design

The following are the assembly precautions required to achieve noise reduction:
- Correct preload on the bearings
- Correct grade and quantity of lubricant (for lubrication guideline refer to Annexure 6.1)
- Cleaning the path of the lubricant to prevent metallic particles and swarf from getting mixed up with lubricant
- Correct procedure for assembly and usage of right assembly tools and aids to prevent any damage to the bearing rings and the cage. Any damage to raceways and cage will result in abnormal noise and vibration in high speed running.
- Correct running clearances and fitting allowances properly adjusted to ensure that there is no rubbing of any components in the assembly. The metallic rubbing causes fine metal dust which can enter the bearings and damage the raceways.

The following noise reduction measures are also being adopted in the bearing design by the manufacturers:

(a) New internal design involving favourable kinematics of ball travel.
(b) Improved dust shields
(c) Optimised cage
(d) Computer aided lubricant testing
(e) Selective ball surface strength
(f) Modern noise testing methods

The noise testing instrument from FAG is shown in Fig. 6.51.

## Preventive Measures for Bearing Failures

Annexure 6.2 gives the list of failures, the identification, cause and preventive measures.

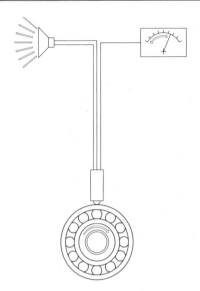
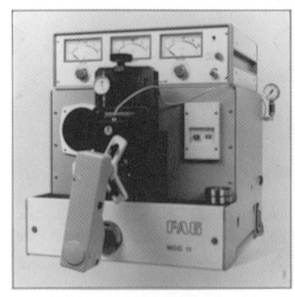

**Fig. 6.51** Measuring principle and view of anderometer noise tester

## Preloading of Bearings

The preload is the initial load applied to the bearings, which will cause axial or radial deflection of the rings in relation to each other after initial clearance has been taken up. This deflection is caused by the elastic deformation of the raceways and the loading elements. Figure 6.52 shows the deflection curve as a function of the load. From the graph it is clear that deflection $d_1$ for a bearing without preload is much more than deflection $d_2$ having an initial preload $P$. Therefore, the preload increases the axial rigidity of the bearings.

The other advantages of preload are:

- Preloading prevents the rolling elements from disengaging the raceways which otherwise result in a sliding action of the rolling elements
- Reduction of noise and vibration
- Increased damping characteristics

Figure 6.53 shows the practical method of preloading and Fig. 6.54 indicates the basic patterns, objectives and characteristics of

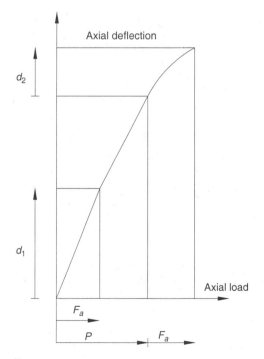

**Fig. 6.52** Deflection curve as a function of the load

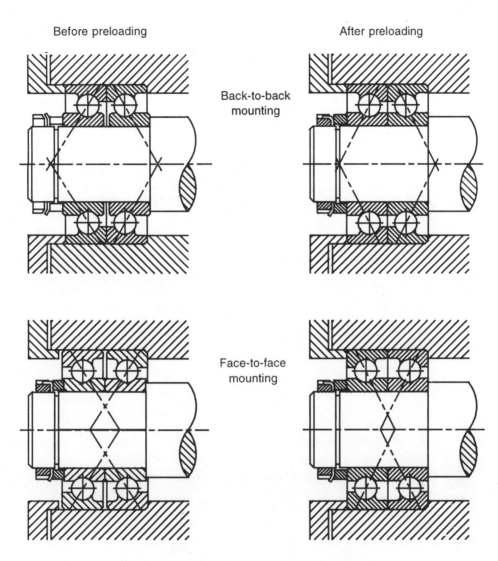

**Fig. 6.53** Practical method of preloading

bearing preload. Bearing preload includes fixed position preload and constant pressure preload. The fixed position preload is effective in fixing the position of each bearing and increases the rigidity. Springs are used to obtain constant pressure preload even if the positions of each bearing change due to the influence of thermal expansion and load while in operation, the amount of preload is still maintained the same.

| Basic Pattern | Applicable bearings | Objectives | Characteristics | Applications |
|---|---|---|---|---|
| Flexible position preload: 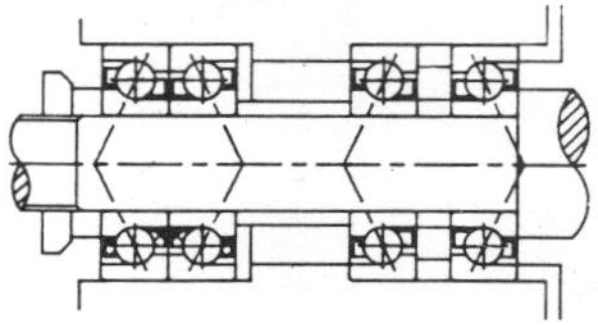 | Precision angular contact ball bearings | Maintaining accuracy of rotating shaft, preventing vibration, increasing rigidity | Preloading is accomplished by a pre-determined offset of the rings or by using spacers | Grinding machines, lathes, milling machines, measuring instruments |
| 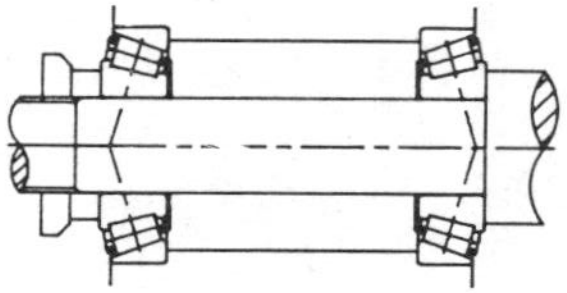 | Tapered roller bearings, thrust ball bearings, angular contact ball bearings | Increasing bearing rigidity | Preload is accomplished by adjusting a threaded screw. The amount of preload is set by measuring the starting torque or axial displacement. Relationship between the starting torque $M$ and preload $T$ is approximately given by: For duplex angular contact ball bearings: $$M = \frac{d_{pw} T}{3200 - 4200} \text{ N m}$$ For duplex tapered roller bearings: $$M = \frac{d_{pw}^{0.8} T}{345 - 690} \text{ N m}$$ | Lathes, milling machines, differential gears of automotives, printing machines, wheel axles |

**Fig. 6.54** (Contd)

| Basic Pattern | Applicable bearings | Objectives | Characteristics | Applications |
|---|---|---|---|---|
| Constant pressure preload: 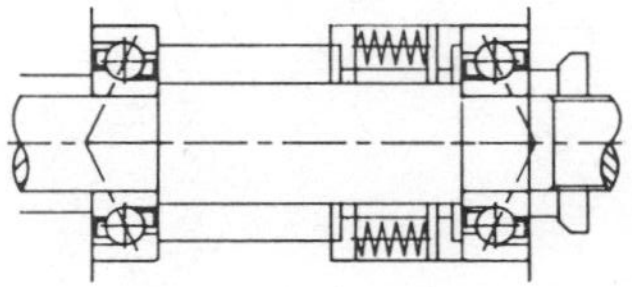 | Angular contact ball bearings, deep groove ball bearings, precision tapered roller bearings | Maintaining accuracy & preventing vibration & noise with a constant amount of preload without being affected by loads or temperature. Preventing smearing on non-loading side of race-way under axial loads | Preloading is accomplished by using coil or belleville springs. Recommended preloads are as follows: *For deep groove ball bearings*: $T = (4\text{-}8)$ dN | Internal grinding machines, electric motors high speed shafts in small machines, tension reels |
| 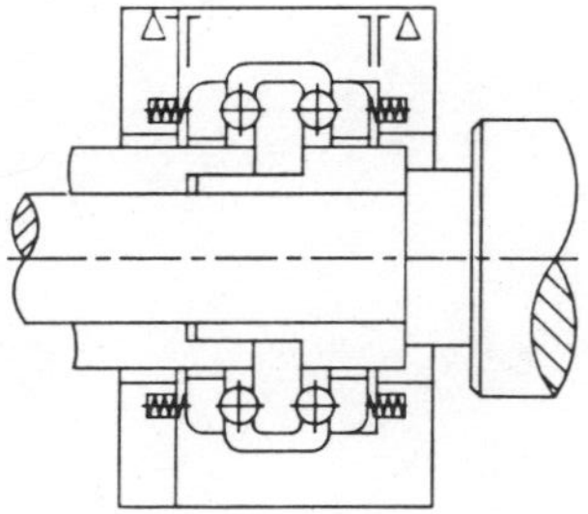 | Tapered roller bearing with steep angle, spherical roller thrust bearings, thrust ball bearings | Preventing smearing on raceway of non-loaded side under axial loads | Preload is accomplished by using coil or belleville springs. Recommended preloads are as follows: *For spherical roller thrust bearings:* $T_1 = 36 \times 10^{-12} (n\, C_{oa})^{0.9}$ N; $T_2 = 0.01\, C_{oa}$N (whichever is greater) | Rolling mills, extruding machines |

*Note*: In the above formulae,

$d_{pw}$ = pitch diameter of bearing, mm (bore + outside diameter)/2
$T$ = preload, N
$d$ = bearing bore, mm
$n$ = number of revolutions, rev/mm
$C_{oa}$ = static axial load rating, N

The starting torque $M$ is, however, greatly influenced by lubricants and a period of run-in time.

**Fig. 6.54** Basic patterns, objective and characteristics of bearing preload

## 6.5 SHOP TOOLS AND EQUIPMENTS FOR ASSEMBLY (TABLE 6.3)

TABLE 6.3

| Tools and Equipment | Application |
|---|---|
| Mandrels and fixtures | • To check alignment of slides with reference to headstock<br>• To check alignment of ballscrew mounting flange or bracket<br>• To align turret tool axis with reference to spindle axis<br>• To check indexing accuracy of C-axis |
| Lapping mandrels | • For improving surface finish of bores, suiting the mating component |
| Torque wrenches | • To preload ballscrew end bearing nut, spindle bearing covers |
| Adjustable tap wrenches | • For tapping in assembly |
| Bearing pullers | |
| —Mechanical | • For assembling bearings |
| —Hydraulic | • For removing bull gears<br>• For removing drive pulley which is shrink fit in assembly |
| Heating equipment | • For enlarging the bore of bearings before assembling on to spindle |
| —Liquid bath | |
| —Induction heating equipment | |
| Length measurement | • Steel scale, tape, vernier, micrometer, dial indicator |
| Temperature | • Thermometer, digital thermometer |
| Boundary measuring equipment (SKF, Sweden make) | • For accurate preloading of cylindrical roller bearings with tapered bore for machine tool spindle (NNK bearings, Japan) of a precision turning centre |
| Surface roughness measuring equipment | • To check surface roughness of flat, round surface both internal and external (e.g., shaft and bore surface) |
| 3D coordinate measuring machine (CMM) | • For component inspection<br>• To check centre distance of bores<br>• Circularity, cylindricity, etc. |
| Autocollimator | • Used for checking straightness of guideways in horizontal or vertical plane<br>• Indexing accuracy of C-axis by using polygon mirror |
| Digital minilevel | • For checking guideway level, flatness |
| Laser equipment | • For measurement of positioning accuracy and repeatability of slides |
| Flow meters | • To check flow rate |
| Force measurement | • Dynamometers 0–2000 kgf |
| Electrical | |
| —Programming unit | • For programmable logic control (PLC) programming |
| —Digital multimeters | • Measurement of current ($A$), voltage ($V$) and resistance ($R$) |
| —Storage oscilloscope | • Drive tuning |
| —Crimping tools | • For wire-end preparation |

## 6.6 HYDRAULICS

- Check completeness of equipment as per circuit or drawing
- Check as per checklist for acceptance of hydraulic power pack
- Check for neatness of assembly, pipe bends, connections and space for operation of connections
- Also refer to available checklists on hydraulics

### 6.6.1 Trouble Shooting and Remedy (Table 6.4)

TABLE 6.4

| Trouble | Causes/Problems | Suggested remedy |
|---|---|---|
| Noise in hydraulic pump | • Oil level is low<br>• Suction stainer blocked<br>• Loose fitting of suction pipe | • Check up oil level<br>• Clean suction stainer<br>• Check up correct fitting of pipe |
| Pressure not developing | • Rotation of pump wrong<br><br>• Shut-off valve open<br>• Sticky movement of plunger in direction control valves | • Check up the direction of rotation of pump as per identification mark made thereon<br>• Close the shut-off valve<br>• Clean and make the plunger move freely |
| Chuck cylinder vibrating while running | • Cylinder is not properly locked | • Tighten the flange against main spindle rear face and lock the chuck nut against the chuck flange |
| Clamp/declamp is not effective | • Chuck pressure not sufficient<br>• Leakage of oil in clamping cylinder | • Set the correct chuck pressure<br>• Replace oil seal and O-rings |
| Jerky movement of slides | • No lubrication<br><br>• Loose gibs | • Check lubrication level and check connecting hose pipes<br>• Readjust gib adjusting screws |
| Slides not moving according to specific cycle | • Sticky movement of plungers in direction control valves<br>• Wrong connection of hydraulic pipe | • Clean the plungers<br><br>• Check up as per identification letters |
| Hydraulic oil leakage through pipe connections and control elements | • Connections are loose or overtight<br><br>• Defective O-rings/oil seals | • Ensure proper fitting of all hydraulic connections<br>• Replace defective or damaged O-rings/oil seals |

## ANNEXURE 6.1

**Lubrication Guidelines**

| Type of bearings | Lubrication guidelines |
|---|---|
| Bearings for screw drives | • Can be operated either with grease or with oil<br>• Maximum temperature in machine tool $\not> 50°C$<br>• ZKLN 2RS bearings are supplied with grease lubricated for life<br>• With oil lubrication $n_{oil} = 1.5\, n_{grease}$ ($n_{oil}$ and $n_{grease}$ are the speeds with oil and grease lubrication respectively)<br>• ZARN/ZARF, oil lubrication preferred<br>• Oil viscosity in accordance with operating condition. Proven media are of ISOVG 32-100<br>• Proven grease is KZK or KTCPE 2K (DIN 51825)<br>• For continuous duty temperature can reach $+ 70$ °C |
| LM guides (ball guide) | • Recommended grease: KP2K(DIN 51825) oil: ISOVG 32-68 (DIN 51517)<br>• Lubrication period: $3 \times 10^6$ m (stroke 1m, temperature $> +50$ °C $C/P = 10$), where $C/P$ is the ratio of dynamic capacity to operating load |
| LM guides roller track | • Oil or grease may be used<br>• Oil is preferred<br>• Grease life is limited to 3-5 years<br>• Recommended oil: oils with viscosity ISOVG 10-68 at operating temperature below $+70$ °C<br>• Recommended grease: KPZK (DIN 51825) |
| Linear roller bearings/tychoways | • Suitable for grease or oil<br>• Recommended grease : KPZK (DIN 51825)<br>• Rough guide for re-lubrication $3 \times 10^6$ m travel (conditions $C/P > 5$, linear velocity, $V < 20$ m/min, $T < 50°C$),<br>• Oil preferred to grease, recommended oil: ISOVG 10–68 |
| Shaft guidance system | • Oil lubrication is preferred<br>• Under low loads $C/P >15$, oils of ISOVG 10-22 are recommended<br>• Under high loads $C/P >8$, oils of ISOVG 68-100 are recommended<br>• Recommended grease: KZK for $C/P > 8$ and KP2K grease for $C/P <8$<br>• Under normal operating conditions, $C/P > 10$, room temperature and 0.6 V limit, service life can be achieved with initial grease filling |
| Track roller guidance | • For guideways and wipers lubrication units are to be used. Lubrication is via oil soaked felt. Oil with viscosity of 200 mm²/s at $T = 40$ °C is to be used |
| Turn table bearings | Can be lubricated either through inner ring or outer ring<br>Supplied initially greased with KPZK (DIN 51825).<br>Suitable for $-30°C$ to $+120°C$ if speeds exceed $n_{grease}$. If speeds exceed $n_{grease}$, then oil should be used. |
| General needle/cylindrical bearings | Grease lubrication<br>• Grease should be capable of $\pm 20°$ beyond operating temperature range<br>• Grease with consistency NLGI 1, 2 or 3 are preferred<br>• Vertical or inclined axis and for sealing grease, high consistency of NLGI 3 should be used<br>• NLGI 2nd range 200,000 to 1 million rpm mm<br>• Lithium based grease suitable up to $+90°C$<br>• Service life of grease suitable up to three years<br>Oil lubrication<br>• Mineral oils – up to $+90°C$ (undoped) |

*(Contd)*

- Doped mineral oil – up to +130°C
- Synthetic oils – up to +200°C
- Oil bath—suitable for $nd \leq 200{,}000$ mm where n is the operating speed, rpm and $d$ is the mean diameter of the bearing
- Oil type—depends on operating temperature and operating speed
- Flowrate—depends on frictional horsepower and oil temperature difference between inlet and outlet
- $\Delta T = 15°$ to $30°C$ for radial bearings
- $\Delta T = 50°$ to $150°C$ for axial bearings

## ANNEXURE 6.2

### Preventive Measures for Bearing Failures

| Type of failure | Identification | Cause | Preventive measure |
|---|---|---|---|
| Fatigue failure | Flaking or spalling of raceway, noisy running of bearings | Normal usage, due to overloading | Bearing life is over, replace; lubrication; avoid overloading |
| Contamination | Scoring, pitting, scratching | Dirty or damp surroundings, abrasive wear materials | Use proper seal or shields in such dirty environments to avoid abrasives from entering inside the bearings. Clean the working surroundings |
| Brinelling | Mounting indication appearing on shoulder of the race and balls | Excessive pressure exerted on rings during assembly, shock loads during transportation | Use proper procedure for assembly, proper mounting during transportation |
| Thrust failure | Counter bored bearings | Improper mounting (hitting in wrong direction on angular contact bearings) | Follow correct assembly procedures |
| Misalignment | Excessive run-out, heating of bearing assembly vibration | Shaft misalignment, housing misalignment, bowing of the shaft | Check the accuracy of the shaft, housing and bowing shaft |
| Electric arcing | Granular race surfaces pitting or cratering | Static electricity emanating from charged belts or from manufacturing process, electric leakage short circuit | Check wiring, installation, etc., grounding, insulating bearings |
| Lubrication failure | Grease appearance, noise, abnormal temperature rise, bearing decolourisation, retainer/seal failure | Dirty lubrication, excessive lubricant, inadequate lubrication, wrong lubricant | Avoid dirty lubrication and check the lubricant path for metallic/abrasive particles, correct the quantity of lubricant. |
| Cam failure | Wobble of bearing | Undersized shafting | Use correct size shaft |

CHAPTER 7

# Drives and Electricals

## 7.1 DRIVES

In a CNC machine tool there are three major groups of elements (Fig. 7.1).

(a) Control and electronics
(b) Electric drives (electromechanical drives)
(c) Mechanical elements (table, slide, tool holder, etc.)

In addition, there can be hydraulic and pneumatic systems which are integrated with the CNC machine tool.

The primary function of the drive is to cause motion of the controlled machine tool member (spindle, slide, etc.) to conform as closely as possible to the motion commands issued by the CNC system. In metal cutting machines, the metal is removed as a result of the movement of the workpiece and the cutting tool as shown in Fig. 7.2.

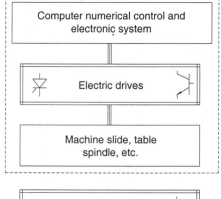
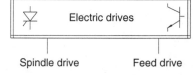

**Fig. 7.1** Elements of a CNC machine tool

In order to maintain a constant material removal rate, the spindle and the tool movements have to be coordinated such that the spindle has a constant power and the slide has a constant torque.

Modern CNC machines are built with higher control accuracies. In order to ensure a high degree of consistency in production, variable speed drives are necessary.

Most of the drives used in machine tools are electrical. Figure 7.3 gives the configuration of an electric drive for a CNC lathe, where both the spindle and feed motors have the facility to vary the speeds infinitely (zero to maximum) in both the directions. Table 7.1 gives the various types of electric motors used in machine tools.

Drives and Electricals **241**

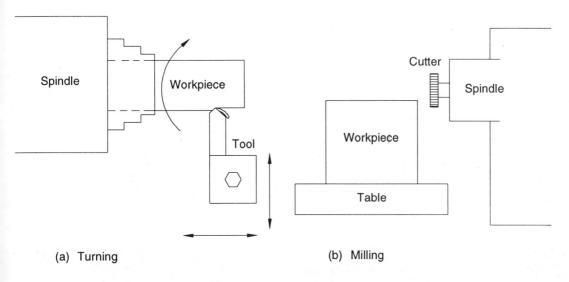

**Fig. 7.2** Drive in a metal cutting machine

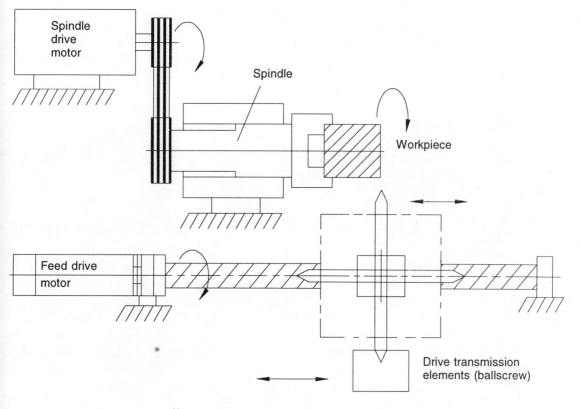

**Fig. 7.3** Electric drives on a CNC lathe

With the developments in power electronics and microprocessor systems, variable drive systems have been developed. These are smaller in size, very efficient, highly reliable and meet all the stringent demands of the modern automatic machine tools.

**TABLE 7.1**

*Types of electric motors used in machine tools*

| Spindle motors | Feed motors |
|---|---|
| DC shunt motor (separately excited) | DC servomotor |
| Three phase ac induction motor | AC servomotor |
|  | Stepper motor |

An electric drive consists of a motor and its associated control electronics. Depending on their characteristics, machine tool drives can be classified as follows.

- Spindle drives—(constant power)
- Feed drives—(constant torque)

## 7.2 SPINDLE DRIVES

The requirements of a spindle drive motor are:
(a) High rotational accuracy
(b) Wide constant power band
(c) Excellent running smoothness
(d) Compactness
(e) Fast dynamic response
(f) Range of rated output from 3.7–50 kW
(g) Maximum speed up to 9000–20,000 rpm (high speed application)
(h) High overload capacity
(i) Large speed range of at least 1 : 1000. A typical speed power requirement of spindle drive is shown in Fig. 7.4.

### 7.2.1 Types of Spindle Drives (Fig. 7.5)

The dc spindle drives are commonly used in machine tools. However, with the advent of microprocessor-based ac frequency inverter, of late, the ac drives are being preferred to dc drives as they offer many advantages. One of the main advantages with the microprocessor-based frequency inverter is the possibility of using the spindle motor for C-axis applications for speed control in the range of 1 : 10,00,000 with positioning.

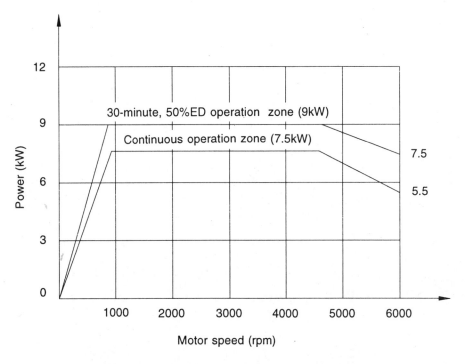

ED : Efficiency duty

**Fig. 7.4** Speed power requirement of spindle drive

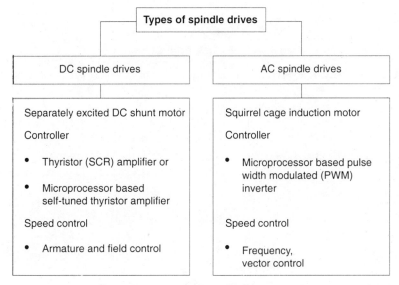

**Fig. 7.5** Types of spindle drives

## 7.3 FEED DRIVES

A feed drive consists of a feed servomotor and an electronic controller. Unlike a spindle motor, the feed motor has certain special characteristics, like constant torque and positioning. Also, in contouring operations, where a prescribed path has to be followed continuously, several feed drives have to work simultaneously. This requires a sufficiently damped servo-system with high band width, i.e. fast response and matched dynamic characteristics for different axes.

### 7.3.1 Requirements of CNC Feed Drive

(a) The required constant torque for overcoming frictional and working forces must be provided (during machining).
(b) The drive speed should be infinitely variable with a speed range of at least 1 : 20,000 which means that both at a maximum speed, say of 2000 rpm, and at a minimum speed of 0.1 rpm, the feed motor must run smoothly and without noticeable waviness.
(c) Positioning of smallest position increments like 1–2 µm should be possible. For a feedmotor this represents an angular rotation of approximately 2–5 angular minutes.
(d) Maximum speeds of up to 3000 rpm.
(e) Four quadrant operation—quick response characteristics.
(f) Low electrical and mechanical time constants.
(g) Integral mounting feedback devices.
(h) Permanent magnet construction.
(i) Low armature or rotor inertia.
(j) High torque-to-weight ratio.
(k) High peak torque for quick responses.
(l) Total enclosed non-ventilated design (TENV)

Figure 7.6 shows the speed-torque characteristics for a typical feedmotor.

### 7.3.2 Types of Feed Drives (Table 7.2)

Variable speed dc feed drives are very common in machine tools because of their simple control techniques. However, with the advent of the latest power electronic devices and control techniques ac feed drives are becoming popular due to certain advantages.

*Advantages of ac servos over dc servo-systems are*

(a) ac servomotor has an almost constant torque output from zero to maximum speed
(b) Fast response
(c) Low rotor inertia
(d) Excellent temperature resistance
(e) Brushless and maintenance-free operations
(f) Increased power density

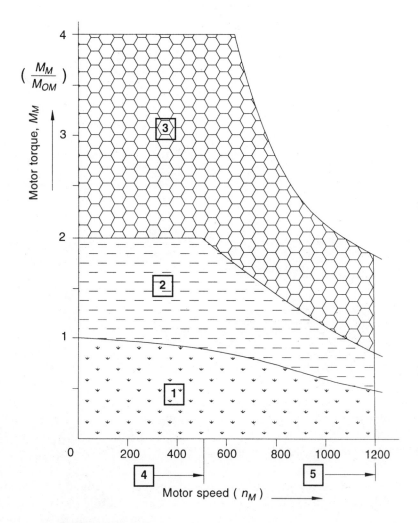

| 1 | : | Continuous duty S1 |
|---|---|---|
| 2 | : | Intermittent duty S3 |
| 3 | : | Dynamic limit range with commutating limit curve |
| 4 | : | Feed range |
| 5 | : | Rapid traverse range |
| $n_M$ | : | Motor speed |
| $M_M$ | : | Motor torque |
| $M_{OM}$ | : | Rated motor torque of dc servomotor |

**Fig. 7.6** Operating ranges of the permanent magnet-excited dc servomotor

## TABLE 7.2
### Types of servo feed drives

| dc servo-drive | ac servo-drive |
|---|---|
| Motor | Motor |
| • Permanent magnet dc motor | • Synchronous three phase ac motor with permanent magnet rotor |
| Controller | Controller |
| (a) Thyristor dc amplifier | (a) Transistor PWM frequency inverter (variable frequency controller), analog drive amplifier |
| (b) Transistor PWM dc chopper | (b) Transistor PWM frequency inverter (variable frequency controller) digital drive amplifier |
| Speed control | Speed control |
| • Armature voltage | • Frequency control |

## 7.4 DC MOTORS

The following section describes the basic principles of electromagnetism which is essential for understanding the working of dc motors.

### 7.4.1 Magnetic Field due to a Straight Current Carrying Conductor

When current is passed through a conducting wire, a magnetic field is produced around it. The direction of the magnetic field due to a straight current carrying wire can be mapped by means of small a compass needle or by iron filings as in the case of a bar magnet. A small experiment can be conducted to see the effect of a current carrying conductor.

Take a sheet of smooth cardboard with a hole at the centre. Place it horizontally and pass a wire vertically through the hole. Sprinkle some iron filings on the cardboard and pass an electric current through the wire. Gently tap the cardboard. We find that the iron filings arrange themselves in concentric circles around the wire as shown in Fig. 7.7. If a small magnetic compass needle is kept anywhere on the board near the wire, the direction in which the north pole of the needle points gives the direction of magnetic field (i.e. magnetic lines of force) at that point.

In this experiment we note that

(i) the magnetic lines of force form concentric circles near the wire, with their plane perpendicular to the straight conductor and with their centres lying on its axis.

(ii) if the direction of current in the wire is reversed, the direction of lines of force is also reversed.

(iii) on increasing the strength of current in the wire, the lines of force become denser and iron filings are arranged in circles up to a larger distance from the wire showing that the magnetic field strength has increased.

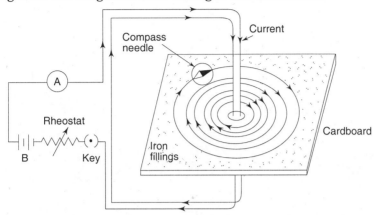

❧ **Fig. 7.7** Magnetic field around a current carrying wire

## Force on a Current Carrying Conductor in a Magnetic Field

When a current carrying conductor is kept in a magnetic field (not parallel to it), a force acts on it. This force is created due to overlap of magnetic field of current in the conductor and the external magnetic field on the conductor. As a result of this superposition, the resultant magnetic field on one side of the conductor is weaker than that on the other side. Hence the conductor experiences a resultant force in one direction.

Since current is due to the flow of charge, a moving charge in a magnetic field also experiences a force (direction of motion not parallel to the field direction). Figure 7.8 represents the direction of force acting on a conductor $A-C$ carrying a current $I$ kept in a magnetic field $B$. The conductor is kept normal to the magnetic lines of force. The force $F$ acts perpendicular to both the current and the magnetic field. The conductor begins to move in the direction of this force.

When no current flows in the conductor, no force acts and the conductor does not move. When the direction of current in the conductor or the direction of the magnetic field is reversed, the direction of force (i.e. direction of movement of conductor) is reversed. Note that if the conductor is kept parallel to the direction of magnetic field, no force acts on it.

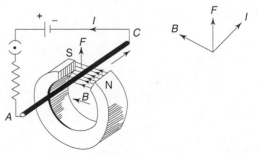

❧ **Fig. 7.8** Force on a current carrying conductor

## Magnitude of Force

Experimentally it is found that the magnitude of the force acting on a current carrying conductor kept in a magnetic field, in a direction perpendicular to it depends on the following factors.

(i) the force $F$ is directly proportional to the current $I$ flowing in the conductor, i.e. $F \propto I$
(ii) the force $F$ is directly proportional to the intensity of the magnetic field $B$, i.e. $F \propto B$
(iii) the force $F$ is directly proportional to the length of the conductor $L$, i.e. $F \propto L$
Combining these,

$$F \propto IBL \quad \text{or} \quad F = KIBL$$

where, $K$ is a constant whose value depends on the choice of the units.

## Direction of Force

The direction of force is obtained by the Fleming's left hand rule, i.e. stretch the forefinger, middle finger and the thumb of your left hand mutually perpendicular to each other as shown in Fig. 7.9. If the forefinger indicates the direction of magnetic field and the middle finger indicates the direction of current, then the thumb will indicate the direction of motion (i.e. force) on the conductor.

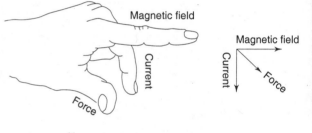

**Fig. 7.9** Fleming's left hand rule

## Simple dc Motor

An electric motor is a device which converts electrical energy into mechanical energy. It works on the principle that when an electric current is passed through a conductor kept normally in a magnetic field, a force acts on the conductor as a result of which the conductor begins to move. The direction of force is obtained with the help of Fleming's left hand rule.

Figure 7.10 shows the construction of a typical electric motor. The main parts of an electric motor are:

(i) the armature coil $ABCD$ mounted on an axle
(ii) the commutator (i.e. a slip ring divided into two parts, $S_1$ and $S_2$ as split rings)
(iii) a pair of brushes $B_1$ and $B_2$ and
(iv) a horseshoe electromagnet $NS$.

The coil $ABCD$ is wound round a soft iron core and is kept in between the pole pieces of a powerful horseshoe magnet. The coil is free to rotate about its axis. The ends of the coil $A$ and $D$ are connected to split parts of the ring $S_1$ and $S_2$ respectively. Two brushes

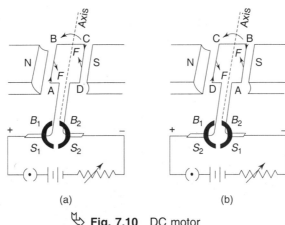

**Fig. 7.10** DC motor

$B_1$ and $B_2$ made of carbon or copper touch the split rings $S_1$ and $S_2$ respectively. A dc source is connected across the brushes $B_1$ and $B_2$. When the coil rotates, the split rings rotate but the brushes do not move. A wheel can be mounted on the axle placed along the axis of the coil, so as to drive the desired parts of the machine.

*Working*   In Fig. 7.10(a) the plane of the coil is horizontal and split ring $S_1$ touches the brush $B_1$, while the split ring $S_2$ touches the brush $B_2$. The current flows through the coil in the direction *ABCD*. According to Fleming's left hand rule, a force $F$ acts on the arm *AB* in the downward direction (into the plane of the paper) and a force $F$ acts on arm *CD* in the upward direction (out of the plane of the paper). The arms *BC* and *DA* being parallel to the magnetic field experience no force. The forces on arms *AB* and *CD* form an anticlockwise couple due to which the coil begins to rotate.

When the coil comes into the vertical position (i.e. perpendicular to the plane of the paper with arm *CD* up and the arm *AB* down), the couple becomes zero (forces on the arms are along same line). But due to inertia of motion, the coil passes through this position and the split ring $S_1$ comes in contact with the brush $B_2$ while the split ring $S_2$ comes in contact with the brush $B_1$ (Fig. 7.10(b)). Now the current flows through the coil in the direction *DCBA* and the forces acting on the arms of the coil again form an anti-clockwise couple due to which the coil remains rotating in the same direction. Thus whenever the coil comes in the vertical position, the direction of current through the coil reverses and the coil continues to rotate in the same direction.

The speed of rotation of the coil can be increased by

(i) increasing the strength of the current
(ii) increasing the number of turns in the coil
(iii) increasing the area of the coil and
(iv) increasing the strength of the magnetic field.

A mechanical switch—called commutator—is connected to the moving conductors and always supplies that conductor with current which is then placed in the magnetic field (exciting field). For constructional reasons, the conductors are placed at the moving part (rotor) and the commutator is developed as a collector ring.

In the conventional dc machine the conductors move in the magnetic field and the total motor current is conducted via the commutator.

### Electronic Commutation

The mechanical commutator has many disadvantages (e.g. wear, etc.). Therefore, it is reasonable not to conduct the motor current via the commutator but to switch electronically.

Instead of a mechanical switch, a non-contact (Hall-) sensor called rotor position sensor is fixed at the machine that signals which conductors are at that moment in the magnetic field and, therefore, generate a torque.

The conductors are individually connected to electronic switches. The sensor signal is evaluated and the converter switches on the current for the adequate conductor. It is reasonable to place the conductors on the non-rotating part called stator because an incoming cable is necessary for each conductor. Then on, the motor is the exciting field which can, for example be permanently excited.

In an electronically commutated dc machine the exciting field moves. The motor current is switched by electronic switches to inflexible conductors, depending on the rotor position.

This motor develops a homogenous torque and is no stepping motor. Direct operation on the line is not possible.

For a homogenous torque, a minimum of three conductors is necessary. The term "3-phase drive" is based on this fact but often leads to misunderstandings concerning the real function.

### Induced Voltage

Voltage is induced in each conductor (emf, $U_q$). Concerning the individual conductor it is an alternating voltage because the conductor runs through different poles ("changing polarity") during one rotation. In a dc machine, however, this alternating voltage is rectified by the commutator and appears as a direct voltage at the output.

The conductors are led outside in a discrete way in an electronically commutated machine—the emf is an alternating voltage.

Figure 7.11 shows a two-pole elementary dc machine (dc motor or generator). A dc machine runs as a dc motor when a direct current voltage is applied to it but if a dc machine is rotated mechanically, it functions as a dc generator. It can be seen in the figure that a dc motor has a field winding (coil) and an armature coil (rotor). The two ends of the armature coil are connected to two conducting (copper) segments which cover slightly less than a semicircular arc. The segments are electrically insulated from each other and from the shaft on which they are mounted. This arrangement is called the commutator. Current is passed to the commutator segments by means of carbon brushes. The armature in a dc motor is always a rotor and the field windings are placed on the stator.

$V_f$ : Field voltage
$I_f$ : Field current
$R_f$ : Field resistance
$\phi$ : Field flux
$E_b$ : Back emf
$V$ : Applied voltage
$I_a$ : Armature current
$R_a$ : Armature resistance

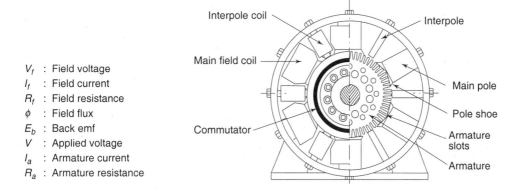

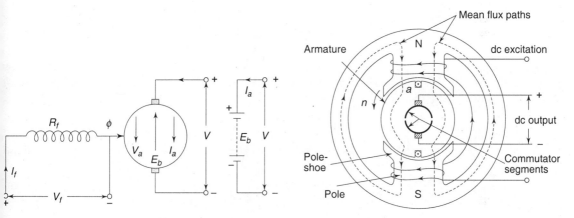

**Fig. 7.11** Two-pole elementary dc machine

## 7.4.2 Speed Control of a dc Motor

The speed of a dc motor can be represented as

$$N = \frac{K_1(V_a - I_a R_a)}{\phi} \qquad (1)$$

Neglecting armature voltage drop, the above equation can be written as

$$N = \frac{K_1 V_a}{\phi} \qquad (2)$$

where  $N$ = speed (rpm)
  $I_a$ = armature current
  $V_a$ = armature voltage
  $\phi$ = field flux
  $K_1$ = proportionality constant

### 7.4.3 DC Servomotors

Direct current servomotors are used as feed actuators in many machine tool industries. These motors are generally of the permanent magnet (PM) type in which the stator magnetic flux remains essentially constant at all levels of the armature current and the speed-torque relationship is linear.

The torque of a dc motor is expressed by the following equation

$$T = K_2 I_a \tag{3}$$

where  $T$ = torque
$K_2$ = proportionality constant
$I_a$ = armature current

From Eq. (1), (2) and (3) we can draw the following conclusions.

(a) The speed of a dc motor is directly proportional to the applied voltage and inversely proportional to the field magnetic flux (field current). In case of a permanent magnet dc servomotor, the field is constant and the speed of such a motor can only be varied by varying the armature voltage. This method of speed control is known as armature control.
(b) By applying the rated armature voltage to the motor and weakening the field current (by reducing the field excitation voltage) in the case of separately excited motors, the speed can be increased above the rated speed. This method of speed control is known as field control.

Direct current servomotors have a high peak torque for quick accelerations. A cross-sectional view of a typical permanent magnet dc servomotor is shown in Fig. 7.12.

Salient features of a typical dc servomotor are:

(a) Smooth rotation at speeds less than 1 rpm.
(b) Brush life of more than 4000 hours.
(c) Tacho generator directly built into the rotor.
(d) Rotor of the tacho generator directly coupled to the motor shaft.

## 7.5  SERVO-PRINCIPLE

The servo-mechanism in a machine tool control can be illustrated through a simple connection of dc servomotor with tacho feedback as shown in Fig. 7.13 (assume that the tacho generator and motor are matched for their current and voltage ratings).

The requirement is that the set speed $n_a$ should be proportional to $V_a$ so that the speed selected is regulated. Without the tacho connection the dc motor will rotate at a speed proportional to $V_a$. But due to circuit losses or load fluctuations, the actual speed may vary beyond the required speed. Therefore, in order to control this the actual speed is sensed through a tacho generator. The tacho generator generates a voltage $V_f$ proportional to the actual motor speed but in the opposite direction. At $t = 0$, when the switch SW1 is closed, the applied voltage to the armature of the motor is

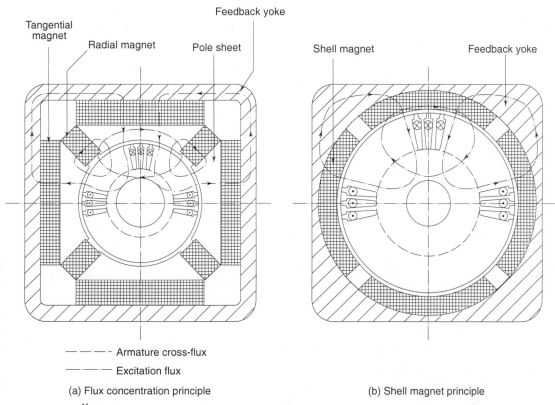

- - - - Armature cross-flux
—·—·— Excitation flux

(a) Flux concentration principle  (b) Shell magnet principle

**Fig. 7.12** Cross-section of a permanent magnet-excited dc servomotor

$$V = V_a - V_f$$

i.e. $V = V_a$ (because at $t = 0$ the motor is not rotating; therefore $V_f = 0$)

Hence, the motor picks up speed. At time $t = t_1$ the motor has attained a speed $n_2$ which is more than the required speed.

Now, the applied voltage is

$$V = (V_a - V_f)$$

Here, if $V_f > V_a$, then $V$ is negative. Hence the motor reduces its speed. Similarly if the actual speed of the motor reduces below the set value, then $V_f$ is lesser than $V_a$ in magnitude. Then the difference is a positive voltage. Hence this positive voltage is applied to increase the actual speed. This process continues and the speed is regulated within the band.

If the parameters of the system are optimised properly, it is possible to reduce the band $(n_2 - n_3)$ to a minimum so that the set speed is maintained irrespective of load conditions or any other influence.

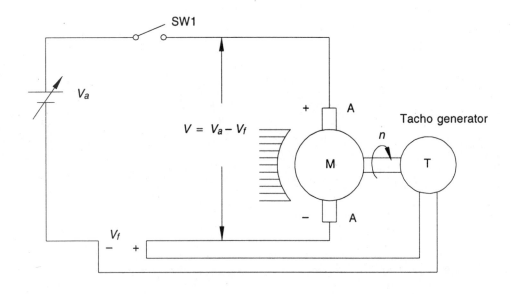

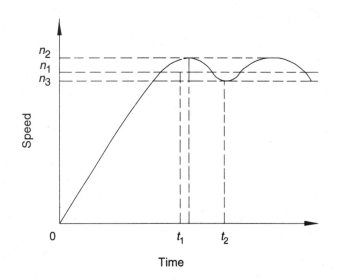

**Fig. 7.13** Servomechanism in a machine tool control

This system with feedback is called closed loop system whereas any system without feedback is called an open loop system. These are shown in Fig. 7.14. A close loop system (with negative feedback) is stable and the outputs are controlled (regulated). On the other hand an open loop system has no control over the output deviations. Therefore, close loop systems (with feedback) are used in CNC machine tools for high degree of positioning accuracy and feed rate control.

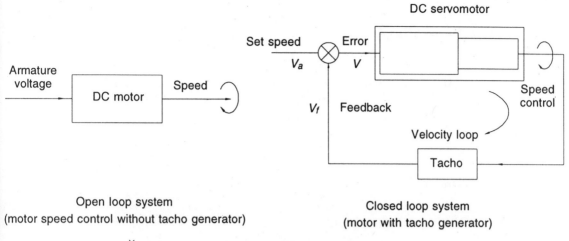

**Fig. 7.14** Open-and closed-loop system
(a) Motor speed control without tacho-generator
(b) Motor with tacho-generator

### 7.5.1 Drive Amplifier

In Fig. 7.15 it is clear that we need a variable dc power supply source which can be adjusted to any voltage, from 0 volts to 100 volts dc. With such an arrangement the motor speed can be varied from zero to its maximum rated speed.

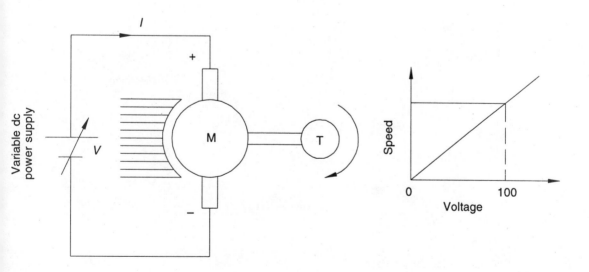

**Fig. 7.15** DC motor speed control (armature)

A proper amplifier whose dc output voltage will vary (from zero to ± maximum of armature voltage) with reference to a low voltage set value is called a drive amplifier or converter.

The basic requirements of a drive amplifier are:
(a) Ability to supply the required voltage of both polarities for bi-directional operation.
(b) The dc output should be free from ripple voltages or ac components, i.e. the form factor should be good.
(c) Regenerative braking must be available for braking the drive quickly, i.e. four-quadrant operation.

### 7.5.2 SCR-DC Drives

Silicon controlled rectifier (SCR) power supplies (also called SCR converters) provide a variable dc supply from single phase or three phase ac mains by controlled rectification.

The SCR (thyristor) is a solid state device with three terminals: an anode, a cathode and a gate. The conduction of current through the device with anode more positive than cathode, cannot take place until positive firing pulses of appropriate voltage and pulse width are applied to the gate with respect to the cathode. At this time the SCR fires or triggers and it will continue to conduct until the anode voltage is removed or reversed in polarity. A very small gate power of the order of milliwatts is sufficient to control kilowatts of power through this device.

Figure 7.16 shows a simple half-wave SCR converter supplying a resistive load. The waveforms are also shown in Fig. 7.17.

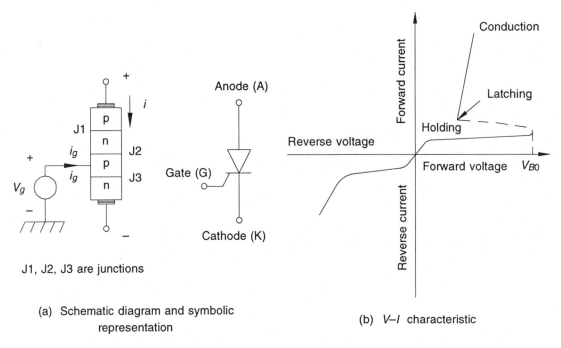

(a) Schematic diagram and symbolic representation

(b) V–I characteristic

**Fig. 7.16** Silicon controlled rectifier (thyristor)

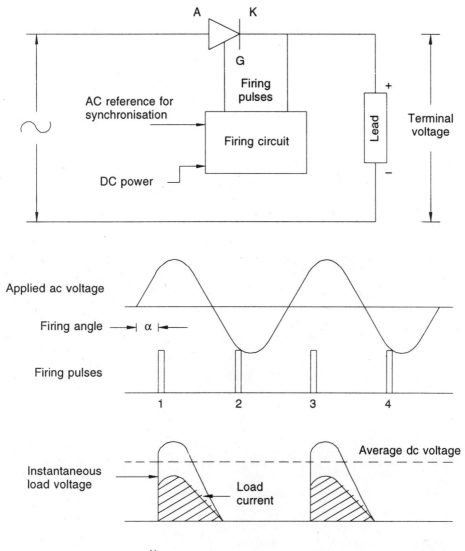

**Fig. 7.17** Waveforms in SCR

By advancing the firing angle, the average dc at the load terminals will keep on decreasing. It should also be noted that the firing pulses (2, 4, etc.) do not fire the SCR since the anode is more negative than the cathode. Other types of SCR drive amplifiers (or converters) are:
(a) Single phase full-wave converter
(b) Three phase half-wave converter

### 7.5.3 Transistor PWM dc Converter

In case of transistor PWM (pulse width modulated) amplifiers (also called transistor choppers), the power switching device is a transistor instead of an SCR. Transistor dc drives are ideal for controlling dc servomotors. The transistors are commonly used in the switching mode at frequencies between 1 kHz and 10 kHz (pulse width modulated). In the PWM technique the average dc voltage is proportional to the pulse width.

Figure 7.18 shows the four-quadrant operation of a dc drive. Four-quadrant operation of an amplifier means the drive is capable of:

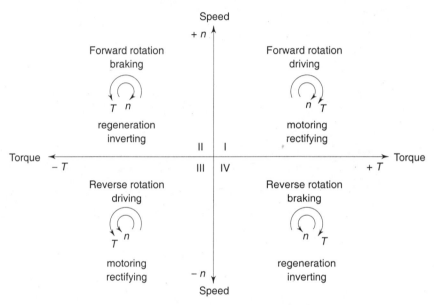

**Fig. 7.18** Four-quadrant operation of a dc drive

(a) Forward running—quadrant I
(b) Reverse running—quadrant III
(c) Forward braking—quadrant II
(d) Reverse braking—quadrant IV

The circuit diagram (Fig. 7.19) shows the single quadrant operation. In such a system motor can rotate in one direction only and hence can develop torque only in the direction of rotation. As can be seen from the associated wave forms shown in Fig. 7.19(b) neither the motor voltage nor the current reverses its direction during the period transistor $Q_1$ is OFF. The average dc voltage applied to the motor depends entirely on the ON-OFF ratio of switching of $Q_1$ and increases proportionately as the ON period increases. The diode provides a free wheeling path and conducts during the time when the motor mean voltage is less than the applied voltage.

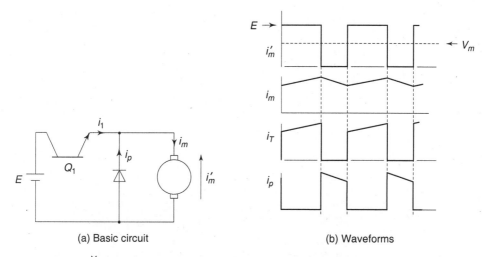

**Fig. 7.19** Chopper control, one-quadrant operation

The circuit diagram (Fig. 7.20) explains the working of the the two-quadrant drive. In such a system, motor current can reverse to allow the motor to generate reverse torque feeding power back to dc source. However the motor voltage and direction does not reverse, which provides explanation for the two quadrant operation involving forward motoring and forward braking.

## Four-Quadrant Operation

Figure 7.21 shows the basic diagram of a transistor dc four-quadrant amplifier. A rectifier feeds from the three phase ac line into a dc bus. The buffer capacitor $C$ can supply stored energy for acceleration and can accept energy as long as the motor absorbs mechanical energy during braking. The buffer capacitor is thus working as generator and supplies electrical energy. The capacitor is so chosen that the dc bus voltage changes only by little. The motor can be controlled selectively for clockwise or counterclockwise rotation and can be accelerated or braked by controlling two diagonally opposite transistors ($Q_1 - Q_3$ or $Q_2 - Q_4$).

The magnitude of the armature voltage $V_A$ and thus the speed $n_M$ is determined by pulse width modulation of transistors acting as switches for a switched transistor chopper. Two energy storages are necessary to operate a transistor controller in all four quadrants:

- a large buffer capacitor $C$ which maintains the voltage $U_C$ constant and is capable of accepting energy to store and to deliver.
- a load inductance, which smoothens the motor current and acts as an energy buffer. This is especially important during braking mode. At high working frequencies of the controller, the armature inductance $L_M$ of the motor is generally sufficient.

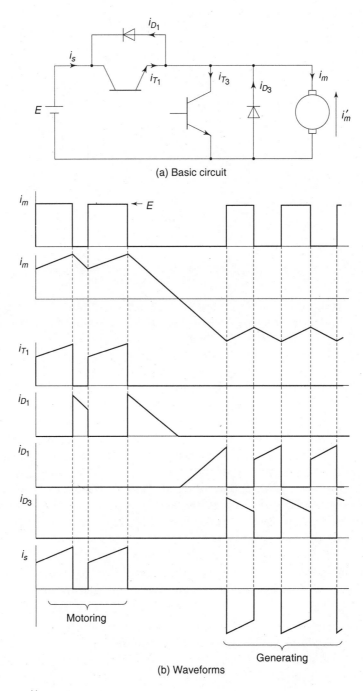

**Fig. 7.20** Chopper control, two-quadrant operation

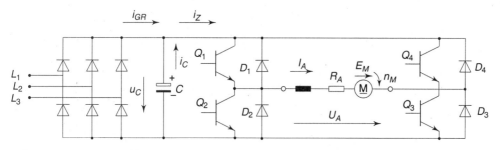

**Fig. 7.21** Transistor dc four quadrant amplifier

### Driving, Clockwise (CW), I Quadrant

The voltage and the current wave forms shown in Fig. 7.22 designate this mode of operation.

At the time $t_0$, both transistors $Q_1$ and $Q_3$ are ON. Logic '0' represents the switched on transistor. The armature voltage of the motor is positive, i.e. $U_A = U_C$ and current flows through the motor via $Q_1$, $L_A$, $R_A$ and $Q_3$.

At time $t_1$, switch $Q_1$ is opened. The motor current commutes from $Q_1$ to diode $D_2$, and is not flowing through the dc bus any more, but circulates in the lower half of the bridge from $Q_3$ through $D_2$, $L_A$, $R_A$, motor and back to $Q_3$ (free wheeling). At this point the motor voltage $U_A$ and the dc bus current $i_2$ jump to zero.

At time $t_2$, the situation is the same as it was at time $t_1$. At time $t_3$, switch $Q_3$ is open. The motor current now commutes through diode $D_4$ and circulates in the upper half of the bridge circuit via $Q_1$, $L_A$, $R_A$, motor $D_4$ and $Q_1$. The motor voltage and dc bus current jump back to zero again. At time $t_4$, switch $Q_3$ is closed and a new switch cycle begins.

The mean value of the motor voltage $U_A$ depends on the ratio between the switch ON time $t_E$ and the switch OFF time $t_A$. During the switch ON time $t_E$, the energy is derived from the dc bus while during the period $t_A$, the current is driven by the energy stored in the inductance. The motor thus maintains a positive product from voltage $U_A$ and current $I_A$ during both the periods $t_E$ and $t_A$ and thus converts electrical energy into mechanical energy. For simplicity the drop across the motor ($I_A R_a$) is neglected since it is very small as compared to the induced voltage $E_M$ in the motor.

### Driving, Counter Clockwise (CCW), III Quadrant

Explanation for driving CCW direction is the same as driving CW; only the direction of current is reversed by switching transistor $Q_2$ and $Q_4$. During free wheeling period the load current is also circulated alternately through the lower and upper halves of the bridge circuit. The period $t_E$ corresponds to motoring in opposite direction by suitably switching of the transistors $Q_2$ and $Q_4$ and the period $t_A$ corresponds to free wheeling phase.

### Braking, Clockwise, IV Quadrant

The mean value of the motor current $I_A$ and the armature voltage $U_A$ must have opposite signs for a back flow of actual power out of the motor circuit. For the purpose of stopping

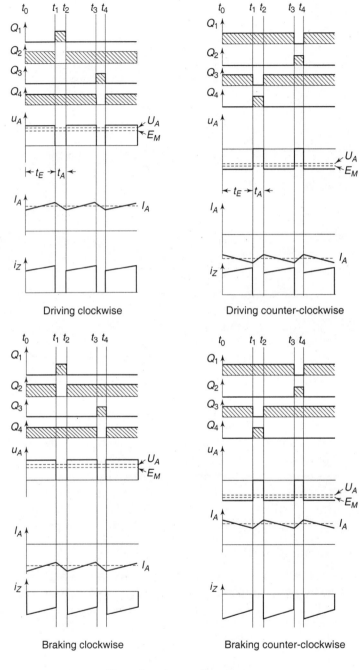

**Fig. 7.22** Voltage and the current wave forms

$U_A$ : Armature voltage
$U_c$ : DC link voltage
$I_A$ : Armature current

the motor, the mean value of the armature voltage $U_A$ is reduced as compared to voltage induced in the motor $E_M$. The direction of current within the motor circuit is reversed. The transition from driving to braking is the result of decrease in $t_E$ and an increase in $t_A$. At time $t_1$, switch $Q_2$ is closed. The voltage $E_M$ drives a current through $R_A$, $L_A$, $Q_2$ and $D_3$. The energy is stored in the $L_A$ inductance which is given by

$$W_{LA} = \frac{1}{2} I_A L_A^2$$

where  $W_{LA}$ = energy stored
$L_A$ = inductance and
$I_A$ = current through the motor.

At time $t_2$, switch $Q_2$ is OFF and current $I_A$ commutes over to diode $D_1$ and flows into the dc bus charging the capacitor and then returns through $D_3$ to the motor. The voltage induced in the inductance and the induced motor voltage are in series. Their sum is larger than the voltages delivered from the dc bus. Energy is thus fed back into dc bus and stored in the capacitor as given by

$$W_C = \frac{1}{2} C U_C^2$$

where  $W_C$ = energy stored
$C$ = capacitance and
$U_C$ = dc link voltage

The voltage of the dc bus increases. If now at time $t_3$, the switch $Q_4$ is turned ON, the armature current will flow in the upper circuit through $R_A$, $L_A$, $D_1$ and $Q_4$ and energy will again be stored in $L_A$. This energy in turn is fed into the dc bus at time $t_4$ via diodes $D_1$ and $D_3$. This cycle is repeated periodically.

The time period $t_4$ is for storing the energy in the inductance and period $t_E$ is for feeding the energy back into dc bus. The mean values of the motor current and motor voltage have opposite polarity. The motor is braked with mean constant torque because the actual power is fed back to the source.

### Braking, Counter Clockwise, II Quadrant

The cycle of operation during CCW braking is similar to the one described for CW braking. The energy is stored in inductance $L_A$ during time $t_4$ by switching the transistor $Q_3$ ON. As soon as, at time $t_2$, $Q_3$ is turned OFF, the current commutes through the free wheeling diodes $D_2$ and $D_4$ and flows back into the dc bus. Here too, the motor current and voltage are of opposite polarity and thus the generator mode of the drive is established.

### 7.5.4 Braking Methods in Servo-drive

Braking of a motor is a normal requirement of CNC machines to stop the slide/spindle to the programmed position or within a definite distance in case of power failure or emergency conditions. There are two types of braking employed in servo-drive.

**Dynamic Braking (Fig. 7.23)** Braking is realised by shorting the armature leads through contactor and dissipating the kinetic energy stored in the motor into the Dynamic Braking Resistor (DBR) in the form of heat. During this period, reverse torque will be generated which will bring the motor to a stand still faster. This type of braking is a fail safe braking and finds application particularly during mains failure and emergency situations.

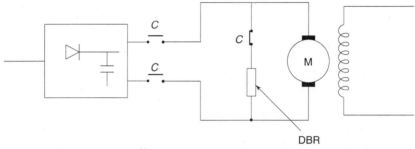

**Fig. 7.23** Dynamic braking

**Regenerative Braking** The motor direction can be reversed by changing the sequence of firing pulses in the power section. To avoid sudden changes within the motor, a ramp network is included in the control circuit. Regenerative braking is possible if the motor is overhauled by load and energy is returned to the source, i.e. dc link or the mains. The feeding of power back to dc source raises the dc link voltage. Depending on the load conditions and speed, this can reach dangerous levels unless the additional energy is returned to ac mains by using the converter in the inverter mode. Regenerative braking is possible only with fully controlled drives. The block diagram of regenerative braking is shown in Fig. 7.24.

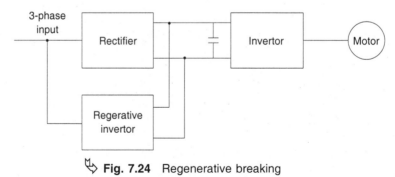

**Fig. 7.24** Regenerative breaking

### Advantages of Transistor PWM dc Drive Over Thyristor Drives

(a) The PWM drive has a high form factor of approximately 1.
(b) Less heating of the motor and an increased torque output (about 20% more than thyristor drives).

(c) Possesses a high frequency response resulting in better surface finish, low machining stress and no resonance problems.
(d) Increased reliability in operation (transistors are used instead of SCRs).

Therefore transistor PWM amplifiers are used advantageously as feed drives for fast and accurate turning and drilling machines. Figure 7.25 gives a schematic block diagram of a typical transistor PWM dc feed drive.

The speed command is given out from the CNC system. The command voltage will be 0 to ± 10 V dc proportional to the required speed of the motor.

*Example* If the motor's maximum speed is 2000 rpm, then 10 V corresponds to the maximum speed (2000 rpm clockwise); –10 V command corresponds to 2000 rpm counter-clockwise. Let the required speed be 500 rpm clockwise, then the system command output will be + 5 V dc. Therefore 5 V dc at the input of the drive amplifier will give the output + 100 V dc (typical value) to the servomotor which corresponds to 500 rpm.

### 7.5.5 AC Servo-drives

Though dc motors are commonly used for variable speed applications, they have certain inherent disadvantages like:

(a) Regular maintenance is required by dc motors
(b) Bulky in size
(c) High inertia of the rotor limits the maximum permissible acceleration
(d) Commutator brushes produce sparking
(e) Additional cooling fans are required to cool the armature
(f) Commutator limits overload and overspeed.

AC servo-drives eliminate all these disadvantages and are best suited for high dynamic response and maintenance-free operation.

### 7.5.6 AC Servo Feed Motor

These are three-phase permanent magnet synchronous motors with built-in brushless tacho and position encoder. The rotor consists of a permanent magnet and the stator contains the three-phase winding. On the rotor, little magnetic tiles are created in order to create three-pole pairs. This means that there are six poles on the circumference. The characteristic features of an ac servomotor are:

(a) High power density with low weight
(b) Low rotor inertia
(c) Constant continuous torque and constant overload capacity over the full speed range
(d) Additional cooling of the motor is not required

### 7.5.7 AC Servo-drive Amplifier (Inverters)

A conventional ac induction motor can be operated by direct connection to a three phase supply. The motor speed is directly proportional to the frequency of the ac mains. Since

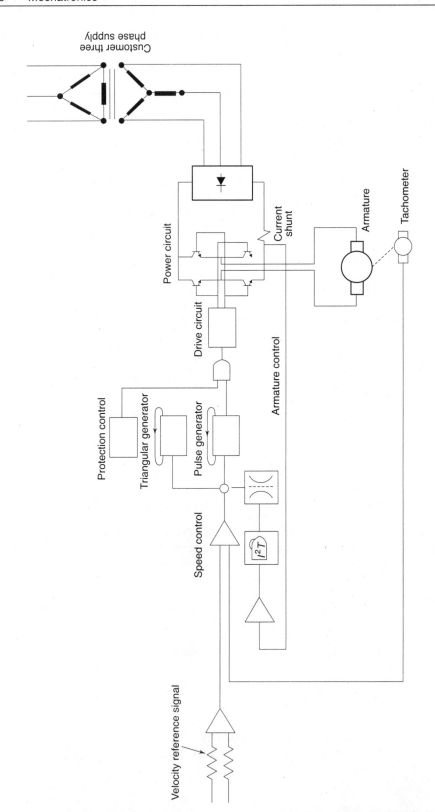

**Fig. 7.25** Typical transistor pulse width modulated dc feed drive

the frequency of the ac mains is constant, say 50 Hz, the motor with a fixed number of poles will always run at a fixed speed given by

$$n = 120 \frac{f}{p}$$

where
$n$ = motor speed
$f$ = frequency of the supply
$p$ = number of poles

Thus, if the number of poles of a motor is constant, then the speed of the ac motor is directly proportional to the frequency. Hence by varying the frequency, speed control can be achieved. However, in addition, the V/F ratio (voltage to frequency), needs to be maintained constant in order to keep the torque constant over a given speed band, where $V$ is the ac rms voltage applied to the motor and $f$ is the frequency of the output voltage.

An ac servo-drive consists of a converter section to convert the ac to dc and an inverter section which inverts the dc to ac with required frequency. The intermediate stage between the converter and the inverter is the dc link (Fig. 7.26).

### Elements of an ac PWM Transistor Module (Fig. 7.27)

(a) Input rectifier unit—rectification of the incoming supply voltage.
(b) DC link capacitor—provides smoothening of energy during motoring and stores energy during braking conditions.
(c) Power supply/monitoring section—generation of control and auxilliary supplies. Also monitoring and enabling functions.
(d) Output power section—switching of output power transistors and monitoring the power section for short circuit.
(e) Controller/trigger section—controlling the ac output voltage by means of speed controller with secondary current control and phase width controlled trigger set.
(f) Coordinator—distribution of pulses to power transistors and processing of current actual value and speed actual value.

### 7.5.8 Brushless dc Drive (Advanced Variable Speed ac Drive)

The brushless dc motor control provides smooth and efficient power and speed control which is not measured in per cent of rpm but in relation to physical shaft position within a single revolution. All these come in a compact package which exceeds the performance of a conventional brush type dc motor control.

***Mode of operation*** The mode of operation of power section of three phase ac servomotor is similar to that of a four-quadrant operation of a dc servomotor (Fig. 7.28).

The three-phase ac power is converted to dc by the bridge converter which charges a bank of storage capacitors called the dc bus. The size of the bus varies depending upon the connected load.

The rectification is accomplished by six diodes or thyristors which may be in a single package or in several modules. The diodes are protected by input fuses. The selection of

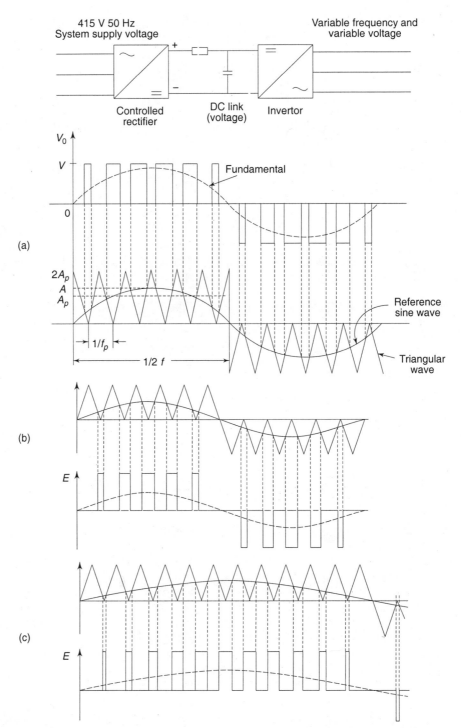

**Fig. 7.26** Formation of firing instants for pulse width modulated wave (a) at maximum output voltage (b) half maximum output voltage (c) at half voltage and half frequency

Drives and Electricals  269

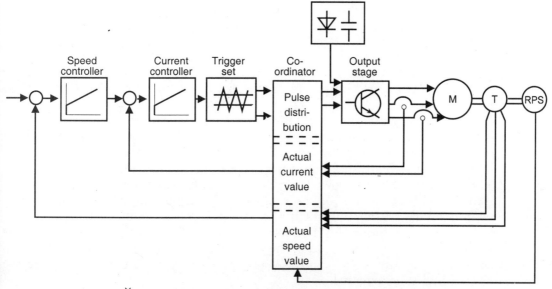

Fig. 7.27  AC pulse width modulated transistor feed drive

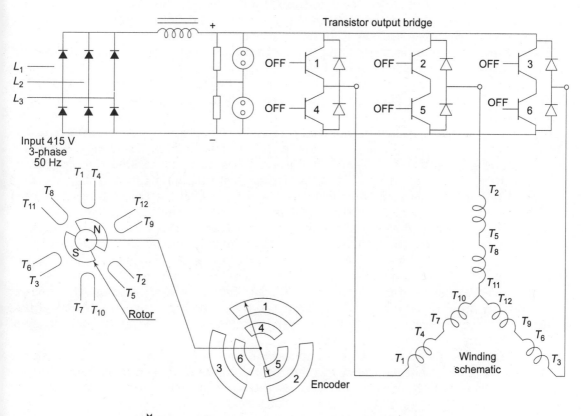

Fig. 7.28  Basic power circuit of brushless dc drive

incoming fuses is based on speed and interrupting capacity. An input choke in the dc side of the rectifier bridge protects against the line transients and limits the rate of change of current.

The input section of this controller is self regulating. The maximum voltage possible across the capacitors is $\sqrt{2}$ times the incoming line voltage. Initially, before the motor is switched ON, the capacitors are charged to this peak value which forms the bus voltage. When the motor is started, it uses power from the bus to perform the work required. The effect of this is to partially discharge the capacitors lowering the bus voltage. With three-phase input power, there are six periods in each cycle of ac when the line-to-line voltage is greater than the capacitor voltage. The capacitors will only draw current from the power lines when the bus voltage is lower than the instantaneous line-to-line voltage and then it will draw enough power to replenish the energy consumed by the motor since the last time the line-to-line voltage was greater than the capacitor voltage.

The torque in a motor is a function of current. Power is a function of speed and torque. Even though the current required by the motor to develop the torque may be large, the actual power used is small at low speeds. The energy drawn from the capacitor bank is the actual power used by the motor and the energy drawn from the supply mains is the actual power supplied to the motor.

The brushless dc motor control is capable of running at very low speeds at very high torques while drawing very low current from the ac mains. The result is that the rms current at the input of the brushless dc motor control is directly proportional to the output power of the motor rather than being proportional to the motor's load.

A brushless dc motor is wound like an ac induction motor, but it uses permanent magnet on the rotor instead of shorted rotor bars. There are three power carrying wires going to the motor. Each of these wires has to be, at synchronised times, connected to either side of the dc bus. This is accomplished with a six transistor power bridge as shown in Fig. 7.28.

Applying power to the motor requires turning ON one transistor connected to the positive side of the bus and one transistor to the negative side of the dc bus, but never the two transistors in the same leg of the output. When these two transistors are turned ON, the entire bus voltage is applied to the windings of the motor through the two wires connected to these transistors and the current will flow until the transistors are turned OFF, provided the counter emf of the motor is smaller than the bus voltage.

Due to the inductive windings of the motor, the current will not stop instantaneously when the transistors are turned OFF. It will decay quickly but the voltage in the bridge would rise dangerously if a suitable RC snubber network is not provided in the circuit to prevent the rate of rise of current. This action is termed as regenerative action.

If the two transistors were turned ON and left for any length of time, the current would rapidly build to very high levels. Thus the transistors are turned ON for definite intervals for a given speed. The variable speed function is realised as the frequency of switching off the transistors is altered.

Figure 7.28 also shows the schematic representation of the windings of the brushless dc motor. The connection shown in the figure is a single-wye. The explanation for sequential switching of the power transistor is the same as that for any PWM drive as applied to a dc motor.

## Construction of a Three-phase ac Servomotor

The motor consists of a laminated stator in which slots for the armature windings are inserted. Two slots are occupied with one "conductor". On the rotor, little magnetic tiles are fixed, so as to create 3 pole pairs. This means that there are 6 poles on the circumference. Figure 7.29 gives the sectional view of ac feed motor.

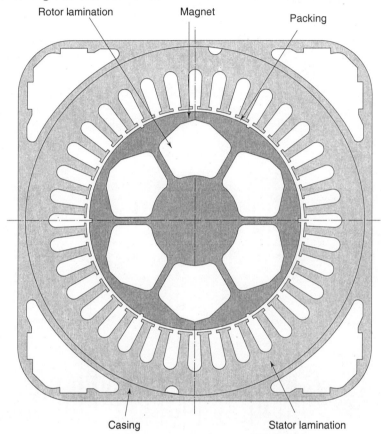

**Fig. 7.29** Sectional view of ac feed motor

## Electronic Commutation of Three-phase ac Servomotor (Brushless dc Motor) (Fig. 7.30)

In an electronically commutated dc machine the exciting field moves. The motor current is switched by transistors to corresponding stator windings depending on the rotor position.

A rotor position sensor, mounted on the motor, signals which stator windings are at that moment in the magnetic field. The sensor signal is evaluated and the converter switches on the current for the corresponding stator windings.

By this the motor develops a homogenous torque and is not a stepping motor. To develop a homogenous torque a minimum of three-stator windings are necessary. The term "three-phase" is based on this fact.

This motor cannot be connected directly to the incoming mains supply.

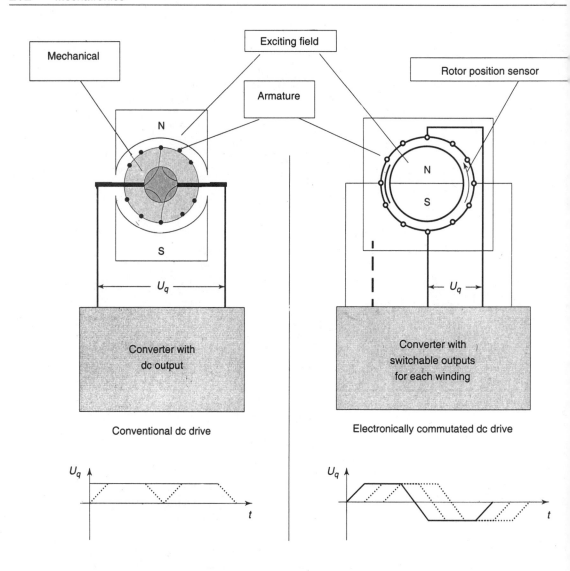

**Fig. 7.30** Electronic commutation of three-phase ac servomotor (brushless dc motor)

### Rotor Position Sensor (Fig. 7.31)

The rotor position is sampled via magnetic fork type barriers. A disc with three soft iron shutters at a distance of 60° is fixed with a cone. The magnetic fork type barriers are connected to the stator with clamp screws. For adjustment, the screws can be opened and the barrier support can be turned. The signals are led to the converter via a plug.

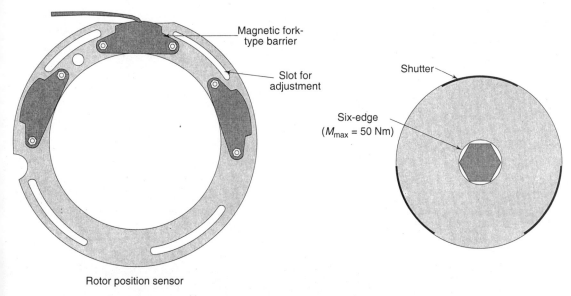

**Fig. 7.31** Rotor position sensor

## Hall Effect (Fig. 7.32)

The hall effect switch is handled by a soft iron shutter which is placed in the air gap between magnet and hall sensor. The barrier short-circuits the magnetic flux before the hall sensor.

The open collector output is current carrying (low), if the barrier is outside the air gap and locks (high) as soon as the barrier penetrates the air gap. As long as the barrier is placed in the air gap, the output remains closed.

Based on the static operation mode, the signal performance is independent of rotation speed. The output signal slopes are independent of the operation frequency.

The circuit has an integrated overvoltage protection against most voltage peaks using a Schmitt trigger type output section. The maximum output current of 40 µA of the open collector causes the most electronic circuits triggered directly.

### 7.5.9 Digital Main Spindle Drives

Microprocessor based digital drive technology provides the advantages of increased accuracy, precise adjustability and reproducibility of set points and controlled parameters. Control variables are no longer adjusted with the aid of potentiometers but are adjusted by using parameters. Further, a higher functional scope can be realised with software which permits a degree of user-friendliness.

A microprocessor controlled transistor PWM converter with an induction motor forms a powerful spindle drive system, which can meet the requirements of modern machine tools. The motor is a separately ventilated induction motor which is designed to obtain variable speeds on spindles. They have the following main features:

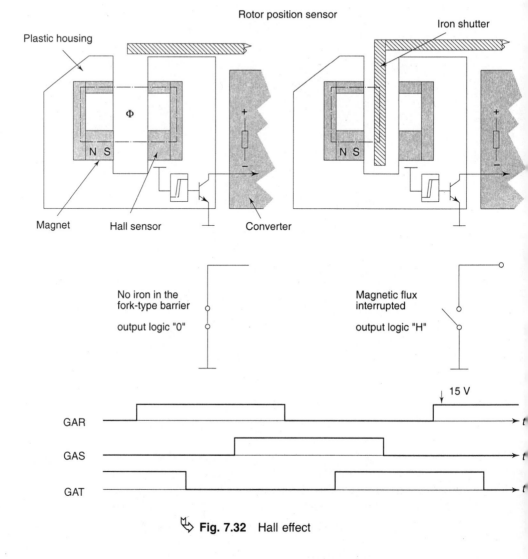

**Fig. 7.32** Hall effect

(a) Rugged, four-pole squirrel-cage rotor
(b) Brushless detection of position and speed
(c) High degree of protection, IP55 (details of IP degree of protection in Annexure 7.1)
(d) High rotational accuracy
(e) Large speed range of at least 1:1000
(f) High maximum speeds of up to 20, 000 rpm
(g) Full rated torque available even at stand-still

Figure 7.33 shows the block diagram of a typical digital spindle drive with a vector control.

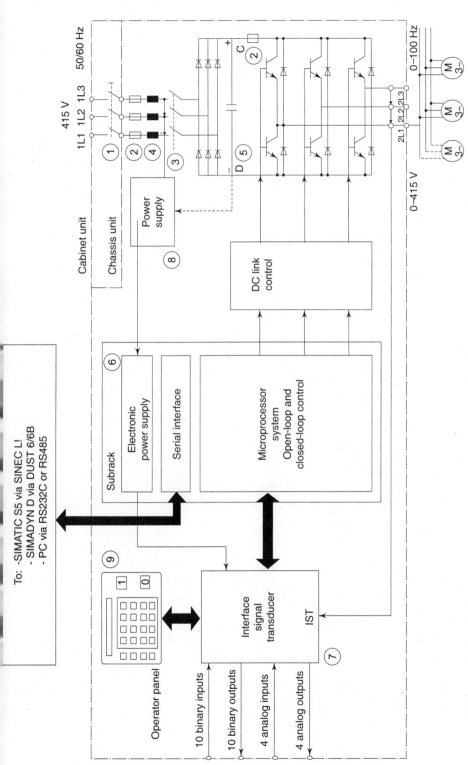

**Fig. 7.33** Digital spindle drive with a vector control

1. Main switch (cabinet units only) 2. Line fuses and dc link fuses for converter protection device 4. Line-side commutating reactor for reduction of noise feedback 5. Power section with line-side rectifier, dc link capacitors and inverter 6. Subrack with control electronics modules and extra slots for expansions and options 7. Interface/signal transducer for set points, actual values and control terminal block for external control 8. Switched-mode power supply with enegy back-up from the dc line (for supply voltages of up to 500 V) 9. Standard operator panel for start-up and local control of the converter

The speed of an ac induction motor is determined by the frequency. The drive converter/inverter provide the motor with a variable frequency, variable voltage three phase ac supply. This is achieved in two steps using line side converter and load side inverter. The line side converter rectifies the line side voltage and generates a constant dc voltage. The constant dc voltage in the dc link generates the three-phase variable ac supply from the dc voltage.

The vector control technique is adopted which enables the preset torque set points to be accurately maintained and effectively limited.

## 7.6 DRIVE OPTIMISATION

The drive optimisation or drive tuning is a procedure to be carried out during commissioning of a drive along with the machine slide. The various servo parameters on the drive amplifier and on the CNC are to be adjusted to get optimised servo characteristics of the slide. Normally, the servo-drive manufacturer will tune the amplifier to match the motor; but, the behaviour of the servomotor is influenced by the mechanical transmission elements coupled to it. Therefore, it is required to re-tune some of the servo-drive parameters to achieve an optimum behaviour.

Optimisation is carried out after the drive is installed onto the machine and after the CNC is established.

### 7.6.1 Measurement of Open-loop Response and Optimising Velocity Loop

In order to adjust any particular drive parameter, it is necessary to measure and record the present behaviour of the motor coupled to the slide. The following measuring procedure for recording the open-loop step response can be adopted. A typical set-up is shown in Fig. 7.34.

- Position control loop is opened (disconnect the CNC command wires)
- Connect a variable dc source through a switch to the reference input of the speed regulator
- Set the velocity command to zero and adjust the drift potentiometer (parameter) such that the motor is not drifting at zero command
- Set the velocity command (dc source) to +0.9 V dc and set the speed adjust pot (parameter) so that the motor rotates at 10% of its required maximum velocity.
- Connect the following signals to the recorder
  $n_s$ – Speed set value (command)
  $n_i$ – Actual speed value (tacho)
  $I_i$ – Actual current value

The following reference value steps (from zero to maximum values) are recommended. 0.1–0.2–0.5–1–2–5–10–20–50–100% of the maximum speed.

Generally, the step responses have to be evaluated between 1 and 10% of the maximum speed. The overshoots are controlled by adjusting the current limit parameter and gain parameters. If required, the damping coefficient parameter is adjusted to achieve the following results.

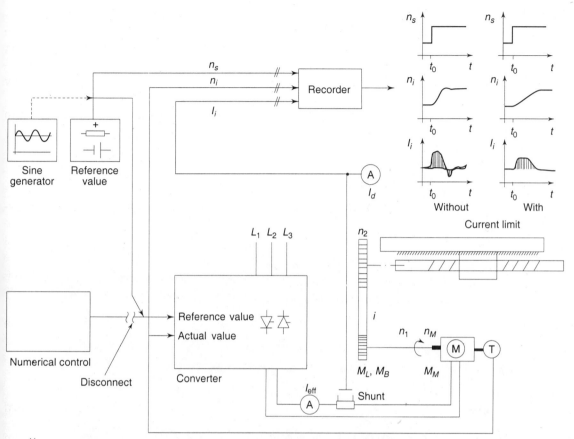

**Fig. 7.34** Measuring arrangement for recording the step response and measuring the current limit, the nominal angular frequency of the feed drive, and the frictional characteristics

$T_a$ – Overshoot magnitude should be 20–30%
$T_{AN}$ – Response time to be within 35 and 8 ms
$T_u$ – Relay time from 10 to 1 ms

In case of small response steps in the range of 0.1–1%, care should be taken to ensure that the response time does not increase too much. Also, for a ± reference change the reversing time should not exceed three times the response time ($T_{AN}$). Depending on the size of the external inertia of the drive, the current limit is reached with a step at approximately 10–20% of the maximum speed.

## 7.6.2 Closed-loop Response and Optimising the Position Loop

The set-up is the same as the open-loop except that the external dc source is disconnected and the velocity command is connected through the CNC system. The position feedback is also connected as shown in Fig. 7.35. The closed-loop measuring procedure and method of optimising the feed drive is as follows:

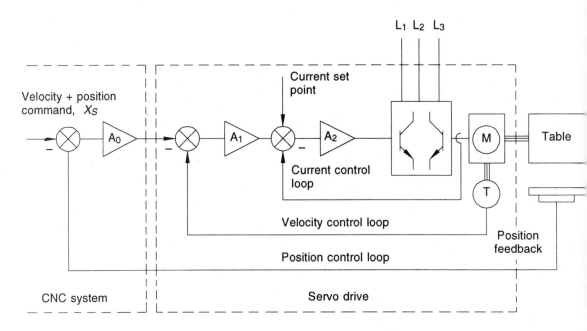

A₀, A₁, A₂ : Amplifiers
L₁, L₂, L₃ : Three phase power supply
T : Tacho generator
M : Motor

**Fig. 7.35** Block diagram of position control loop

Establish the position control loop through CNC. Set the CNC drive parameters like drift, multigain, $K_v$ factor, acceleration and maximum velocity for the particular axis
Signals for the following are recorded.

$n_s$ – Command speed value
$n_i$ – Actual speed value (tacho signal)
$I_i$ – Actual current value

Program the control for 0.1–0.5–10–50–100% of the maximum speed with a particular distance command and record the values of $n_s$, $n_i$ and $I_i$. Figure 7.36 shows the tacho and current waveforms before and after optimisation.

Critically damped characteristics correspond to optimised behaviour of the system. The parameters like $K_v$ factor (CNC), acceleration (CNC), proportional and integral gain (drive), damping factor (drive) are adjusted and set to achieve the optimised response. The procedure is repeated such that the slide behaviour is nearer to the critically damped characteristics at all velocity commands. This procedure is to be repeated and optimised for setting the maximum value of the acceleration and position loop gain parameters.

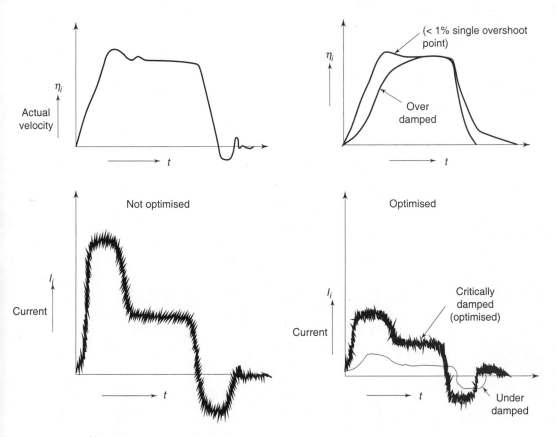

**Fig. 7.36** Tacho and current wave forms before and after optimisation

### Results of a Better Optimised Slide

1. With higher acceleration (values of 1–1.5 m/s² are common in typical CNC machine tools) in the absence of an overshoot, the slide will be able to attain the programmed velocity within a short time which results in reduced cycle times for positioning blocks.
2. With a high position loop gain, the slide will be able to correct at a faster rate for any disturbances in the system; in other words, a better response. Position loop gain in the range of 1-2.5 mm/min/micron will result in better servo response characteristics.

### 7.6.3 Tuning the Drive for the Smallest Position Increments

For a uniform movement and accurate positioning of the slide at the smallest feed rates, it is necessary that the drive should respond to every small position increment command from the CNC system. To ensure this, the following measuring process and tuning is done.

(a) The machine slide is moved by a few millimetres in the positive direction.
(b) A dial indicator with a resolution of 0.001 mm is mounted on the slide
(c) With the help of CNC, the slide is moved in jog (incremental) mode with the smallest position increments. After each command value output, its execution is checked with the dial indicator.
(d) The drive response to minimum increments can be improved by slightly increasing the proportionality gain (P-gain) parameters of the drive amplifier.
(e) The reversal error is measured and compensated as backlash compensation in the CNC system parameter.

After eliminating the reversal error, the single position increments are recognised and executed. No sticking due to static friction should occur.

## 7.6.4 Spindle Drive Optimisation

The spindle drive tuning is similar to that of open-loop feed drive optimising. In addition to the tacho response, the linearity of speed with reference to the command from zero to maximum speed is to be checked and adjusted if required.

Some of the latest advanced microprocessor controlled digital spindle drives have built-in software to tune itself to the motor and the load. The first level of tuning is done automatically by the computer and can be further improved by manual tuning (if needed) (Fig. 7.37).

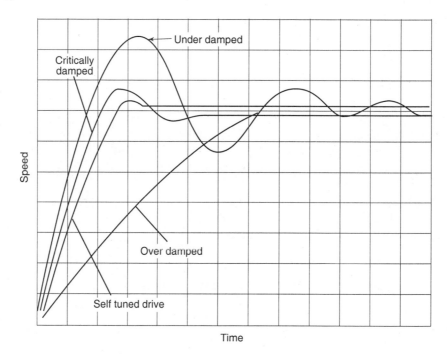

**Fig. 7.37** Spindle drive optimisation

## 7.7 DRIVE PROTECTION

One of the important features of any drive amplifier or a controller is to provide a high degree of protection to the motor and the drive itself. Every drive amplifier consists of a monitoring module to perform protective functions such as:

(a) Voltage monitoring
(b) Fuse failure monitoring
(c) Motor connection continuity check
(d) Tacho feedback circuit fault monitoring
(e) Motor current limiting
(f) Power ($I^2 t$) monitoring
(g) Voltage limiting (regenerative voltage)
(h) Maximum speed monitoring
(i) Motor temperature monitoring
(j) Internal circuit function monitoring
(k) Wrong connection check

In case of any of the above faults, the drive will stop and the status is displayed by light emitting diodes (LED). Also, the fault message is communicated to the CNC system to draw the operator's attention.

In addition to the internal monitoring circuits, drive amplifiers are also protected against overload, short circuits, and no-voltage circuits externally in the control panel by the machine builder.

Feed motors should have IP 64 degree of protection with built-in thermistor to monitor the motor temperature. Spindle motors usually confirm to IP 54 degree of protection with built-in thermistor protection.

## 7.8 SELECTION CRITERIA FOR AC DRIVES

Following factors are to be borne in mind while selecting an ac drive.

- Maximum torque required to start and run the machine and the type of load or duty.
- Maximum speed required.
- If a motorised drive can be used, what electrical service is available and whether open (dripproof), totally enclosed or explosion resistant motor construction is required.

The determination of torque requirement is the most important step in the selection of drive for any application. The torque rating one requires is the product of the maximum torque required to start and run the machine multiplied by the service factor resulting from the type of load or duty. Over 90% of all industrial drive applications other than pumps require constant-torque, although many machines have peak loads that occur occasionally, which must be taken into consideration while choosing the proper drive for the application (Table 7.3).

Torque is a twisting or turning effort which tries to cause, maintain, or change rotation. Torque requirements for driven machines can be classified as:

- Starting
- Accelerating
- Running

Starting torque is the torque needed to start the machine. It is usually greater than the torque needed to maintain motion. Starting torque has two parts, viz. static-friction torque and process-demand torque.

- Static-friction torque is always greater than running friction torque. A simple example of this phenomenon can be best illustrated below. Try pushing a chair or table slowly. It takes more force to start it than to keep it in motion. This is true of all machines.
- Process-demand torque varies with the application. It often determines the size of the drive. It can be seen that while hand-cranking a gasoline engine, one needs a large force to pull the crank over the first compression stroke. Once in motion, the next compression is easier.

TABLE 7.3

**Service factors**

| Type of load | Type of duty | |
|---|---|---|
| | 8 to 10 hours a day | 24 hours a day |
| Uniform | 1.0 | 1.5 |
| Moderate shock | 1.5 | 2.0 |
| Heavy shock | 2.0 | 3.0 |
| Reversing service | | |
| Low inertia | 2.0 | 3.0 |
| High inertia | Not recommended | Not recommended |

*Example*: Starting torque measures 100 inch-pounds, running torque measures 60 inch-pounds, load is uniform, 10 hours a day, so service factor is 1, thus 100 inch-pounds × 1 = 100 inch-pounds torque required for this job.

Accelerating torque is the torque needed to bring the machine to running speed. High inertia machines need special consideration.

Running torque is the torque needed to drive the machine after it accelerates to running speed.

The torque can be determined by calculation if the horsepower and speed are known. If several speeds are required, use the lowest, for that is generally where the torque will be greatest. The formula for converting horsepower and speed to inch-pounds of torque is

$$\frac{\text{hp} \times 63025}{\text{rpm}} = \text{torque (inch pounds)}$$

For easy reference, 1 hp at 1800 rpm equals approximately 36 inch-pounds torque.

## Estimating Starting Torque

The starting torque may be estimated after running torque is known. From Table 7.4 the type of application similar to the one under consideration may be determined. Then, the running torque is to be multiplied by the respective multiplier.

For best results, one should select a drive that has adequate torque for the job and has a maximum speed that most closely matches the maximum speed required.

## 7.9 ELECTRIC ELEMENTS

The schematic diagram in Fig. 7.38 shows a typical CNC system. Other than drives there are several electrical and electromechanical devices used as control elements. These elements form the input or output devices for the CNC system.

### 7.9.1 Input Elements

Some of the commonly employed input elements are push-button, foot switch, proximity switch, float switch, relay contact, photo transistor switch, selector switch, pressure switch, limit switch and flow switch.

Limit switch (LS) (Fig. 7.39) or a push-button (PB) switch is a control which makes or breaks an electric circuit.

Proximity switches are non-contact type switching devices. There are two types of proximity switches.

(a) Inductive proximity sensors—operated on inductance principle
(b) Capacitive proximity sensors—operated on capacitance principle

### Operating Principle of Inductive Proximity Sensor

An alternating electromagnetic field is developed at the sensing face of a proximity switch. When a metallic object enters this field, energy is absorbed. The amplitude of the electromagnetic field is consequently reduced. This reduction in amplitude is processed internally, amplified and an output is generated.

Proximity switches are used where reliable operation even under extreme conditions (insensitive to water, lubricants, coolants, dust, vibration, etc.) is required. Proximity switches can operate up to 5000 times per second (operating frequency of the switch) and have an exceptionally long life.

**284** Mechatronics

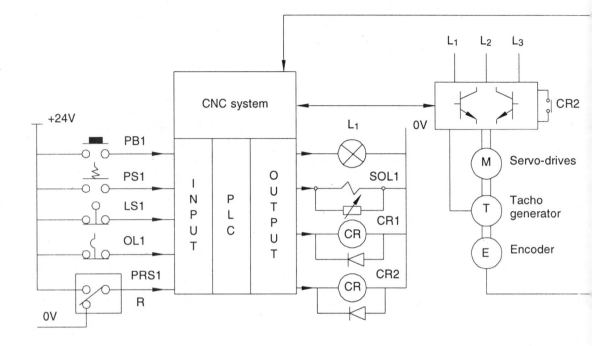

CR : Control relay
LS : Limit switch
OL : Over load
PB : Push button
PRS : Proximity switch
PS : Pressure switch
SOL : Solenoid

**Fig. 7.38** Schematic of a CNC system with an electrical interface

### 7.9.2 Output Elements

Output elements that are commonly used are:

- Indicating lamps
- dc control relays (electromagnetic)
- Power contactors
- dc and ac solenoids
- Electromagnetic clutch and brake
- Solid state relay

Drives and Electricals **285**

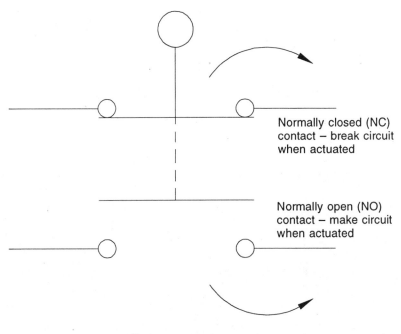

**Fig. 7.39** Limit switch

**TABLE 7.4**

| Types of applications | Running torque multiplier |
|---|---|
| General machines with ball or roller bearings | 1.2–1.3 |
| General machines with sleeve bearings | 1.3–1.6 |
| Conveyors and machines with excessive sliding friction | 1.6–2.5 |
| Machines that have "high" load spots in their cycle like some printing and punch presses and machines with cam- or crank-operated mechanisms | 2.5–6.0 |

A control relay (CR) is an electromagnetic device excited through an ac or dc electric coil. The dc relays are used as interface between the CNC-PLC and the ac or dc power switching devices (Fig. 7.40).

Contactors (C) are also electromagnetic devices which are excited with ac voltages (110 V or 220 V or 440 V) and are used to switch large current circuits. Usually contactors are used for ON/OFF functions of induction motors, induction coils, drive power circuits, etc. Power contactors are designed to switch currents up to several hundreds of amperes at 440 V ac three-phase.

The electrical power circuit diagram in Fig. 7.41 shows the connections of a typical power contactor used for controlling the ON/OFF function of the induction motor.

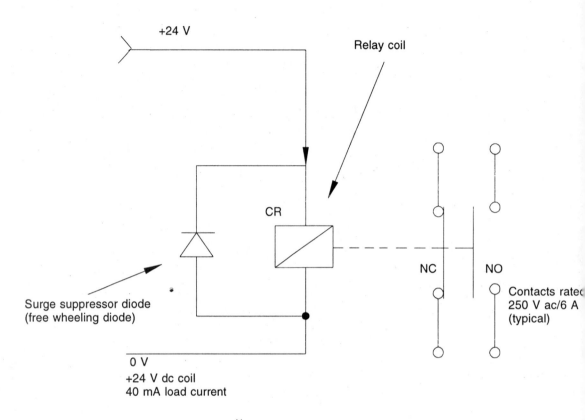

Fig. 7.40 DC control

When using power contactors or switching relays in an electric circuit for a CNC machine, care has to be taken to suppress the electrical switching noise (electromagnetic interference) so that the electronic circuitry of the drives or the CNC system is not affected.

Any electromagnetic switching device is like a generator of electromagnetic interference. Hence, incorporating surge suppressors (Fig. 7.42) to the switching coils will drastically reduce the switching surges.

### 7.9.3 Overload (OL) Relays

Bi-metallic thermal overload relays are very commonly used as overload protection devices for various ac motors such as hydraulic pump motor, coolant pump motor, lubrication motor, blower induction motor, or any other power ac circuit. The overload (OL) relay, when connected in series with the power circuit, will open out when the current increases beyond a preset value.

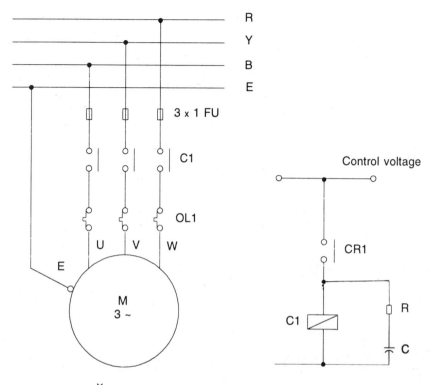

**Fig. 7.41** Electrical power circuit diagram

### 7.9.4 Miniature Circuit Breaker (MCB)

An MCB is a protective device which will provide both overload and short circuit protection when connected in a circuit. Hence an MCB will replace a fuse and a bi-metallic overload relay. When an MCB trips, it has to be reset manually. Compact MCBs with auxiliary trip contact are used in the electrical control panel for CNC machines.

## 7.10 WIRING OF ELECTRICAL CABINETS

The design and wiring of an electrical cabinet for a CNC machine requires certain special guidelines. Unlike a conventional panel, the CNC panel may house the CNC system, servo-drive amplifiers, position measuring modules, or any other sensitive electronic circuits. The cabinet may also house power transformers, relays, switching contactors, inductors, power supplies, and other ac power circuits.

Great care has to be taken in the design of the sheet metal panel and the electrical layout, and also while wiring of the CNC panel. Following are some of the guidelines.

- Electrical panel should be as compact as possible, with provision for enough space between the elements.

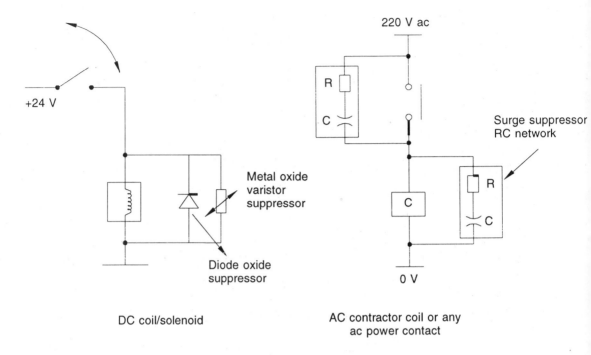

**Fig. 7.42** Surge suppressor circuits

- Power switching devices like contactors, transformers, chokes and power resistors are to be grouped and separated from the CNC system and drive amplifiers. Under special cases, instructions from CNC and drive manufacturers can be obtained to maintain minimum space around the electronic modules.
- Running ac power cables and the signal cables together should be avoided in a single raceway. To the extent possible, these cables should not run parallel for long distances.
- Too many terminations of power and signal cables should be avoided.
- Extreme care has to be taken to design the earth pattern of various devices, cables, and connectors. Ground loops should be avoided.
- Earth wire should never be used as neutral for single phase loads.
- It is preferable to use plugs and sockets as disconnecting means. Terminals should be avoided.
- While routing the CNC signal cable, care has to be taken to clamp every cable properly such that no strain comes onto the connector or connecting terminal.
- The layout and wiring methods and spacings should be as per electrical standards.
- The cabinet temperature should be maintained within 27°C and the humidity controlled accordingly.
- Cabinet should be of high quality and protected against external entry of smoke or dust.

If the design and wiring of the CNC panel is not taken care of, it may result in:

- Electromagnetic interference resulting in a malfunction or failure of electronic circuits.
- Heating up of microprocessor modules because of power devices in the vicinity.
- Problems on CNC display, e.g. cathode ray tube (CRT) display with floating characters.
- Difficulty in servicing.

## 7.11 POWER SUPPLY FOR CNC MACHINES

Indian standards for industrial power supply is based on three-phase, four-wire system, i.e. three phase + earth at 415 V ac, 50 cycles. The electrical system of the machine is designed to suit this specification. The CNC control panel consists of highly sensitive electronic circuits. However, because of very poor regulation of the power supply and too much variation in supply voltages, it is necessary to use a voltage stabiliser for the machine control.

A servo-controlled voltage stabiliser with the following features is preferred:

- Three-phase unbalanced input correction
- Input voltage: 320 to 450 V ac
- Output voltage: 415 V ac to within ± 1%
- Servo-controlled high rate of correction; min. 20 V/s per phase
- No waveform distortion
- Built-in filter for spike suppression
- kVA capacity suitable for machine rating

## 7.12 ELECTRICAL STANDARD

Standards pertaining to the interface between CNC systems and the electrical equipment of machine tools, which are controlled by CNC are given in IMTMA (Indian Machine Tool Manufacturers' Association) standard 12-1993: Interface between CNC system and machine tools (Annexure 7.2).

## 7.13 ELECTRICAL PANEL COOLING (AIR CONDITIONING)

As the electronic circuits are sensitive to voltage fluctuations, they are also sensitive to temperature changes. For example, the CNC system is designed to function up to 45 °C. The electrical panel which houses the other power devices also generates additional heat. In case of servo-amplifiers, the current ratings are specified at 20 °C. Hence an increase in temperature will change the amplifier's maximum current output. Therefore, it is necessary to maintain the temperature inside the cabinet of the electrical/electronic system within the normal value. Electronic panel coolers are specifically designed for this application. The following are the basic requirements of an electronic panel cooler:

- Continuously operating compressor—because start/stop type compressor will induce vibrations and electrical noise onto the electronic panels.
- While running the compressor, running, should generate very low level of vibration.
- Compact design for ease in mounting onto the panel.
- Closed-loop temperature control.
- Humidity controlled by incorporating heater coils.

In order to select a panel cooler, the cooling capacity is decided on the basis of the amount of heat generated by various electrical elements within the cabinet and the ambient temperature. Accordingly, the correct functioning of the panel cooler is sensed by a thermostat placed within the electrical cabinet and is interlocked with the operation of the machine.

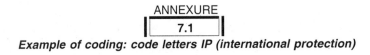

ANNEXURE 7.1

*Example of coding: code letters IP (international protection)*

A system of classification of types of enclosure for electric motors has been agreed internationally in International Electric Commission, UK (IEC) in its standard IEC 34-5 (also DIN IEC 34, VDE 0530, part 5). The system accounts for protection by the enclosure in three ways:

(a) to persons against contact with live and moving parts;
(b) to the motor against ingress of solid foreign bodies;
(c) to the motor against harmful ingress of liquid.

The first two of these are inter-related and the system groups them as one. The classification does not include special systems of protection as suitable for example for hazardous atmospheres or corrosive vapours. The classification consists of the letters IP followed by two numbers. The first number indicates the degree of protection regarding contact with live or moving parts and regarding ingress of solid foreign bodies. The second indicates the degree of protection against harmful ingress of water. The two numbers may be followed by a letter (S or M) indicating whether the motor was tested against ingress of water with the motor stationary or running respectively. The absence of a letter indicates that the motor was tested under both conditions. The first number has the following significance:

0   non-protected
1   protected against solid bodies greater than 50 mm diameter
2   protected against solid bodies greater than 12 mm diameter
4   protected against solid bodies greater than 1 mm diameter
5   protected against dust

The second number indicates the following degree of protection against ingress of water:

0   non-protected
1   dripping water
2   drops of water 15° from vertical
3   spraying water
4   splashing water
5   water jets
6   conditions on ship's decks
7   immersion of limited duration
8   submersible

In addition, the letter W between IP and the number indicates weather protection, i.e. when the motor can withstand operation in rain or snow and in the presence of airborne particles. The possible combinations of first and second numbers are extensive in theory, but in practice only a few combinations are necessary. As examples, the following equivalents relate to some of the more common enclosures:

| | |
|---|---|
| IP 11 S | ventilated, drip-proof |
| IP 44 | totally enclosed |
| IP 54 | dust-proof |

Discussion took place during 1974 in IEC and British Standards Institution (BSI) concerning the need for an intermediate degree of protection between IP4 and IP5 to cover totally enclosed machines. The argument is that IP allows ingress of solid bodies of 1 mm diameter, while IP5 is a dust-tight enclosure. A totally enclosed machine can be said to lie between the two and a classification IP4-5 giving protection against solid bodies greater than 0.1 mm diameter has been suggested.

## ANNEXURE 7.2

### Indian Machine Tool Manufacturers Association Standard*
### Interface Between CNC System And Machine Tools

## 1. GENERAL

### 1.1 Scope

This standard applies to the interface between CNC system and the electrical equipment of machine tools which are controlled by CNC (For definitions, refer Clause 2).

### 1.2 Object

The purpose of this standard is to provide specifications for an interface between CNC and the other electrical equipment associated with the machine with the following objectives:-
(a) to ensure the safety of personnel and equipment;
(b) to standardize all mandatory and other features to the extent necessary for the safe use of equipment from different manufacturers;
(c) to define certain signals going through the interface, to make recommendations for standardizing their electrical characteristics and also to recommend methods of connection and installation.

## 2. Definitions

### 2.1 Computer Numerical Control (CNC)

The CNC is a computer based electronic equipment that receives commands in digital form from perforated tape or other types of input, as well as positional information of certain elements of the machine. The CNC interprets certain of these digital data as requirements for new positions of the machine elements and gives appropriate commands of direction and velocity. The CNC also interprets certain other digital data as command of velocity, of discrete functions, of actions etc.

### 2.2 Interface

The interface consists physically of the connections between the CNC and the machine including its electrical control equipment.
Functional areas of interface
Figure A 7.1 shows the area of interface between the CNC, the machine and the machine control equipment that are defined as 'machine/CNC interface'.

*Source*: IMTMAS: 12-1990

The interface is divided functionally into four groups of connections (see Fig. A7.1). The composition of the groups does not depend on the arrangement of enclosures; they are valid in both cases, when machine control and servo drives are physically separate, and when either or both are in enclosure of the CNC.

2.2.1 Group I: Drive Commands

Connections frm the CNC to the drive controllers including servo error signals (Refer Fig. A 7.2 and the motor connections.

This group includes in particular the connections to hydraulic servo valves and to servo drives which are used for driving CNC axes or machine spindles. It also includes the connection for velocity commands and outputs to the servo actuators.

Approved & accepted on : 1990-12-04    Total No of
Date of issue            : March 1991   Pages : 16

2.2.2 Group II: Interconnections with Measurement Systems and Measuring Transducers.

Connections for measurement of position on liner and rotary axes and of spindle orientation; also for other measuring transducers associated with the CNC such as adaptive control signals. This group includes the connections for associated excitation and supply voltages, also tachogenerator signals when these are connected directly to the CNC.

2.2.3 Group III: Power Supply and Earthing Circuit

Connections and recommended procedures for connecting the CNC and the various parts to the earthing circuit and to the main power supply system.

2.2.4 Group IV: ON/OFF and Coded Signals

ON/OFF signals, both mandatory and optional, interchanged between the CNC and the machine (see Fig. A 7.3 and A 7.4 and other signals not belonging to Groups I, II, and III.

## 3. General Requirements of Interface Connections

### 3.1 Connection of Shielding

The shielding of cable shall be connected to the frame or to the earthing circuit at the CNC. The CNC builder shall be responsible for these connections or shall indicate exactly their location if they are, as an exception, to be made outside the CNC.

The user shall not connect the shielding at any other point to any metallic structure, nor by any alternative route to the frame or to the earthing circuit, except by agreement with the CNC builder.

The shielding shall have electrical continuity from the CNC to the remote termination and the length of unshielded cable connection shall be minimum.

The shielding shall not be used as a connection nor as a current-carrying element between the devices at each end of the cable.

All shielded cables shall be externally insulated to prevent accidental contact with any other metallic parts. This outer insulation shall either afford mechanical protection of the cable or it shall be additionally covered with a mechanically protective outer layer or layers.

### 3.2 Routing and Spacing of Cables

Where conductors from different groups run parallel to each other in the same raceway and also at their points of connection to the CNC, they shall be separated by a sufficient distance and/or additional shielding shall

Drives and Electricals **293**

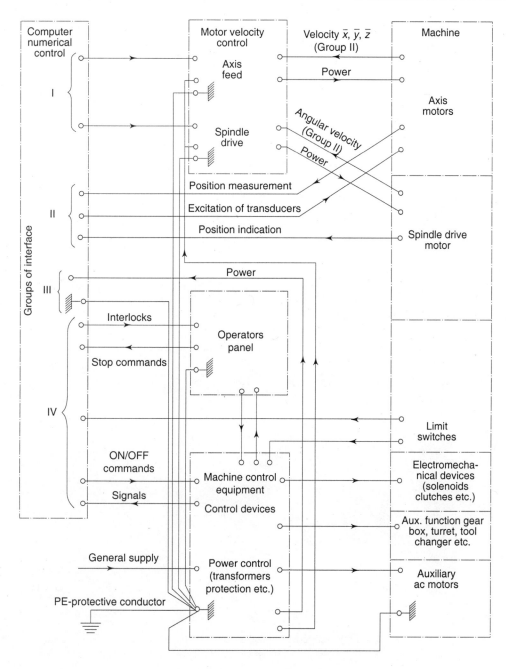

**Fig. A7.1** Relations between the computer numerical control, the control equipment and the machine

294  Mechatronics

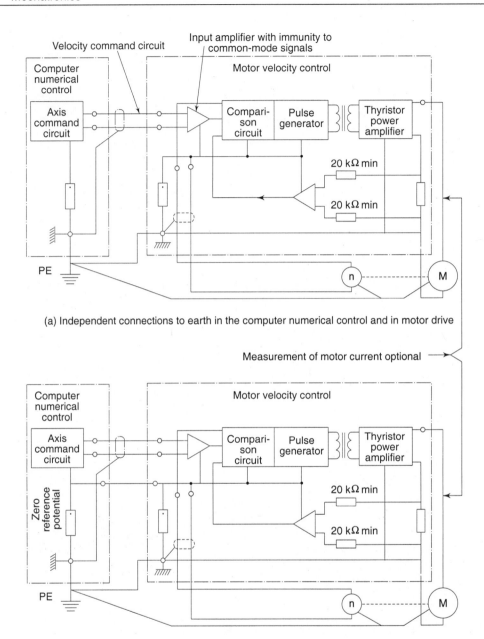

**Fig. A7.2** Group I of the interface velocity controls

reduce mutual interference effects to an insignificant level. Alternatively, each group shall have its own shield. It shall be the responsibility of the CNC builder to give the advice necessary to fulfill this requirement.

The CNC builder shall advise if it is necessary to group separately and shield the inputs and outputs of the Group IV connections.

In the case of Group II cables, connections to each measuring device shall be shielded separately.

Exceptions to this sub-clause are permitted in relation to connections of Groups I and II when their terminations at the machine are in proximity and they are functionally related and are low level signals.

Specific examples are velocity command signals, such as voltage signals to motor drives and current signals to hydraulic servo-valves. Either of these may be associated with connections to measuring transducers. However, such connections to measuring transducers associated with related low level signal connection shall be shielded accordingly.

### 3.3 Specification of Shielded Cables

The CNC builder shall recommend types of shielded cable suitable for use with the CNC. These shall have an insulating cover over the shield and have suitable resistance to mechanical damage. Unless otherwise specified, the interior wires shall be in twisted pairs or groups with a recommended pitch of one full turn in 15 mm or less. Insulation and stranding shall be suitable for the purpose and the mechanical duty.

## 4. Recommendations and Requirements for Group I Interface
(Velocity Command Signals, see Fig. A 7.2)

### 4.1 Isolation Between CNC and Motor Drives

Inside a motor control there shall be no electricity conductive connection between the power circuit of the motor and the CNC except as specified in Sub-clause 4.2. Connections to the earthing conductor shall be in accordance with Sub-clause 6.7.

Connections for the velocity command signals from the CNC to the drives for axis motors may or may not comprise a zero-reference potential connection (see Fig. A7.2a and A7.2b). The drive shall always be insensitive to common mode inputs between its input terminals and the frame.

This sub-clause does not apply to transistor switching devices.

### 4.2 Connections Between Control and Power Circuit

The manufacturer of the motor drive shall ensure that no intentional electrical path under 20 K Ohms resistance exists between the control and power circuits, except for any common connections to the earthing circuit. This includes any means for measuring motor current.

Any breakdown in the drive shall not have any damaging effect on the CNC.

### 4.3 Hydraulic Servo Valves

The CNC builder is recommended to provide for output current proportional to machine velocity or alternatively to advise the machine builder of appropriate voltage to current conversion devices.

### 4.4 Velocity Command Voltage Signals

This sub-clause applies when the axes of the machine are driven by proportional servo systems and the power controller for the motor is physically external to the CNC. The CNC output of 'position error' then forms a command of machine velocity.

This sub-clause also applies to commands of velocity of rotation of the spindles of the machine.

The maximum allowable length of cable between CNC and drive controller for velocity commands is as per the recommendations by CNC system builder.

4.4.1. It is recommended that the motor rated speed for a ±10 V signal should be chosen so that the voltage command for maximum motor rpm under the conditions of use lies between 7 V and 10 V. The CNC shall provide an output that is linearly proportional to the speed commanded, in the range of velocity corresponding to contouring on axis drives, and throughout the range on spindle drives. The manufacturer of the motor drive shall make appropriate provision for control of field energization where necessary.

4.4.2 The output impedance of the CNC for voltage signals shall be less than 100 Ohms.

4.4.3 The input impedance of the motor controller for voltage signals shall be more than 2 K Ohms.

## 4.5 Servo Energization

Depending on the presence of brakes, clamps or detents, on the accuracy and on the rigidity required, etc., the CNC builder shall agree with the machine builder* whether or not the servo loops for the machine axis drives remain fully excited when the axes are in position. It should be recommended that servo loop for machine axis to be removed after the axis is clamped wherever mechanical clamps are used.

## 5. Recommendations and Requirements for Group II Interface (measuring circuits)

### 5.1 Shielding of Connections

Sub-clauses 3.1 and 3.2 shall be compiled with and materials in accordance with Sub-clause 3.3 shall be used. In addition, where more than one shielded cable is used for connections to a measuring device, the shielding shall not be connected together at the device.

### 5.2 Routing of Connections

The machine builder shall follow the requirements of Sub-clause 3.2, particularly concerning the location of the cables from the measuring transducers with regard to connections of inductive load that are interrupted during the machine cycle and of armatures of motors.

### 5.3 Information from CNC Builder

The CNC builder shall specify the types of transducer that may be used, and he shall make recommendations on methods of connection.

## 6. Recommendations and Requirements for Group III Interface

(Power Supply and Earthing Circuits)

### 6.1 Power Supply for the CNC

#### 6.1.1 Rated Voltages

The rated voltages are of 24 V dc, 110 V or 220 V, 50 Hz and 115 V or 230 V, 60 Hz, normally single phase. It is recommended that in case of ac voltages, the CNC should be able to operate with either of the two frequencies 50 Hz and 60 Hz and to be adopted to the voltages 110/115 V and 220/230 V for example by changing the connections. If different equipment is necessary for these changes, then this shall be clearly stated by the CNC builder and the machine builder* shall inform the CNC builder of the frequency specified by the user.

---

* The expression 'machine builder' as used in this standard includes the builder or manufacturer of the machine control electrical equipment

If one of these standard voltages is used, the machine builder is responsible for supplying the necessary power for the CNC at that voltage.

### 6.1.2 Other Voltages

By agreement between machine and, CNC builders, the CNC may be supplied directly with the voltage of the supply network at the frequency specified by the user. In this case, the CNC may be equipped with an additional transformer or otherwise adapted to the required voltage.

### 6.1.3 Power Consumption of the CNC

The CNC builder shall indicate the nominal power consumption of the CNC in volt-amperes (excluding that of the drives) and includes this information on the CNC nameplate.

## 6.2 Connection to the Supply Network

The transformer supplied by the machine builder* (according to Sub-clause 6.1.1) or incorporated in the CNC (according to Sub-clause 6.1.2) is normally connected between two (or possibly three) phases. Connection between phase and neutral is permitted only with the written agreement of the user.
In the CNC, neither pole of the supply shall be connected to the frame or to the earthing circuit.

## 6.3 Independent Supply for Drives

6.3.1 The power for operating drives (for the axes and spindles) shall not be supplied from the same transformer as used for the CNC. It is recommended that, as far as possible, separate conductors be used for supplying such drives from the load terminals of the supply disconnecting device.

6.3.2 The power consumption of the motor drives shall be indicated separately from the consumption of the CNC, even if the drives are placed in the same control enclosure.

## 6.4 Protection Against Short Circuit and Overload

The protection against short circuit and overload shall be in conformity with Clauses 5.2 and 5.3 of IEC 204/1-1981.

## 6.5 Immunity to Externally Generated Electrical Interference

The following deviations in electrical interference found in supply voltage shall not cause the CNC system to malfunction.

| | |
|---|---|
| Voltage | 85% to 110% of rated voltage |
| Frequency | 50 Hz ±6% |
| Low harmonics | Total sum of 2nd, 3rd, 4th of 5th harmonics up to 10% rms of rated voltage. |
| Higher harmonics (6th to 30th) | Total sum up to 2% rms. of rated voltage |
| Radio-frequency voltages | 2% rms of rated voltage in the range 10 kHz to 10 MHz. Th CNC shall be immune to ratio frequency interference when applied either in common mode or in differential mode. |
| Impulse Voltage | 200% peak voltage up to 1 rms duration with a rise time between 0.5 us and 500 µs. |
| Voltage drops and interruptions of the supply | Reduction of 50% for one entire supply cycle, and as a separate event, supply disconnected or at zero voltage for a complete half-cycle, with intervals between successive events of at least one second. |

## 6.6 Earthing

Connections to the earthing shall be in accordance with Clause 5 of IEC 204/1 - 1981.

## 7. General Recommendations and Requirements for Group IV Interface

(ON/OFF Signals) (Refer Fig. A 7.3)
These recommendations and requirements apply to all input and output signals of the Group IV interface.

### 7.1 Terminations and Connectors

For each signal, the CNC shall be provided with at least one termination and in addition a sufficient number of terminations for the common connections, but it is recommended that a pair of terminations should be provided for each signal. These terminations may be terminals of suitable type or alternatively plug connectors with a multiplicity of connections. When the CNC has multipin plug connectors, the CN builder shall provide the mating part of the plug. Such connectors shall not be interchangeable with those of higher voltage circuits.

### 7.2 Inputs

A signal voltage of either 24 V dc or 110/115 V ac may be used, but only one and the same voltage for the equipment. The CNC builder shall state which type of input signal is used.

One side of each input circuit shall be connected to the live side of the supply, the other side to an individual termination leading into the CNC. Inside the CNC, the other end of the input circuits shall be connected to the low potential side of the supply, which is connected to the earth terminal.

#### 7.2.1 DC Inputs

It is recommended that inputs be connected between the 24 V dc supply and input terminations at the CNC. This/dc/potential is assumed to be positive referred to the common reference potential (or to earth) unless of CNC builder clearly states that the polarity is negative to earth.

The CNC builder shall ensure that the current passing through the input terminations is sufficient for reliable operation of contacts, for example at least 10 mA where contacts are exposed to the atmosphere. When 24 V dc is used, not more than five such contacts exposed to the atmosphere shall be connected in series.

#### 7.2.2 AC Inputs

The CNC builder may accept signals of 110/115 V ac between the input terminations. A minimum current of 10 mA rms shall be drawn. Not more than eight contacts exposed to the atmosphere shall be connected in series.

#### 7.2.3 Isolation

The part of such input signal circuits outside the CNC shall have no conductive connection to other circuits, or to the earthing conductor (except for the connection to the supply if ac, see Sub-clause 7.4.2).

#### 7.2.4 Contact Bounce

The CNC and the machine control shall be arranged so as to be unaffected by intermittent contact in any input circuit occurring in the time interval immediately after a change of any signal. This interval of insensitivity shall be atleast 5 ms but not longer than 10 ms.

If all control devices such as push-buttons, position sensors etc., effect their command by non-mechanical means (such as Hall effect), the requirements of this sub-clause need not be fulfilled, subject to agreement between the builders of the machine and of the CNC.

### 7.3 Outputs

The recommended output voltages of the CNC are 24 V dc or 110/115 V ac. The use of other voltages such as 220/230 V ac is permitted by agreement between the CNC builder and the machine builder. The CNC builder shall state voltage and current ratings for each output in accordance with Sub-clauses 7.3.1 and 7.3.2. One side

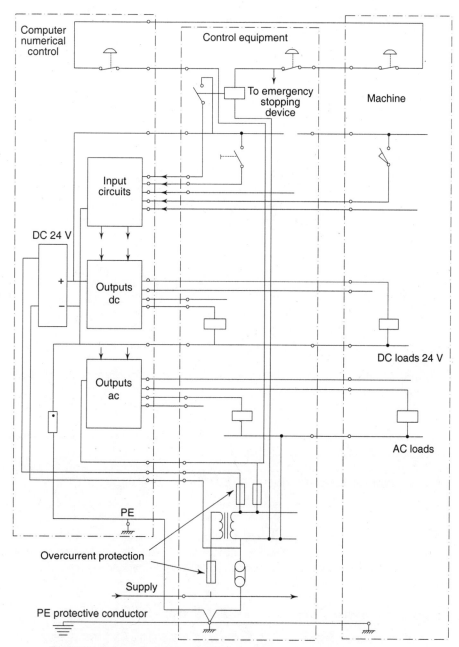

Fig. A7.3 Grooup IV of the interface ON/OFF signals

of each output circuit, i.e. of each output load, shall be connected to the common (low) potential side of the supply.

### 7.3.1 Electromagnetic Switching Devices

Outputs of switching devices using contacts (e.g. relays) shall be suitable for both 24 V dc and 110/115 V ac. The CNC builder shall specify ratings for resistive and inductive loads at both voltages. These shall be atleast 100 mA at 24 V dc and 1 A at 110/115 V ac, when used with inductive loads with suppression of transient voltages as in Sub-clauses 7.5 and 7.5.1.

### 7.3.1 Static Switching

When outputs with static switching are provided, the CNC builder shall state for each output the operating voltage (either 24 V dc or 110/115 V ac) and the minimum and maximum continuous current ratings. He shall also provide data on permissible peak in-rush currents (e.g. for lamps or magnetic valves).

The CNC builder is also responsible for recommendations on suppression devices for use with inductive loads (see Sub-clause 7.5).

## 7.4 Power Supply for Input and Output Circuits

### 7.4.1 Low Voltage dc

All input and output circuits operating at low voltage dc shall be supplied by the CNC. The CNC builder is responsible for the connection of the common (low) potential side of the supply to the earthing circuit. This connection shall be independent of any connection from the internal data processing system of the CNC to the earthing circuit.

It is recommended and it will be assumed unless the CNC builder states otherwise that the negative side of the low-voltage dc is connected to the earthing circuit.

The over current protection and the connections to the earthing shall be in accordance with Sub-clause 6.2 of IEC 204/1–1981.

The CNC builder shall give information on current ratings of the low voltage dc supply system and shall make recommendations on the correct use of inputs and outputs.

The low-voltage dc supplied by the CNC can only be used by the machine builder for purposes other than Group IV signals after agreement with the CNC builder.

### 7.4.2 AC Voltage

Input and output circuits operating on ac shall be supplied from the machine control equipment. The machine builder shall provide this supply by means of a transformer with separate windings and he shall be responsible for the short circuit protection and the appropriate connection(s) to the earthing in accordance with Sub-clause 6.2 of IEC 204/1–1981.

## 7.5 Suppression of Transient Voltages on Outputs of the CNC

The machine builder connecting magnetic devices to the output of the CNC shall be responsible for providing suppression of transient voltages at interruption of current into an inductive load, in accordance with the recommendations of the CNC builder. Suppression shall also be provided across the windings of motors that are started and stopped while the CNC is energised.

The suppression device shall, whenever practical, be connected to the load by the minimum wiring length. Additional suppression may be needed in special circumstances, on the recommendation of the CNC builder.

### 7.5.1 Proposed Specifications for Suppression Devices

Unless otherwise specified by the CNC builder, devices chosen from the following list shall be connected across inductive loads. These devices should be connected as close to the inductive load as possible.

| | |
|---|---|
| 24 V dc | Upto I A Varistor 40 V; zener 40 V; 220 Ohms and 0.22 µF diode 50 V PIV |
| 24 V dc | 1 to 3 A Varistor 40 V; zener 40 V; 100 Ohms and 1 µF diode 50 V PIV |
| 110/115 V ac | 1 to 3 A Varistor 300/350 V; 100 Ohms 0.47 µF |
| 220/230 V ac | 1 A Varistor 500/600 V; 220 Ohms and 0.22 µF |
| 380/415 V ac | for motors; suitable resistor-capacitor networks or varistors |

*Notes*:
1. 'Varistor' denotes resistor with inherent non-liner variability, voltage dependent.
2. Diodes slow down the release of magnetic devices to a greater extent than the other devices.
3. The capacitor rating should be suitable for this duty.
4. In certain cases of operation from ac with solid-state devices the CNC builder may allow the omission of suppression devices.

## 8. Mandatory Signals in Group IV Interface

This clause describes the input and output signals of the CNC which shall be provided by the machine builder. The voltage and current levels of all signals shall be in accordance with Sub-clauses 7.2 and 7.3

Three categories of signals shall be provided.
(a) Input signals to the CNC to inhibit operation and to protect the operator and the machine (Sub-clause 8.1)
(b) Input signals to the CNC necessary for operation (Sub-clause 8.2)
(c) Output signals of the CNC to the machine, necessary for operation (sub-clause 8.3)

### 8.1 Mandatory Input Signals for Inhibiting Operations

An inhibiting operation shall be signaled to CNC by interrupting the input circuit connected to the terminations provided for this signal.

The CNC shall always accept these signals within 10 ms and respond correctly to them with priority over all other signals at any time it is operating.

When more than one STOP or INHIBIT button is provided for the same function on a machine with a CNC, these buttons shall be either connected in series or connected via normally energized "fail safe" devices.

#### 8.1.1 Emergency Stop (E-Stop)

The emergency stopping device according to Sub-clause 5.6.1 of IEC 204/1–1981 shall stop the machine independently of the CNC. It shall override any command of motion from the CNC.

Note: Only upon special agreement between CNC and machine builders, the CNC may command a controlled and rapid deceleration of moving (rotating) parts. In this case, the direct stop command (independent of the CNC) may be deferred to allow the deceleration command over the CNC to be effective, but this delay shall be for only as long as necessary and in no case longer than one or a few seconds.

The EMERGENCY STOP signal shall result in a response of the CNC which should preferably be the same as on the input of FEED HOLD, i.e. it is not recommended to switch off the CNC.

The CNC shall cease to give further commands of motion even after the machine energization has been restored. The normal operation of the machine is restored by the devices provided by the machine builder to start the machine. If the CNC is not in the MANUAL mode (see Sub-clause 8.3.3), the machine shall remain immobilized until the signal CYCLES START has been given to the CNC.

If the operation of the EMERGENCY STOP has taken place at a time or in conditions such that the CNC has lost the machine reference (Home position), the cycle start of the CNC shall depend on the completing of a HOMING Cycle of the CNC to restore the correct relationship of the CNC and the measured position of the machine.

The signal EMERGENCY STOP shall be given to the CNC by interrupting an input circuit of the CNC in accordance with Sub-clause 7.2. This circuit shall be electrically insulated from the circuit of the emergency stopping device of the machine (See Fig. A7.3). For additional safety it is preferable to connect a contact of emergency stop relay in series with power supply line to the outputs.

A red mushroom head push-button with a (mechanical) latch shall be mounted on the CNC, provided for connection into the circuit of the emergency stopping device of the machine.

### 8.1.2 Feed Hold

The object of this signal is to stop the movement of the axes as quickly as possible, together with the ability to restore operation without loss of data.

The minimum requirement is that the CNC gives out commands of stop for the motions of all axes of the machine which are controlled by the CNC and that the CNC ceases to give out any further commands.

Note:

By special agreement with the machine builder, a limited continuation before stopping may be permitted, for example for a controlled separation of the tool from the work. In automatic operation, interlocking shall be provided between the spindle rotation and the feed motion so as to ensure the rotation of the spindles before the tool is in contact with the workpiece and to create and ensure feed hold safe when rotation of spindle is stopped intentionally or unintentionally. In some cases, this interlocking may only involve the corresponding contractors.

A push-button with this function shall be provided in each installation, normally by the CNC builder. The CNC shall always provide terminations for remote operation.

The CNC builder shall specify other results of this command in each mode of operation of the CNC and he shall provide information on the method of restarting the CNC.

An output to indicate that the input FEED HOLD has been operated as recommended, but is not compulsory.

### 8.1.3 Over Travel Limit

The minimum requirement is that a signal from any limit of travel of any CNC axis on the machine shall cause the action of FEED HOLD.

The machine builder preferably provides and installs the limit switches required. He shall continue independently to conform with Sub-clause 6.2.4.6 of IEC 204/1–1981 and provide overrun limit protection operating on the machine control or on the power switching devices of the axis drives. (If there is no mechanical over travel safety device, any limit switch having its normal function of limiting the travel shall be duplicated by an E-Stop limit switch stopping all motions of the machine.)

The IN CYCLE signal remains energised in feed hold condition.

The CNC builder shall provide information on the method of restarting the CNC. It is not mandatory to provide an output for the over travel condition.

### 8.1.4 Permit Entry of New Data

The CNC builder shall provide this input when auxiliary functions are used (for example M, S, T see Clause 9). For machine not using auxiliary functions, it is optional.

When this input signal is interrupted, the CNC shall cease to give out any further commands at the end of the current block of instructions.

It is recommended that the machine builder should remove this signal on change of the auxiliary functions (such as M,S,T) and restore this signal only after these auxiliary functions have been executed on the machine.

## 8.2 Input Signal to the CNC Necessary for Operation

The input signal initiates the operation of the CNC.

### 8.2.1 Cycle Start

This command initiates operation in MDI or automatic mode (see Sub-clause 8.3.3). The CNC starts to read and give out the instructions contained in the next block or blocks of information of a program source (tape or memory) or in the data from a manual data input.

When in AUTO/MDI mode, the movement of any accessible part of the machine shall be allowed only after CYCLE START command has been operated.

The CNC builder shall normally provide a push-button for this function and when requested, terminations for an additional push button on the machine.

The CYCLE START signal shall be a momentary closure of the input circuit for at least 10 ms.

## 8.3 Output Signals from the CNC to the Machine Necessary for Operation

It is recommended that the actuated state of a signal be transmitter to the machine by energizing the terminations provided for that signal.

### 8.3.1 CNC is Ready

The object of this signal is to allow the power drives on the machine to reach the operating conditions. The CNC output shall be actuated when the essential internal supply voltages and measurements signals of Group I have all reached their final operating condition. A failure of any of these voltages capable of impairing correct operation shall deenergize this output.

Together with interruption of this signal, the CNC shall react in the same way as on reception of the input signal FEED HOLD or E STOP.

### 8.3.2 Cycle on

The object of this signal is to provide remote indication of the operating condition of the CNC and to allow certain machine functions.

The signal CYCLE ON shall be actuated when the input CYCLE START has been given and the CNC is executing instructions in one of the CNC modes (see Sub-clause 8.3.3).

The CNC builder shall provide a visible indication of this signal.

### 8.3.3 Mode Indication

The principal object of this signal is to permit the machine builder to provide interlocking by differentiating MDI/AUTO modes of the machine.

The output signal MANUAL-MODE authorizes the machine and the CNC for such functions as 'Jog', 'Power feed', 'Zero shift' etc. It inhibits the CNC from giving out other than manual commands.

When the output MANUAL-MODE is not energized, the selection of other modes, in which commands to the machine are given out from the CNC only, shall be authorized and operative. Direct manual commands, at the machine shall be inhibited. Examples of such modes are 'automatic', 'single block', 'manual data input' etc.

Note: It is recommended that an output signal complementary to the signal MANUAL, i.e. an output which is energized always when MANUAL is not energized authorizes the above mentioned modes and inhibits direct manual commands at the machine.

## 9. Auxiliary and Miscellaneous Signals in Group IV Interface

These signals and commands are read by the CNC from a tape or other program source or given by manual input, typically addressed by "M" for miscellaneous functions, by "S" for spindle speed functions or by "T" for tool functions, as defined in the ISO Standards 1056, 1057, 1058 and 1059. Where additional addresses are used for auxiliary functions, the interface shall still generally conform with the recommendations of this Sub-clause shall still generally conform with the recommendations of this Sub-clause. When any of these functions are used by the machine builder, he shall provide an input in accordance with Sub-clause 7.3 (outputs of the CNC). The voltage, current and duration of all signals, when these are of the type Group IV, shall conform with Sub-clauses 7.2 and 7.3.

It is recommended that the CNC builder provide either pulsed or maintained command outputs, as required by the machine builder; however, in the case of S Signals, preferably maintained outputs should be provided.

If the output commands are a Group IV, they are permitted to be in binary coded decimal format, or decimal format, or even partially or fully coded into the format required by the machine builder. The CNC builder shall state the format provided.

### 9.1 Pulsed Mode (Refer Figure A7.4a)

It is mandatory that the group of signals representing the new command (for example, a coded command group) be validated by a separate output signal of the CNC. This 'validate' or 'Strobe' signal shall commence after a time interval T2 from initiation of the new command group and it shall have the duration T3. The command group shall terminate after a time interval T4.

It is permitted to provide an additional signal 'M, S or T change' which commences before the new command group by a time interval T1 and has a duration of T1. Its function is to allow time for disengagement of the machine control equipment from the previous command group.

### 9.2 Maintained Mode (see Fig. A7.4b)

In this mode the output signals of the command group are maintained.

Changes in output occur as a group and shall be completed by the commencement of a time interval T2.

The validate signal (strobe) is the same as in Sub-clause 9.1.

### 9.3 Time interval values         (Minimum recommended)

T1 preparatory delay              20 ms
T2 pre-validate interval          20 ms
T3 validate signal                40 ms
T4 Command-group duration         80 ms

It is recommended that the indicated minimum time should be used. These may be longer, but any reduction shall be permissible only by agreement between the CNC builder and the machine builder.

## 10. Examples of Optional Signals in Group IV Interface

These examples are illustrative only and given as an indication.

However, all inputs and outputs provided shall be in accordance with all the relevant requirements of Clause 7.

### 10.1 Axis Enable (Output)

This signal indicates that the CNC is ready to give a command to move the indicated axis. For each axis controlled by the CNC, thee shall be an independent output. If these output are provided, the CNC shall also be provided with the corresponding inputs according to Sub-clause 10.2 (Permission for axes to move).

It is recommended that the energized state of the output should correspond to the 'ready to move' condition. When the axis is in its new position, the output shall change back to the de-energized state. An example of this use is to release side clamps. The signal from the machine 'clamps are released' is then used as an input according to Sub-clause 10.2.

Drives and Electricals   **305**

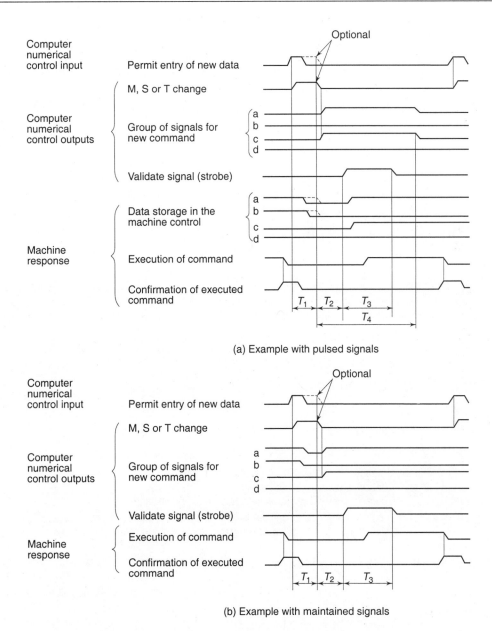

**Fig. A7.4** Examples of timing for miscellaneous commands such as those addressed by M, S or T

## 10.2 Axis Enable Input

When this type of signal is provided, the CNC shall provide for independent inputs for each axis. Where the movements of different axes are coordinated as in contouring control, the absence of any relevant permission shall inhibit all coordinated axes.

This input signal is associated with the output according to Sub-clause 10.1. It shall always be effective in the AUTO modes, and the CNC builder shall define its status in MANUAL/AUTO modes. The signals to the CNC indicate that the machine has fulfilled safety or operating requirements, e.g. slide clamps are free or some physical obstacle has been removed.

If the CNC provides such inputs, individual terminations shall be provided for each axis controlled by the CNC. The signal shall be a circuit closure. The effect of the signal shall be to allow the CNC to initiate move on that axis. An interruption of the signal before the move is complete is not necessarily to interrupt an axis motion already in progress.

## 10.3 Block Stop/Operation Stop (Single Block)

The objective of BLOCK STOP is to enable the user of the machine to terminate operation at the end of the current block.

This is commanded by actuating a selector switch or a push-button. When actuated it causes the CNC to terminate operation after the current block of instructions has been given out by the CNC and executed at the machine. It may also be connected in the circuits of the machine for example as safety interlock during a manual tool change.

The CNC builder shall provide and locate the BLOCK STOP push button or alternatively provide a selector switch with, in addition a position "single block". If required by the machine builder, he shall also provide terminations for remote operation.

When the CNC has executed BLOCK STOP, the signal CYCLE ON (see Sub-clause 8.3.2) is extinguished. The CNC shall be able to be restarted by activating the command CYCLE START.

## 10.4 Referencing of Axis (Homing)

This input command sets the CNC to the initial measuring positions on all axes of a machine. Homing is strongly recommended when incremental measuring systems are used.

The method of HOMING CNC shall be one of the following:
 (a) all axes are set to zero in the CNC at the present position of the machine:
 (b) all axes are set to a predetermined value in the CNC at the present position of the machine.
 (c) all axes are set in the CNC to the null point or reference point of the measuring system and this is a predefined position on the machine.

The CNC builder shall state which method is used. The axes may be referenced in any sequence or simultaneously.

When the command is given, the homing starts. The CNC shall inhibit CYCLE START until the homing is completed.

## Related Standards

The following is the list of various National and International Standards that are referred to in this standard.

1. IEC 204/1–1981                Electrical Equipment of Machine tools. Part I—Electrical Equipment of Machines for General use.

2. IEC 550           Interface between Numerical Controls and Industrial Machines.
3. ISO 1056          Numerical Controls of Machines. Punched tape block formats. Coding of preparatory functions G and miscellaneous functions M.
4. ISO 1057          Numerical Controls of Machines. Interchangeable punched tape variable block format for positioning and straight-cut machining.
5. ISO 1058          Numerical Controls of machines. Punched tape variable block format for positioning and straight-cut machining.
6. ISO 1059          Numerical Controls of Machines. Punched tape fixed block format for positioning and straight-cut machining.

CHAPTER **8**

# CNC Systems

### 8.1 INTRODUCTION

Numerical control (NC) is a method employed for controlling the motions of a machine tool slide and its auxiliary functions with an input in the form of numerical data. A computer numerical control (CNC) is a microprocessor based system to store and process the data for the control of slide motions and auxiliary functions of the machine tools. The CNC system is the heart and brain of a CNC machine which enables the operation of the various machine members such as slides, spindles, etc. as per the sequence programmed into it, depending on the machining operations.

The main advantage of a CNC system lies in the fact that the skills of the operator hitherto required in the operation of a conventional machine is removed and the part production is made automatic.

The CNC systems are constructed with an NC unit integrated with a programmable logic controller (PLC) and sometimes with an additional external PLC (non-integrated). The NC controls the spindle movement and the speeds and feeds in machining. It calculates the traversing paths of the axes as defined by the inputs. The PLC controls the peripheral actuating elements of the machine such as solenoids, relay coils, etc. Working together, the NC and PLC enable the machine tool to operate automatically. Positioning and part accuracy depend on the CNC system's computer control algorithms, the system resolution and the basic mechanical machine inaccuracies. Control algorithms may cause errors while computing, which will reflect during contouring, but they are very negligible. Though this does not cause point-to-point positioning errors, but when mechanical machine inaccuracies are present, it will result in a poorer part accuracy.

This chapter gives an overview of the configuration of the CNC system, interfacing and introduction to PLC programming. Annexure 8.1 gives the general specifications of a CNC system and Annexure 8.2 gives the basics of binary system, the numbering system with which all the computers work with.

### 8.2 CONFIGURATION OF THE CNC SYSTEM

Figure 8.1 shows a schematic diagram of the working principle of an NC axis of a CNC machine and the interface of a CNC control.

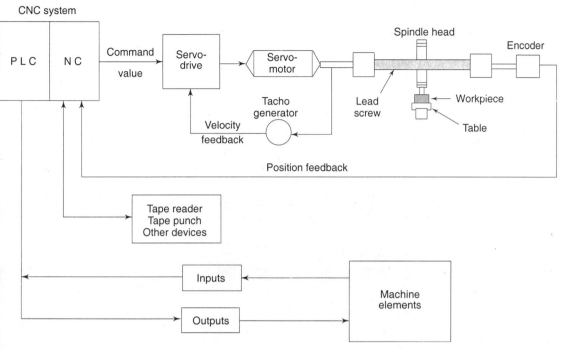

**Fig. 8.1** Schematic diagram of a CNC machine tool

A CNC system basically consists of the following:

- Central processing unit (CPU)
- Servo-control unit
- Operator control panel
- Machine control panel
- Other peripheral devices
- Programmable logic controller

Figure 8.2 gives the typical numerical control configuration of Hinumerik 3100 CNC system.

### 8.2.1 Central Processing Unit (CPU)

The CPU is the heart and brain of a CNC system. It accepts the information stored in the memory as part program. This data is decoded and transformed into specific position control and velocity control signals. It also oversees the movement of the control axis or spindle and whenever this does not match with the programmed values, a corrective action is taken.

All the compensations required for machine accuracies (like lead screw pitch error, tool wear out, backlash, etc.) are calculated by the CPU depending upon the corresponding inputs made available to the system. The same will be taken care of during the

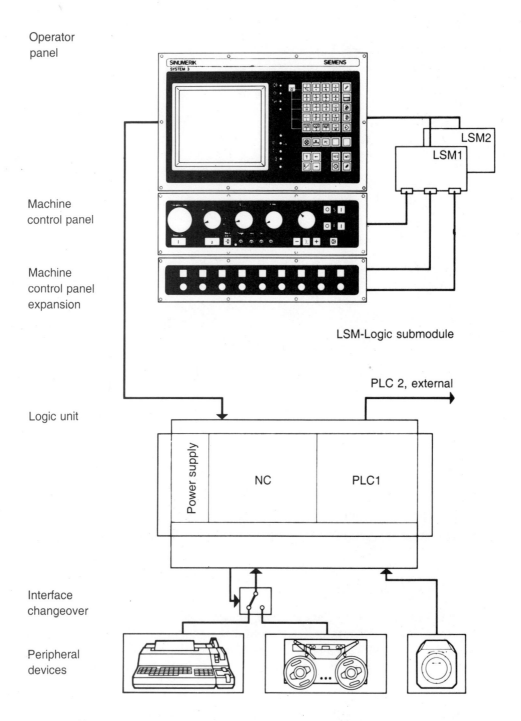

**Fig. 8.2** Typical numerical control configuration of Hinumerik 3100 CNC system

generation of control signals for the axis movement. Also, some basic safety checks are built into the system through this unit and continuous necessary corrective actions will be provided by the CPU unit. Whenever the situation goes beyond control of the CPU, it takes the final action of shutting down the system and in turn the machine.

## Speed Control Unit

This unit acts in unison with the CPU for the movement of the machine axes. The CPU sends the control signals generated for the movement of the axis to the servo-control unit and the servo-control unit converts these signals into a suitable digital or analog signal to be fed to a servo-drive for machine tool axis movement. This also checks whether machine tool axis movement is at the same speed as directed by the CPU. In case any safety conditions related to the axis are overruled during movement or otherwise they are reported to the CPU for corrective action.

### 8.2.2 Servo-control Unit

The decoded position and velocity control signals, generated by the CPU for the axis movement forms the input to the servo-control unit. This unit in turn generates suitable signals as command values. The command values are converted by the servo-drive unit which are interfaced with the axes and the spindle motors (Fig. 8.1). (For more details on drives refer to Chapter 7.)

The servo-control unit receives the position feedback signals for the actual movement of the machine tool axes from the feedback devices (like linear scales, rotary encoders, resolvers, etc.). The velocity feedback are generally obtained through tacho generators. The feedback signals are passed on to the CPU for further processing. Thus, the servo-control unit performs the data communication between the machine tool and the CPU.

As explained earlier, the actual movement of the slides on the machine tool is achieved through servo drives. The amount of movement and the rate of movement are controlled by the CNC system depending upon the type of feedback system used, i.e. closed-loop or open-loop system (Fig. 8.3).

## Closed-loop System

The closed-loop system is characterised by the presence of feedback. In this system, the CNC system sends out commands for movement and the result is continuously monitored by the system through various feedback devices. There are generally two types of feedback to a CNC system—position feedback and velocity feedback.

**Position Feedback**  A closed-loop system, regardless of the type of feedback device, will constantly try to achieve and maintain a given position by self-correcting. As the slide of the machine tool moves, its movement is fed back to the CNC system for determining the position of the slide to decide how much is yet to be travelled and also to decide whether the movement is as per the commanded rate. If the actual rate is not as per the required rate, the system tries to correct it. In case this is not possible, the system

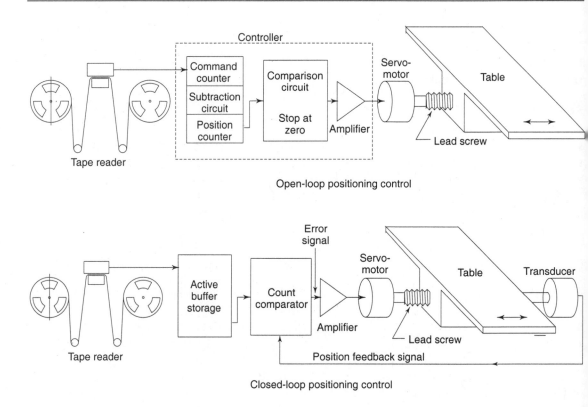

**Fig. 8.3** Open- and closed-loop positioning systems

declares a fault and initiates action for disabling the drives and if necessary, switches off the machine.

***Velocity Feedback*** In case no time constraint is put on the system to reach the final programmed position, then the system may not produce the required path or the surface finish accuracy. Hence, velocity feedback must be present along with the position feedback whenever CNC systems are used for contouring, in order to produce correct interpolations and also specified acceleration and deceleration velocities. The tacho generator used for velocity feedback is normally connected to the motor and it rotates whenever the motor rotates, thus giving an analog output proportional to the speed of the motor. This analog voltage is taken as speed feedback by the servo-controller and swift action is taken by the controller to maintain the speed of the motor within the required limits.

### Open-loop System

The open-loop system lacks feedback. In this system, the CNC system sends out signals for movement but does not check whether actual movement is taking place or not. Stepper motors are used for actual movement and the electronics of these stepper motors is run on digital pulses from the CNC system. Since system controllers have no access to any real-time information about the system performance, they cannot counteract disturbances appearing during the operation. They can be utilised in point-to-point system, where loading torque on the axial motors is low and almost constant.

*Servo-drives*

As shown in Fig. 8.1, the servo-drive receives signals from the CNC system and transforms it into actual movement on the machine. The actual rate of movement and direction depend upon the command signal from the CNC system. There are various types of servo-drives, viz., dc drives, ac drives and stepper motor drives. A servo-drive consists of two parts, namely, the motor and the electronics for driving the motor. (For more details on drives refer to Chapter 7).

### 8.2.3 Operator Control Panel

Figure 8.4 shows a typical Hinumerik 3100 CNC system's operator control panel. The operator control panel provides the user interface to facilitate a two-way communication between the user, CNC system and the machine tool. This consists of two parts:

- Video display unit
- Keyboard

*Video Display Unit (VDU)*

The VDU displays the status of the various parameters of the CNC system and the machine tool. It displays all current information such as:

- Complete information on the block currently being executed
- Actual position value, set or actual difference, current feed rate, spindle speed
- Active G functions, miscellaneous functions (refer to Chapter 9, Sections 9.6.7 and 9.6.8)
- Main program number, subroutine number
- Display of all entered data, user programs, user data, machine data, etc.
- Alarm messages in plain text
- Soft key designations

In addition to a CRT, a few LEDs are generally provided to indicate important operating modes and status.

Video display units may be of two types:

1. Monochrome or black and white displays
2. Colour displays

*Keyboard*

A keyboard is provided for the following purposes:

- Editing of part programs, tool data, machine parameters.
- Selection of different pages for viewing.

# 314 Mechatronics

**Operator's and machine control panel**

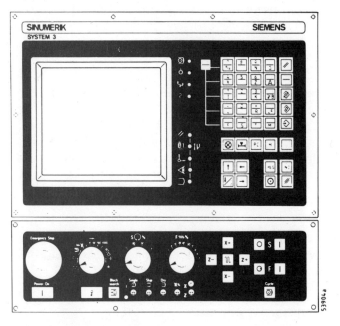

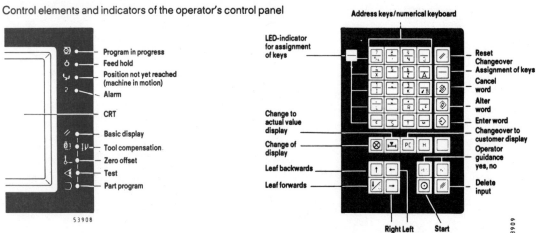

👆 **Fig. 8.4** Operator control panel of Hinumerik 3100 system

- Selection of operating modes, e.g. manual data input, jog, etc.
- Selection of feed rate override and spindle speed override
- Execution of part programs
- Execution of other tool functions

## 8.2.4 Machine Control Panel (MCP)

It is the direct interface between the operator and the NC system, enabling the operation of the machine through the CNC system. Figure 8.5 shows the MCP of Hinumerik 3100 system.

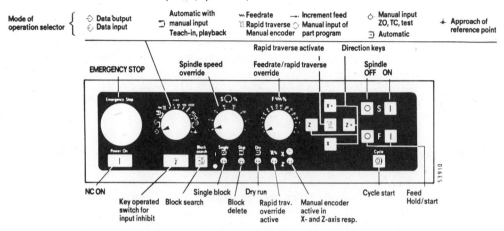

**Fig. 8.5** Machine control panel of Hinumerik 3100 system

During program execution, the CNC controls the axis motion, spindle function or tool function on a machine tool, depending upon the part program stored in the memory. Prior to the starting of the machining process, machine should first be prepared with some specific tasks like,

- Establishing a correct reference point
- Loading the system memory with the required part program
- Loading and checking of tool offsets, zero offsets, etc.

For these tasks, the system must be operated in a specific operating mode so that these preparatory functions can be established.

## Modes of Operation

Generally, the CNC systems can be operated in the following modes:

- Manual mode
- Manual data input (MDI) mode
- Automatic mode
- Reference mode
- Input mode
- Output mode, etc.

***Manual Mode*** In this mode, movement of a machine slide can be carried out manually by pressing the particular jog buttons (positive or negative). The slide (axis) is selected through an axis selector switch or through individual switches (e.g., X+, X–, Y+, Y–, Z+, Z–, etc.). The feed rate of the slide movement is prefixed. Some CNC systems allow the axis to be jogged at a high feed rate also. The axis movement can also be achieved manually using a handwheel interface instead of jog buttons. In this mode, slides can be moved in two ways:

- Continuous
- Incremental.

*Continuous Mode* In this mode, the slide will move as long as the jog button is pressed.

*Incremental Mode* Hence, the slide will move through a fixed distance which is selectable. Normally, systems allow jogging of axes in 1,10,100,1000,10000 increments. Axis movement is at a prefixed feed rate. It is initiated by pressing the proper jog+ or jog– key and will be limited to the number of increments selected even if the jog button is continuously pressed. For subsequent movements the jog button has to be released and once again pressed.

***Manual Data Input (MDI) Mode*** In this mode the following operations can be performed:

- Building a new part program
- Editing or deleting of part programs stored in system memory
- Entering or editing or deleting of:
  — Tool offsets (TO)
  — Zero offsets (ZO)
  — Test Data, etc.

### Teach-in

Some systems allow direct manual input of a program block and execution of the same. The blocks thus executed can be checked for correctness of dimensions and consequently transferred into the program memory as part program.

### PlayBack

In setting up modes like jog or incremental, the axis can be traversed either through the direction keys or via the electronic hand wheel, and the end position can be transferred into the system memory as command values. But the required feed rates, switching functions and other auxiliary functions have to be added to the part program in program editing mode.

Thus, teach-in or playback operating method allows a program to be created during the first component proveout.

### Automatic Mode (Auto and Single Block)

In this mode the system allows the execution of a part program continuously. The part program is executed block by block. While one block is being executed, the next block is read by the system, analysed and kept ready for execution. Execution of the program can be one block after another automatically or the system will execute a block, stop the execution of the next block till it is initiated to do so (by pressing the start button). Selection of part program execution continuously (*Auto*), or one block at a time (*Single Block*) is done through the machine control panel.

Many systems allow blocks (single or multiple) to be retraced in the opposite direction. Block retrace is allowed only when a cycle stop state is established. Part program execution can be resumed and its execution begins with the retraced block. This is useful for tool inspection or in case of tool breakage.

Program start can be effected at any block in the program, through the BLOCK SEARCH facility.

## Reference Mode

Under this mode the machine can be referenced to its home position so that all the compensations (e.g., pitch error compensation) can be properly applied. Part programs are generally prepared in absolute mode with respect to machine zero. Many CNC systems make it compulsory to reference the slides of the machine to their home positions before a program is executed while others make it optional.

## Input Mode and Output Mode (I/O Mode)

In this mode, the part programs, machine set-up data, tool offsets, etc. can be loaded/unloaded into/from the memory of the system from external devices like programming units, magnetic cassettes or floppy discs, etc. During data input, some systems check for simple errors (like parity, tape format, block length, unknown characters, program already present in the memory, etc.). Transfer of data is done through a RS232C or RS422C port.

### 8.2.5 Other Peripherals

These include sensor interface, provision for communication equipment, programming units, printer, tape reader/punch interface, etc.

Figure 8.6 gives an overview of the system with a few peripheral devices.

### 8.2.6 Programmable Logic Controller (PLC)

A PLC matches the NC to the machine. PLCs were basically introduced as replacement for hard-wired relay control panels. They were developed to be re-programmed without hardware changes when requirements were altered and thus are re-usable. PLCs are now available with increased functions, more memory and larger input/output capabilities. Figure 8.7 gives the generalised PLC block diagram.

In the CPU, all the decisions are made relative to controlling a machine or a process. The CPU receives input data, performs logical decisions based upon stored programs and drives the outputs. Connections to a computer for hierarchical control are done via the CPU.

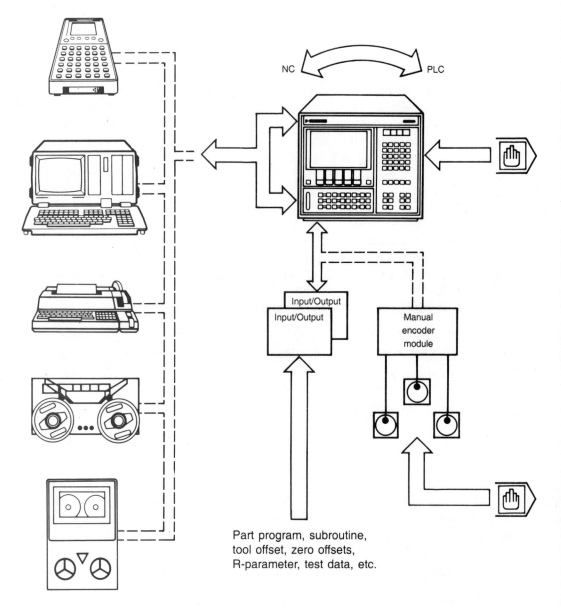

**Fig. 8.6** Peripheral devices

The I/O structure of PLCs is one of their major strengths. The inputs can be push buttons, limit switches, relay contacts, analog sensors, selector switches, thumb wheel switches, proximity sensors, pressure switches, float switches, etc. The outputs can be motor starters, solenoid valves, position valves, relay coils, indicator lights, LED displays, etc.

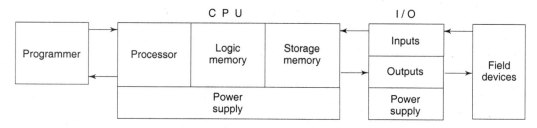

**Fig. 8.7** Generalised PLC block diagram

The field devices are typically selected, supplied and installed by the machine tool builder or the end user. The type of I/O is thus normally determined by the voltage level of the field devices. So, power to actuate these devices must also be supplied external to the PLC. The PLC power supply is designated and rated only to operate the internal portions of the I/O structures, and not the field devices. A wide variety of voltages, current capacities and types of I/O modules are usually available (Refer section 8.7 for PLC programming details).

## 8.3 INTERFACING

Interconnecting the individual elements of both the machine and the CNC system using cables and connectors is called interfacing.

Extreme care should be taken during interfacing. Proper grounding in electrical installations is most essential. This reduces the effects of interference and guards against electronic shocks to personnel. It is also essential to properly protect the electronic equipment.

Cable wires of sufficiently large cross-sectional area must be used. Even though proper grounding reduces the effect of electrical interference, signal cables require additional protection. This is generally attained by using shielded cables. All the cable shields must be grounded at control only, leaving the other end free. Other noise reduction techniques include using suppression devices, proper cable separation, ferrous metal wire ways, etc. Electrical enclosures should be designed to provide proper ambient conditions for the controller. Power supply to the controller should match with the suppliers' specifications. Figures 8.8 and 8.9 show typical examples for interface details of a two-axes machine with the system and other machine elements.

## 8.4 MONITORING

In addition to the care taken by the machine tool builder during design and interfacing, basic control also includes constantly active monitoring functions. This is in order to identify faults in the NC, the interface control and the machine at an early stage to prevent damages occurring to the workpiece, tool or machine. If a fault occurs, first the

**320** Mechatronics

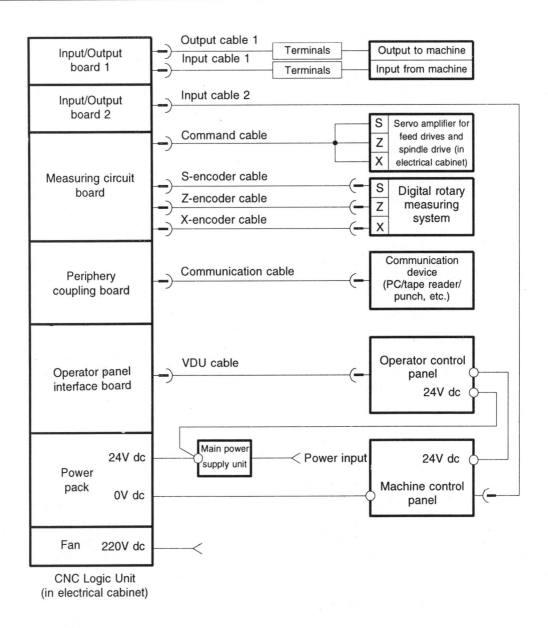

**Fig. 8.8** Typical cabling details of CNC system used in two-axes machine

machining sequence is interrupted, the drives are stopped, the cause of the fault is stored and then displayed as an alarm. At the same time, the PLC is informed that an NC alarm exists. In Hinumerik CNC system, for example, the following can be monitored:

- Read-in
- Format

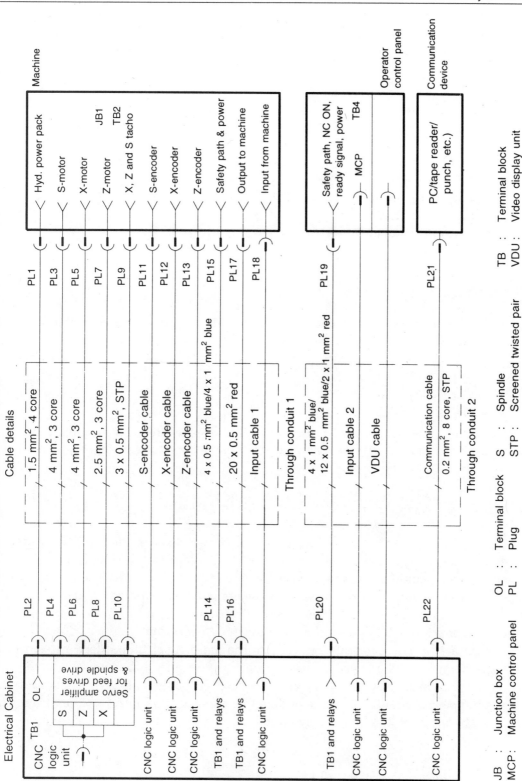

**Fig. 8.9** Typical interface details for two axes CNC machine

- Measuring circuit cables
- Position encoders and drives
- Contour
- Spindle speed
- Enable signals
- Voltage
- Temperature
- Microprocessors
- Data transfer between operator control panel and logic unit
- Transfer between NC and PLC
- Change of status of buffer battery
- System program memory
- User program memory
- Serial interfaces

## 8.5 DIAGNOSTICS

The control will generally be provided with test assistance for service purposes in order to display some status on the CRT such as:

- Interface signals between NC and PLC as well as between PLC and machine
- Flags of the PLC
- Timers of the PLC
- Counters of the PLC

For the output signals, it is also possible to set and generate signal combinations for test purposes in order to observe how the machine reacts to a changed signal. This simplifies trouble shooting considerably.

## 8.6 MACHINE DATA

Generally, a CNC system is designed as a general purpose control unit which has to be matched with the particular machine to which the system is interfaced. The CNC is interfaced to the machine by means of data which is machine specific. The NC and PLC machine data can be entered and changed by means of an external equipment or manually by the keyboard. These data are fixed and entered during commissioning of the machine and generally left unaltered during machine operations.

Machine data entered is usually relevant to the axis travel limits, feed rates, rapid traverse speeds and spindle speeds, position control multiplication factor, Kv factor, acceleration, drift compensation, adjustment of reference point, backlash compensation, pitch error compensation, etc. Also, the optional features of the control system are made available to the machine tool builder by enabling some of the bits of machine data.

## 8.7 COMPENSATIONS FOR MACHINE ACCURACIES

Machine accuracy is the accuracy of the movement of the carriages and tables, and is influenced by:

(a) Geometric accuracy in the alignment of the slideways
(b) Deflection of the bed due to a load
(c) Temperature gradients on the machine
(d) Accuracy of the screw thread of any drive screw and the amount of backlash (lost motion)
(e) Amount of twist (wind up) of the shaft which will influence the measurement of rotary transducers

The CNC systems offer compensations for the various machine accuracies. These are detailed below:

### Lead Screw Pitch Error Compensation

To compensate for the movements of the machine slide due to inaccuracy of the pitch along the length of the ballscrew, a pitch error compensation is required. To begin with, the pitch error curve for the entire length of the screw is built-up by physical measurement with the aid of an external device (like laser). Then the required compensation at predetermined points are fed into the system. Whenever a slide is moved, these compensations are automatically added by the CNC system (Fig. 8.10).

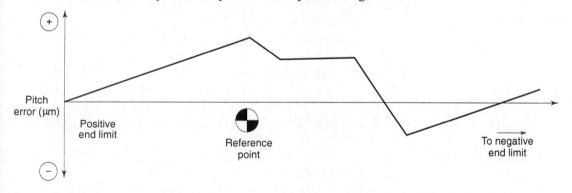

**Fig. 8.10** Typical error curve

### Backlash Compensation

Whenever a slide is reversed, there is some lost motion due to backlash between the nut and the screw; a compensation is provided by the CNC system for the motion lost due to reversal, i.e. extra movement is added into the actual movement whenever reversal takes place. This extra movement is equal to the backlash between the screw and the nut. This has to be measured in advance and fed to the system. This value keeps on varying due to wear of the ballscrews, hence the compensation value has to be updated regularly from time to time (Fig. 8.11).

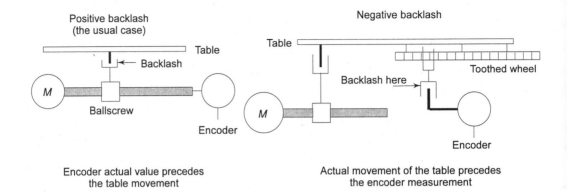

Fig. 8.11 Backlash compensation

### Sag Compensation

Inaccuracies due to sag in the slide can be compensated by the system. Compensations required along the length of the slide have to be physically measured and fed to the system. The system automatically adds up the compensation to the movement of the slide.

### Tool Nose Compensation

Tool nose compensation is normally used on tools for turning centres. While machining chamfers, angles or turning curves, it is necessary to make allowances for the tool-tip radius; this allowance is known as *nose radius compensation*. As shown in Fig. 8.12(a), if this allowance is not made, the edges of the tool tip radius would be positioned at the programmed $X$ and $Z$ coordinates, and the tool will follow the path $AB$ and the taper produced will be incorrect. In order to obtain correct taper, tool position has to be adjusted.

It is essential that the radius at the tip of the tool is fed to the system to make an automatic adjustment in the position and movement of the tool to get the correct taper on the work. In Fig. 8.12(b), the distance $X_c$ is the adjustment necessary at the start of the cut and distance $Z_c$ is the adjustment necessary at the end of the cut.

### Cutter Diameter Compensation

The diameter of the used tool may be different from the actual value because of re-grinding of the tool or due to non-availability of the assumed tool. It is possible to adjust the relative position of cutter size and this adjustment is known as cutter diameter compensation.

### Tool Offset

A part program is generated keeping in mind a tool of a particular length, shape and thickness as a reference tool. But during the actual mounting of tools on the machine,

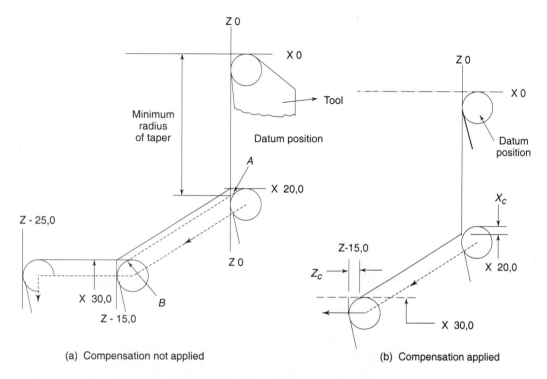

**Fig. 8.12** Tool nose radius compensation

different tools of varying lengths, thickness and shapes may be available. Correction for dimensions of the tools and movements of the workpiece has to be incorporated to give the exact machining of the component. This is known as tool offset. This is the difference in the positions of the centre line of the tool holder for different tools and the reference tool. When a number of tools are used, it is necessary to determine the tool offset of each tool and store it in the memory of the control unit. Figure 8.13 explains the function of the tool offset.

Normally, it is found that the size of the workpiece (diameter or length) is not within tolerances due to wear of the tool; it is then possible to edit the value of offsets to obtain the correct size. This is known as *tool wear compensation*.

## 8.8 PLC PROGRAMMING

The principle of operation of a PLC is determined essentially by the PLC program memory, processor, inputs and outputs.

The program that determines PLC operation is stored in the internal PLC program memory. The PLC operates cyclically, i.e. when a complete program has been scanned, it starts again at the beginning of the program. At the beginning of each cycle, the processor examines the signal status at all inputs as well as the external timers and counters

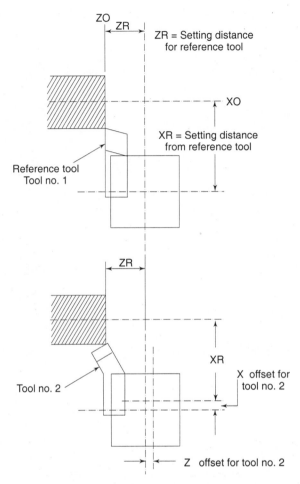

**Fig. 8.13** Tool offsets

and are stored in a process image input (PII). During subsequent program scanning, the processor then accesses this process image.

To execute the program, the processor fetches one statement after another from the program memory and executes it. The results are constantly stored in the process image output (PIO) during the cycle. At the end of a scanning cycle, i.e. program completion, the processor transfers the contents of the process image output to the output modules and to the external timers and counters. The processor then begins a new program scan.

STEP 5 programming language is used for writing user programs for Simatic S5 programmable controllers. The programs can be written and entered into the programmable controller as in:

- Statement list (STL) (Fig. 8.14(a))
- Control system flow chart (CSF) (Fig. 8.14(b))
- Ladder diagram (LAD) (Fig. 8.14(c))

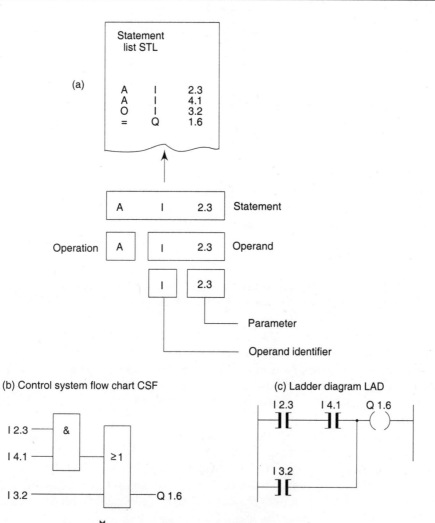

**Fig. 8.14** Programmable controller

The *statement list* describes the automation task by means of mnemonic function designations.

The *control system flow chart* is a graphic representation of the automation task.

The *ladder diagram* uses relay ladder logic symbols to represent the automation task.

The *statement* is the smallest STEP 5 program component. It consists of the following:

Operation, i.e. *What is to be done?*

e.g.
        A = AND operation (series connection)

        O = OR operation (parallel connection)

        S = SET operation (actuation)

Operand, i.e.   *What is to be done with?*
         e.g.    I 4.5, i.e. with the signal of input 4.5

The operand consists of:

- Operand identifier (I = input, Q = output, F = flag, etc.)
- Parameter, i.e. the number of operand identifiers addressed by the statement. For inputs, outputs and flags (internal relay equivalents), the parameter consists of the byte and bit addresses, and for timers and counters, byte address only (see Annexure 8.2 for bits and bytes).

The statement may include absolute operands, e.g. I 5.1, or symbolic operands, e.g. I LS1. Programming is considerably simplified in the latter case as the actual plant designation is directly used to describe the device connected to the input or output.

Typically, a statement takes up one word (two bytes) in the program memory.

## 8.8.1 Structured Programming

The user program can be made more manageable and straight forward if it is broken down into related sections. Various software block types are available for constructing the user program.

*Program blocks (PB)* contain the user program broken down into technologically or functionally related sections (e.g. program block for transportation, monitoring, etc.). Further blocks, such as program blocks or function blocks can be called from a PB.

*Organisation blocks (OB)* contain block calls determining the sequence in which the PBs are to be processed. It is therefore possible to call PBs conditionally (depending on certain conditions).

In addition, special OBs can be programmed by the user to react to interruptions during cyclic program processing. Such an interrupt can be triggered by a monitoring function if one or several monitored events occur.

*Function blocks (FB)* are blocks with programs for recurrent and usually complex functions. In addition to the basic operations, the user has an extended operation set at his disposal for developing function blocks. The program in a function block is usually not written with absolute operands (e.g. I 1.5) but with symbolic operands. This enables a function block to be used several times over with different absolute operands.

For even more complex functions, *standard function blocks* are available from a program library. Such FBs are available, e.g. for individual controls, sequence controls, messages, arithmetic functions, two-step control loops, operator communication, listing, etc. These standard FBs for complex functions can be linked into the user program just like user-written FBs simply by means of a call along with the relevant parameters.

*The sequence blocks (SB)* contain the step-enabling conditions, monitoring times and output conditions for the current step in a sequence cascade. Sequence blocks are employed, for example, to organise the sequence cascade in communication with a standard FB.

*The data blocks (DB)* contain all fixed or variable data of the user program.

## 8.8.2 Cyclic Program Processing

The blocks of the user program are executed in the sequence in which they are specified in the organisation block.

## 8.8.3 Interrupt Driven Program Processing

When certain input signal changes occur, cyclic processing is interrupted at the next block boundary and an OB assigned to this event is started. The user can formulate his response program to this interrupt in the OB. The cyclic program execution is then resumed from the point at which it was interrupted.

## 8.8.4 Time Controlled Program Execution

Certain OBs are executed at predetermined time intervals (e.g. every 100 ms, 200 ms, 500 ms, 1s, 2s or 5s). For this purpose, cyclic program execution is interrupted at a block boundary and resumed again at this point, once the relevant OB has been executed. Figure 8.15 gives the organisation and execution of a structured user program.

## 8.8.5 Examples of PLC Program

Before attempting to write a PLC program, first go through the instruction set of the particular language used for the equipment, and understand the meaning of each instruction. Then study how to use these instructions in the program (through illustration examples given in the manual). Once the familiarisation task is over, then start writing the program.

Follow the following steps to write a PLC program.

- List down each individual element (field device) on the machine as Input/Output.
- Indicate against each element the respective address as identified during electrical interfacing of these elements with the PLC.
- Break down the complete machine auxiliary functions which are controlled by the PLC into individual, self contained functions.
- Identify each individual function as separate block (PBxx/FBxx).
- Once the PBs/FBs for each function are identified, take them one by one for writing the programs.
- List down the preconditions required for the particular function separately.
- List down the respective elements available on the machine.
- Note down the address of the listed elements.
- Write down the flow chart for the function.
- Translate the flow chart into PLC program using the instructions already familiarised.
- Complete the program translation of all individual functions in similar lines.
- Check the individual blocks independently and correct the program to get the required results.
- Organise all the program blocks in the organisation block depending upon the sequence in which they are suppose to be executed as per the main machine function flow chart.

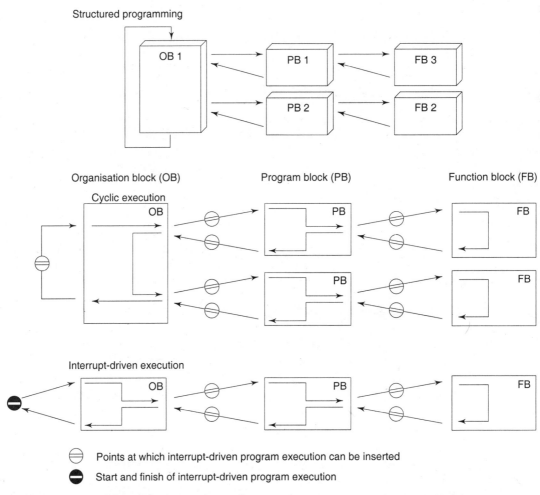

**Fig. 8.15** Organisation and execution of a structured user program

- Check the complete program with all the blocks incorporated in the final program.

## Example 1: Spindle ON (Flowchart 1)

| Preconditions | Feedback elements | Address | Fault indication | Address | Remarks |
|---|---|---|---|---|---|
| Tool clamp | Pressure switch | I 2.4 | Lamp | Q 2.1 | |
| Job clamp | Proximity switch | I 3.2 | Lamp | Q 1.7 | |
| Door close | Limit switch | I 5.7 | Lamp | Q 4.0 | |
| Lubrication ON | PLC output bit | Q 1.0 | Lamp | Q 7.7 | |
| Drive ready | Input signal from drive unit | I 4.6 | Lamp | Q 0.4 | |

PB 12 written above is the individual function module for making spindle ON with all the preconditions checked and found satisfactory. This function is required to be executed only when spindle rotation is requested by the NC in the form of a block in the part program.

 e.g. N 210 S 100 M 03 *  (refer Chap. 9 for part programming details)

Whenever NC decodes this block, it in turn informs the PLC through a fixed buffer location that spindle rotation is requested. Say Flag Bit 100.0 is identified for this information communication. With this data, spindle ON function module can be called in the organisation block OB1 as follows.

```
OB 1
...
...
A  F  100.0
JC PB12
...
...
...
BE
```

Now, Spindle ON function module PB 12 will be executed only when F 100.0 is set. Otherwise the function execution will be bypassed.

*Example 2: Job Loading on to the Machine by Robot (Flowchart 2)*

| Preconditions | Feedback elements | Address | Fault indication | Address | Remarks |
|---|---|---|---|---|---|
| Shutter open | Limit switch | I 3.0 | Lamp | Q 2.1 | |
| Chuck jaw open | Proximity switch | I 3.1 | Lamp | Q 1.7 | |
| Robot in loading position | Limit switch | I 3.2 | Lamp | Q 7.7 | |
| Robot with component held in the gripper | Pressure switch | I 3.3 | Lamp | Q 0.4 | |

Assume that miscellaneous code M80 is identified for the job loading by robot. In part program, loading initiation can be done at an appropriate point as follows.

 e.g.  N 355 M80 *

When this block is decoded by NC, a flag bit reserved for the code M80 will be set (F20.1). This flag bit is used in organisation block OB 1 as shown below.

```
OB 1
...
...
A  F  20.1
JC PB32
...
...
BE
```

## 332  Mechatronics

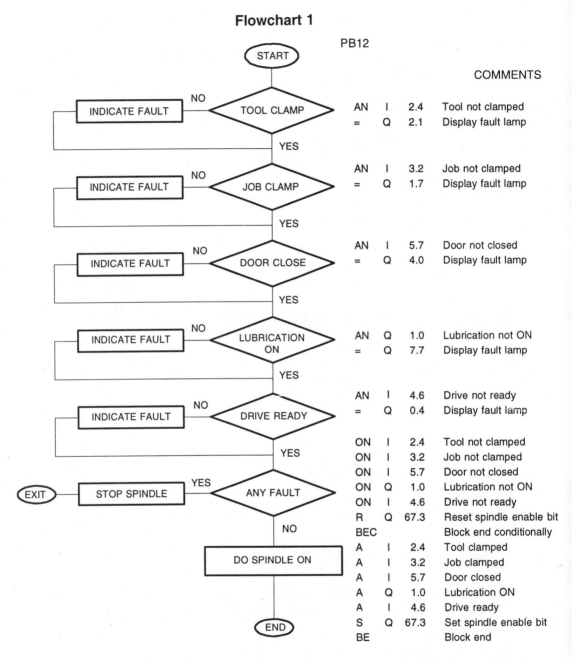

Thus, whenever M80 is programmed in the NC part program, NC informs PLC through the flag bit F20.1. This bit in turn initiates the execution of the program block PB32, which is nothing but the individual function module for job loading on to the machine by robot.

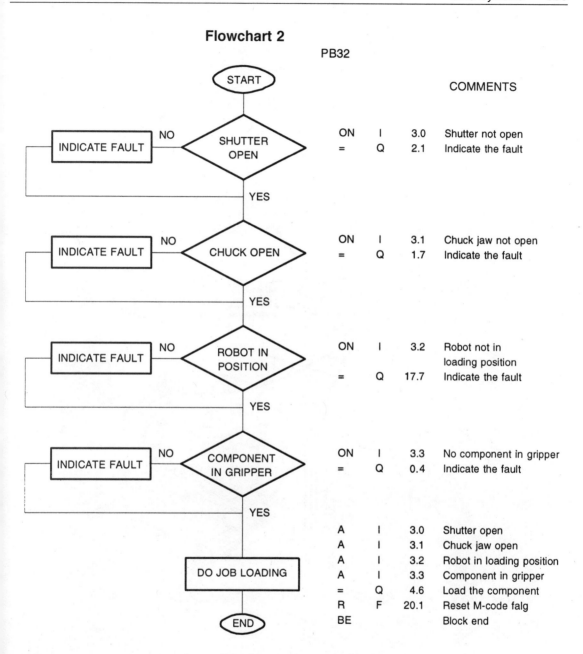

## 8.9 DIRECT NUMERICAL CONTROL (DNC)

In a production situation, when several CNC machines are used, it has become a common practice to use a central computer to connect all the machine CNC systems for preparing part programs at a central place and to transfer the same to the CNC system when required. This feature is called DNC linking. The RS232C, 2 way communication interface option on the system enables the user to immediately access any program from the library of programs in the host computer. Also this link permits transmission of messages and other critical data. The transmission may be effected in the foreground, i.e. operating the system in a specific mode (like input mode or output mode), or it may be done in the background, i.e. without interrupting the machining process (like auto mode). Thus by offloading the preparation of part program and other preparatory work to the central computer, the machine availability for cutting can be enhanced considerably (Fig. 8.16).

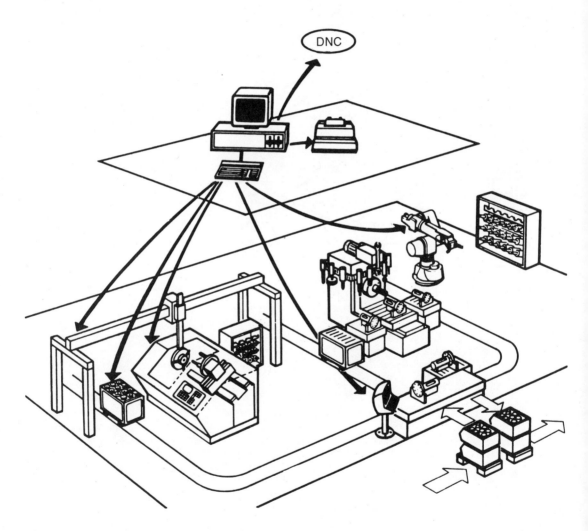

**Fig. 8.16** Direct numerical control (DNC)

## ANNEXURE 8.1

### Specifications of a CNC System

| | | | |
|---|---|---|---|
| 1. | Number of controlled axes | : | Two/Four/Eight, etc. |
| 2. | Interpolation | : | Linear/circular/parabolic or cubic/cylindrical |
| 3. | Resolution | : | Input resolution (feedback) |
| | | : | Programming resolution |
| 4. | Feed rate | : | Feed/min |
| | | : | Feed/revolution |
| 5. | Rapid traverse rate | : | Feed rate override |
| | | : | Feed per minute |
| 6. | Operating modes | : | Manual/Automatic/MDI(editing)/Input/Output/Machine data set-up/Incremental, etc. |
| 7. | Type of feedback | : | Digital (rotary encoders with train of pulses) |
| | | : | Analog (transducers, etc.) |
| | | : | Both |
| 8. | Part program handling | : | Number of characters which can be stored |
| | | : | Part program input devices |
| | | : | Output devices |
| | | : | Editing of part program |
| 9. | Part programming | : | Through MDI |
| | | : | Graphic simulation |
| | | : | Blue print programming |
| | | : | Background editing |
| | | : | Menu driven programming |
| | | : | Conversational programming |
| 10. | Compensations | : | Backlash |
| | | : | Lead screw pitch error |
| | | : | Temperature |
| | | : | Cutter radius compensation |
| | | : | Tool length compensation |
| 11. | Programmable logic controller | : | Built-in (integrated)/External |
| | | : | Type of communication with NC |
| | | : | Number of inputs, outputs, timers, counters and flags |
| | | : | User memory |
| | | : | Program organisation |
| | | : | Programming Languages |
| 12. | Thread cutting/Tapping | : | Types of threads that can be cut |
| 13. | Spindle control | : | Analog/Digital control |
| | | : | Spindle orientation |
| | | : | Spindle speed overrides |
| | | : | RPM/min; constant surface speed |
| 14. | Other features | : | Inch/metric switchover |
| | | : | Polar coordinate inputs |
| | | : | Mirror imaging |
| | | : | Scaling |

(Contd)

- Coordinate rotation system
- Custom macros
- Built-in fixed cycles
- Background communication
- Safe zone programming
- Built-in diagnostics, safety functions, etc.
- Number of universal interfaces
- Number of active serial interfaces
- Direct numerical control interface
- Network interface capability

CNC Systems 337

## ANNEXURE 8.2
### Numbering System

A computer can only understand and compute all the calculations in binary code, i.e. a numbering system that has 2 as base and all the numerical values are composed of '1' and '0'. In electrical terms, 'ON' or 'OFF', sensing the presence or absence of magnetism or voltage, etc. Hence all the information to the computer are fed as binary coded numbers.

To understand a binary system, it is helpful to look at the familiar decimal system. This system uses 10 as a base. Whenever we see a number written in base ten, we seldom think or analyse before we start working with it. The number, for example, 538 should be represented as $(538)_{10}$. Base 10 has digits from 0,1,2,..., to 9 and the numerical values are constructed from multiples of units, tens, hundreds, thousands, etc.

Units has a value of $10^0$, Ten has the value of $10^1$, Hundred has the value of $10^2$, etc.
Thus the number $(538)_{10}$ is made up as follows:

| Hundreds | | Tens | | Units |
|---|---|---|---|---|
| $5 \times 10^2$ | + | $3 \times 10^1$ | + | $8 \times 10^0$ |
| $= 5 \times 100$ | + | $3 \times 10$ | + | $8 \times 1$ |
| $= (538)_{10}$ | | | | |

In general, a number '$N$' can be expressed as :

$$N = d_n R^n + d_{n-1} R^{n-1} + \ldots + d_2 R^2 + d_1 R^1 + d_0 R^0$$

where, $d_n$ = digit of $n^{th}$ position
$R^n$ = base or radix of the $n^{th}$ position

Comparing the decimal and binary bases with their powers, there is no difference in the basic principle.

$10^0 = 1$  $\quad\quad$ $2^0 = 1$
$10^1 = 10$ $\quad\quad$ $2^1 = 2$
$10^2 = 100$ $\quad\quad$ $2^2 = 4$
$10^3 = 1000$ $\quad\quad$ $2^3 = 8$ and so on.

By using the small range of binary values, any decimal digit can be expressed as shown below.

| Decimal Digit | Binary Equivalent | | | | Computation | | | | | | | | |
|---|---|---|---|---|---|---|---|---|---|---|---|---|---|
| | $2^3$ (8) | $2^2$ (4) | $2^1$ (2) | $2^0$ (1) | | | | | | | | | |
| 0 | 0 | 0 | 0 | 0 | $2^3 \times 0$ + | $2^2 \times 0$ + | $2^1 \times 0$ + | $2^0 \times 0$ | = | 0 |
| 1 | 0 | 0 | 0 | 1 | $2^3 \times 0$ + | $2^2 \times 0$ + | $2^1 \times 0$ + | $2^0 \times 1$ | = | 1 |
| 2 | 0 | 0 | 1 | 0 | $2^3 \times 0$ + | $2^2 \times 0$ + | $2^1 \times 1$ + | $2^0 \times 0$ | = | 2 |
| 3 | 0 | 0 | 1 | 1 | $2^3 \times 0$ + | $2^2 \times 0$ + | $2^1 \times 1$ + | $2^0 \times 1$ | = | 3 |
| 4 | 0 | 1 | 0 | 0 | $2^3 \times 0$ + | $2^2 \times 1$ + | $2^1 \times 0$ + | $2^0 \times 0$ | = | 4 |
| 5 | 0 | 1 | 0 | 1 | $2^3 \times 0$ + | $2^2 \times 1$ + | $2^1 \times 0$ + | $2^0 \times 1$ | = | 5 |
| 6 | 0 | 1 | 1 | 0 | $2^3 \times 0$ + | $2^2 \times 1$ + | $2^1 \times 1$ + | $2^0 \times 0$ | = | 6 |
| 7 | 0 | 1 | 1 | 1 | $2^3 \times 0$ + | $2^2 \times 1$ + | $2^1 \times 1$ + | $2^0 \times 1$ | = | 7 |
| 8 | 1 | 0 | 0 | 0 | $2^3 \times 1$ + | $2^2 \times 0$ + | $2^1 \times 0$ + | $2^0 \times 0$ | = | 8 |
| 9 | 1 | 0 | 0 | 1 | $2^3 \times 1$ + | $2^2 \times 0$ + | $2^1 \times 0$ + | $2^0 \times 1$ | = | 9 |
| 10 | 1 | 0 | 1 | 0 | $2^3 \times 1$ + | $2^2 \times 0$ + | $2^1 \times 1$ + | $2^0 \times 0$ | = | 10 |
| 11 | 1 | 0 | 1 | 1 | $2^3 \times 1$ + | $2^2 \times 0$ + | $2^1 \times 1$ + | $2^0 \times 1$ | = | 11 |
| 12 | 1 | 1 | 0 | 0 | $2^3 \times 1$ + | $2^2 \times 1$ + | $2^1 \times 0$ + | $2^0 \times 0$ | = | 12 |
| 13 | 1 | 1 | 0 | 1 | $2^3 \times 1$ + | $2^2 \times 1$ + | $2^1 \times 0$ + | $2^0 \times 1$ | = | 13 |
| 14 | 1 | 1 | 1 | 0 | $2^3 \times 1$ + | $2^2 \times 1$ + | $2^1 \times 1$ + | $2^0 \times 0$ | = | 14 |
| 15 | 1 | 1 | 1 | 1 | $2^3 \times 1$ + | $2^2 \times 1$ + | $2^1 \times 1$ + | $2^0 \times 1$ | = | 15 |

The digits in the binary system are called *bits*, i.e. binary digit. A byte is typically 8 bits. 16 bits form a word and 32 bits is a double word.

# CHAPTER 9

# Programming and Operation of CNC Machines

## 9.1 INTRODUCTION TO PART PROGRAMMING

A program is a set of encoded information giving coordinate values and other details to indicate how a tool should be moved in relation to a workpiece in order to achieve a desired machining form. A part program contains all the information for the machining of a component which is input to the CNC system. The CNC system provides signals at the correct time and in the correct sequence to the various drive units of the machine.

### 9.1.1 What is Programming?

The numerical control machine tool receives information through a punched paper tape or a floppy disk. The tape is prepared in accordance with a program manuscript written for the job or operations to be carried out on the CNC machine. The program is prepared by listing the coordinate values $(X, Y, Z)$ of the entire tool paths as suited to machine the complete component. The coordinate values are prefixed with preparatory codes to indicate the type of movement required (point-to-point, straight or circular) from one coordinate to another. Also, the coordinates are suffixed with miscellaneous codes for initiating machine tool functions like start, stop, spindle movement, coolant on/off and optional stop. In addition to these coded functions, spindle speeds, feeds and the required tool numbers to perform machining in a desired sequence are also given. All these elements represent a line of information and form one meaningful command for the machine to execute and is called a block of information. The number of such blocks written sequentially form a part program for the particular component.

### 9.1.2 What the Programmer has to do? (Fig. 9.1)

1. Study the relevant component drawing thoroughly.
2. Identify the type of material to be machined.
3. Determine the specifications and functions of machine to be used.
4. Decide the dimension and mode—metric or inch.
5. Decide the coordinate system—absolute or incremental.

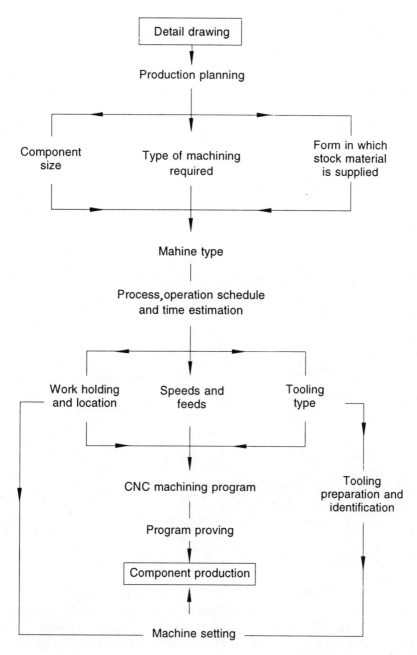

**Fig. 9.1** Procedures associated with part programming

6. Identify the plane of cutting.
7. Determine the cutting parameters for the job/tool combination.
8. Decide the feed rate programming—mm/min or mm/rev.

9. Check the tooling required.
10. Establish the sequence of machining operations.
11. Identify whether use of special features like subroutines, mirror imaging, etc. is required or not.
12. Decide the mode of storing the part program once it is completed.

## 9.2 COORDINATE SYSTEM

The cartesian coordinate system is used to locate any section or point on any drawing, machine or part along three mutually perpendicular axes. As shown in Fig. 9.2, four sections, called quadrants, are formed in the plane where the $X$ and $Y$ axes cross each other.

These quadrants are numbered counterclockwise as I, II, III and IV. The centre, called the origin, is the point at which the horizontal axis $X$ equals zero and the vertical axis $Y$ equals zero. The $Z$-axis (not shown) is the line perpendicular to the $X$ and $Y$ axes that goes through the origin, or zero point.

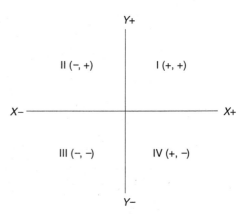

**Fig. 9.2** Cartesian coordinate system

From the origin to the right, all $X$ values are positive and to the left all $X$ values are negative. From the origin upwards, all $Y$ values are positive. Downwards $Y$ values are negative.

Each point is determined by two values, an $X$ value and a $Y$ value, each of which is either positive or negative, depending on its location. For example, if a point is stated as (–2, 3), it will be found in the second quadrant, where $X$ is negative and $Y$ is positive.

Any point in the first quadrant is (+, +) i.e., both $X$ and $Y$ are positive. Similarly, a point in the second quadrant is (–, +), in the third it is (–, –) and in the fourth it is (+, –).

## 9.3 DIMENSIONING

There are two types of dimensioning—absolute and incremental (Fig. 9.3).

Absolute dimensioning is the most commonly used method of dimensioning drawings for part production on CNC machines. In this system, all dimensions are taken from a single point, called a *datum point* or *origin* which is denoted by the preparatory function G90 (see Sub-section 9.6.7).

This method has definite advantages and some disadvantages. For example, in Fig. 9.3, if hole 2 is put in the wrong location, hole 3 would not be affected. This is one reason why absolute dimensioning is used more often than incremental dimensioning.

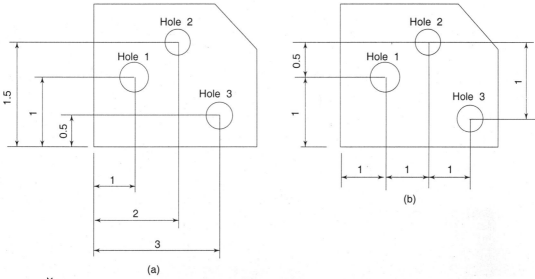

**Fig. 9.3** Dimensioning (a) Absolute dimensioning (b) Incremental positioning

In incremental dimensioning, the measurement is taken from hole to hole. Figure 9.3 shows how this method differs from absolute dimensioning. This system proves advantageous for machining a complicated pocket or recess and is denoted by the preparatory function G91.

Proper care should be taken when using incremental dimensioning because, if one hole, e.g. hole 2 is wrongly located, then all the remaining holes will be incorrectly positioned.

## 9.4 AXES AND MOTION NOMENCLATURE (FIG. 9.4)

The axes and motion nomenclature for computer numerically controlled (CNC) machines is intended to simplify programming and facilitate easy training of the programmers.

The methodology is to use a standard right-handed cartesian coordinate system on the assumption that the cutting tool moves relative to the workpiece. For $X$ and $Y$ axes, the positive direction of movement is that which causes an increasing positive dimension on the workpiece. In case of $Z$-axis, positive movement means the cutting tool moves away from the workpiece. $A$, $B$ and $C$ are the rotary axes of motion around the $X$, $Y$ and $Z$ axes respectively. Positive $A$, $B$ and $C$ are in the direction which advances a right handed screw in the positive $X$, $Y$ and $Z$ axes respectively.

*X-Axis:* The $X$-axis is always horizontal and parallel to the work holding surface. If $Z$-axis is also parallel to $X$-axis as in a horizontal machining centre, the positive $X$-axis is to the right when looking from the spindle towards the workpiece. When $Z$-axis is vertical as in a vertical machining centre, the positive $X$-axis is to the right when looking from the spindle.

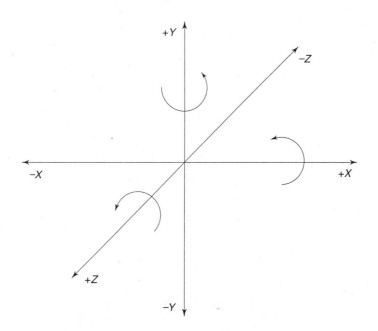

**Fig. 9.4** Axes and motion nomenclature

*Y-Axis:* The Y-axis is perpendicular to both X and Z-axes. The positive direction of Y-axis can be determined by reference to the right hand rule.

*Z-Axis:* The axis of the main machine spindle (or principal spindle) is denoted as the Z-axis. It can either be the axis of the tool spindle (as in machining centres) or the axis about which the workpiece rotates (as in turning centres). The Z-axis is positive in a direction from the workpiece towards the spindle.

If there are several spindles on a machine, one spindle is selected as the principal spindle and its axis is then considered to be Z-axis.

## 9.5 STRUCTURE OF A PART PROGRAM

As mentioned earlier, a part program (denoted by the symbol %) defines a sequence of NC machining operations. The information contained in the program can be dimensional or non-dimensional like speed, feed, auxiliary functions, etc. The basic unit of a part program input to the control is called a block. Each block contains adequate information for the machine to perform a movement and/or functions. Blocks in turn are made up of words and each word consists of a number of characters. All blocks are terminated by the *block end* character. The maximum block length for each CNC is fixed.

A block may contain any or all the following:

- Optional block skip (/)
- Sequence or block number (N)

- Preparatory functions (G)
- Dimensional information (X, Y, Z, etc.)
- Decimal point (.)
- Feed rate (F )
- Spindle speed (S)
- Tool number (T)
- Tool offset function (D)
- Miscellaneous functions (M, H, etc.)
- End of block (EOB/ *)

## 9.5.1 Block Example (Fig. 9.5)

N1234 G.. X.. Y.. S.. M.. F.. T.. D.. LF

The word addressed system is the method generally used for writing part programs for the numerical control of a wide range of machines. In this system there is an identifying letter (alpha character) preceding each numerical data. This alpha character in the word is known as the address, e.g. G2, N2, etc. A control unit recognises a word through its address. A number of words make up a block and two blocks are separated by a marker. Generally, the words need not be programmed in a particular order. Different words may have a different number of digits; some have only one digit and others may have up to seven digits. Whenever a word represents machine movement data, a + or a – may be required between the letter and the number.

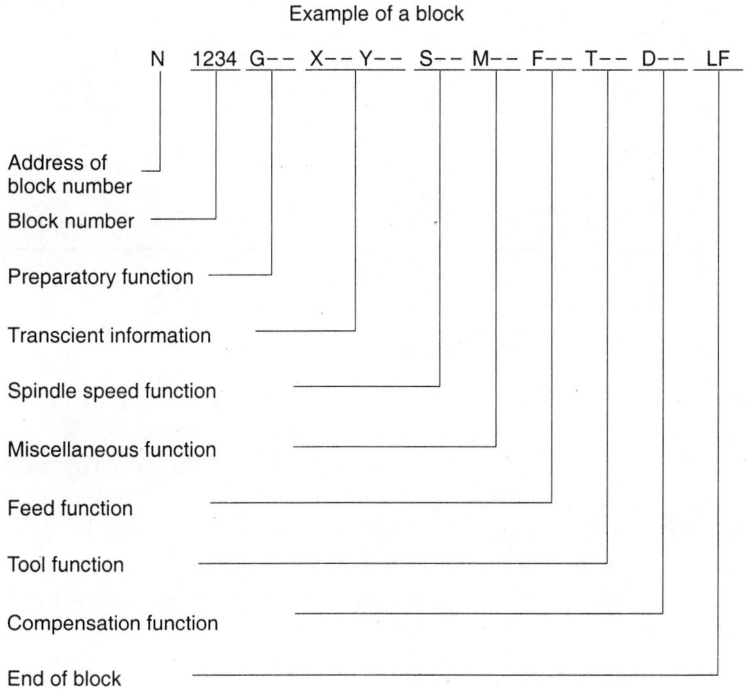

Example of a block

N  1234 G-- X-- Y-- S-- M-- F-- T-- D-- LF

Address of block number
Block number
Preparatory function
Transcient information
Spindle speed function
Miscellaneous function
Feed function
Tool function
Compensation function
End of block

**Fig. 9.5** Example of a block in programming

Designation of some letters for certain addresses has been standardised, for example the letter N always designates block number; other letters in the standardised list are P, G, M, X, Y, Z, I, J, K, L, U, V, W, A, B, C, F, S, T, EOB. A CNC manufacturer may use some letters which are not standardised for special words, e.g. R and H used in controls manufactured by Fanuc or Siemens.

## 9.6 WORD ADDRESSED FORMAT

As said earlier, an operation of a machine may be defined by a group of words or a single word. This is termed as a block. A typical block has the following format:

Metric system/N40G02X+or–4.3Y+or–4.3Z+or–4.3I+or–4.3J+or–4.3K+ or –4.3F4S4 T4M2EOB

### 9.6.1 Optional Block Skip Character(/)

A programmer may program all the operations of a particular family of components. The operator can omit certain operations in a program for a particular component. A particular operation may be required for a particular component and may not be required for another component. These blocks are programmed with the character / (slash) as the first character. Whenever a program is read by the control unit, the blocks preceded by the/character are omitted (not executed) when a switch on the system is activated. The following example explains the functioning of the block skip character.

| Original program | Execution of the program when the switch is not actuated | Execution of the program when the switch is actuated |
|---|---|---|
| / N05 X40.001 Z20 | N05 X40.001 Z20 | — |
| N10 Z-20 | N10 Z-20 | N10 Z-20 |
| / N15 X4.100 | N15 X4.100 | — |
| N20 Z+40 | N20 Z+40 | N20 Z+40 |

(Block numbers 5 and 15 are not executed)

### 9.6.2 Block or Sequence Number (N)

A part program is constructed with a number of blocks. Block number represents the operation number in a program and is usually the first character.

The number of digits in a block number depends upon the control manufacturer (usually it is four digits). Block numbers are mainly used for the convenience of an operator in identifying the different operations (block numbers need not be consecutive and in most of the cases it is not a must to have a block number). Many times it is useful to give a single block number to a set of operations when they are related to each other. For e.g.,

```
N0000    X..........Y........Z..........EOB
         Y....................Z..........EOB
```

```
            S200.......M03...................EOB
            T0012                             EOB
            X..............Y..................EOB
   N0005    M04                               EOB
```

It is convenient for the block numbers within the original program to be in multiples of five so that additional blocks could be inserted during any editing that may be needed.

### 9.6.3  Spindle Speed (S)

This may indicate either the spindle rpm or the constant cutting speed in m/min.

### 9.6.4  Feed Rate Word (F)

The feed rate or the rate at which the cutter travels through the material, is specified in mm/min or mm/rev.

### 9.6.5  Tool Number (T)

For machines having automatic tool changers or turrets, the T-word calls out a particular tool that has to be used for cutting.

### 9.6.6  D-Word

This word activates the cutter radius and length compensations.

### 9.6.7  Preparatory Function (G)

These are the codes which prepare the machine to perform a particular function like positioning, contouring, thread cutting and canned cycling. In general the following preparatory functions can be identified.

### PREPARATORY FUNCTIONS (G CODES)

    G00  - Rapid traverse
    G01  - Linear interpolation
    G02  - Circular interpolation, clockwise
    G03  - Circular interpolation, counter clockwise
    G04* - Dwell time, under address F or X in seconds
    G10  - Polar coordinate programming, rapid traverse
    G11  - Polar coordinate programming linear interpolation
    G17  - Plane selection XY-plane
    G18  - Plane selection XZ-plane
    G19  - Plane selection YZ-plane
    G40  - Cancel cutter radius compensation
    G41  - Cutter radius compensation, left

G42 - Cutter radius compensation, right
G53 - Suppress the zero offsets
G54 - Select zero offset 1
G55 - Select zero offset 2
G56 - Select zero offset 3
G57 - Select zero offset 4
G59* - Programmable additive zero offset
G60 - Exact stop
G62 - Continuous mode, fine tolerance
G64 - Continuous path mode, coarse tolerance
G70 - Input system in inch, reset state
G71 - Input system in metric, machine data
G80 - No boring cycle
G81 - Drilling, centring, boring cycle
G82 - Drilling and countersinking cycle
G83 - Deep hole drilling and boring cycle
G84 - Tapping cycle
G85 - Boring cycle
G86 - Boring cycle
G87 - Boring cycle
G88 - Boring cycle
G89 - Boring cycle
G90 - Absolute dimension programming
G91 - Incremental dimension programming
G92 - Limitation of spindle speed S when using with G96
G94 - Feed rate under address F mm/min.
G95 - Feed rate under address F mm/rev.
G96 - Constant cutting speed

**Problem 1** The part drawing in Fig. 9.6(a) requires the $X$ and $Y$ axes to position first to point 1 (PT1) and then to PT2, PT3 and finally PT4. The following example shows how this can be done using G90 absolute mode.

```
N5  G71           (Metric)
N10 G90           (Absolute)
N15 G00           X100 Y100 (PT1)
N20               X150 Y175 (PT2)
N25               X200 Y125 (PT3)
N30               X250 Y150 (PT4)
```

**Problem 2** The part drawing in Fig. 9.6(b) requires the $X$ and $Y$ axes to position first to PT1 and then to PT2 and finally to PT3. The set-up dimensions are the various distance from the absolute zero position to the part reference points. All dimensions are in mm.

```
N5  G90 X100 Y100 (G90 is used to locate part ref. point)
N10 G91           (Incremental mode)
```

Programming and Operation of CNC Machines  347

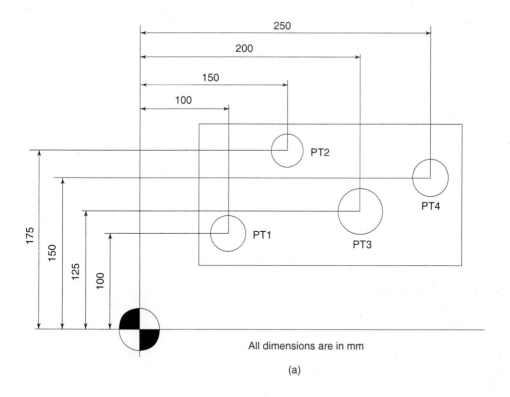

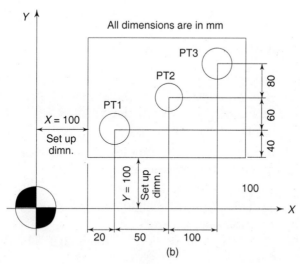

**Fig. 9.6** (a) Absolute programming (b) Incremental programming

```
N15  G00         X20  Y40  (PT1)
N20              X50  Y60  (PT2)
N25              X100 Y80  (PT3)
```

## 9.6.8 Miscellaneous or Auxiliary Function (M)

These are the operations associated with the machine for functions other than positioning or contouring, e.g. coolant on or off, spindle rotation, etc.

The following are the various miscellaneous functions in general.

### MISCELLANEOUS FUNCTION (M-CODES)

M00 - Unconditional program stop
M01 - Conditional program stop
M02 - End of program with return to program start
M03 - Spindle rotation, clockwise
M04 - Spindle rotation, counter clockwise
M05 - Spindle stop
M08 - Coolant ON
M09 - Coolant OFF
M13 - Coolant ON with Spindle ON
M15 - Coolant OFF with Spindle OFF
M17 - End of subroutine, written in the last block of the subroutine
M19 - Spindle orientation
M20 - Cancel mirror image
M21 - Mirror image $X$-axis
M22 - Mirror image $Y$-axis
M23 - Mirror image $Z$-axis
M30 - End of program

## 9.7 G02/G03 CIRCULAR INTERPOLATION

The interpolation parameters together with the axis commands determine the circle or arc. The starting point $KA$ is determined by the previous block. The end point $KE$ is fixed by the axis values of the plane in which the circular interpolation is programmed. The circle centre point $KM$ is determined

(a) either through the interpolation parameters (Sub-section 9.7.1) (Fig. 9.7)
(b) or by specifying the circle radius (Sub-section 9.7.2)

Radii should not be programmed when the angle between the radius vectors drawn from $KM$ to $KA$ and from $KM$ to $KE$ is 0° or 360°. In these cases the full circles must be programmed using the interpolation parameters.

Circular interpolation is possible in two out of four axes. The direction in which the arc is traversed is determined by G02 or G03 (Fig. 9.8).

A right-hand system in the three primary axes is achieved with the following axes combinations:

$$X...\ Y...,\ Z...\ X...\ \text{and}\ Y...\ Z...$$

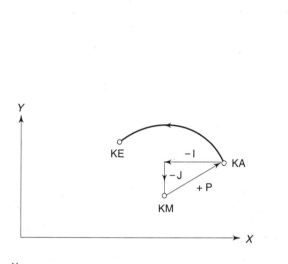
Fig. 9.7 Circular interpolation parameter

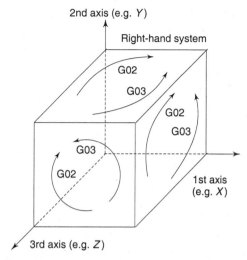
Fig. 9.8 Circular interpolation direction

## 9.7.1 Circular Interpolation Using Interpolation Parameters

The starting point of the circle or arc is determined by the previous block. The end point is fixed by the respective axis values. The circle centre is determined through the interpolation parameters:

Vector $I$ parallel to $X$ axis  The magnitudes $I$, $J$, $K$ are incremental distances
Vector $J$ parallel to $Y$ axis  from circle start point to centre point
Vector $K$ parallel to $Z$ axis

The parameter vectors are positive in the direction of increasing coordinate values with respect to their respective axis, the vectors are negative otherwise (Fig. 9.7).

If only one axis coordinate is programmed, the value belonging to the primary axis of the selected plane (G17, G18, G19) is used as secondary axis coordinate. If the signal *fourth axis = primary axis* is active, then this value is used according to the selected plane.

The fourth axis may be defined as being parallel to the $X$, $Y$, or $Z$ axis using machine parameters.

The address of the circular interpolation parameter for the fourth axis is then equal to the associated parallel primary axis. If an interpolation parameter is not programmed, zero is assumed.

### Example

| | | | | | | |
|---|---|---|---|---|---|---|
| N5 | G17 | G42 | D03 | ... | LF | Plane and tool offset selection |
| N10 | G03 | X17. | Y30. | I-9 J8. | LF | Complete definition of the circle with direction, circle end point coordinates and interpolation parameters |

N25   G03   X17. I-9           LF      Circle programming with missing
                                       addresses. If no other plane and
                                       traverse distance in the Y-axis is
                                       programmed between N10 and N25
                                       the control generates the following:

N25   G17   G03   X17. Y30. I-9J0. LF

\*

## Example of Circular Interpolation Using Interpolation Parameters

### Absolute Dimension Programming (Fig. 9.9)

N5  G02  G90  X45. Y30. I0. J15.  LFN5  G03  X60. Y15. I15. J0. LF
The tool moves from point $P_2$         The tool moves from point $P_1$
to point $P_1$                           to point $P_2$

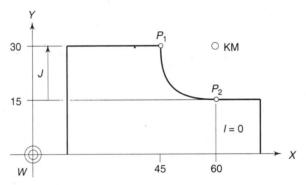

Fig. 9.9  Circular interpolation specifying circular interpolation parameter

### Incremental Dimension Programming

N10  G02  G91  X-15. Y15. I0. J15. LF   N10  G03  X15. Y-15. I15. J0. LF
The tool moves from point $P_2$          The tool moves from point $P_1$
to point $P_1$                            to point $P_2$

### 9.7.2  Circular Interpolation by Specifying the Radius

The starting point of the circle or arc is determined by the previous block. The end point is given by two of the axis values (e.g., X and Y ). The circle centre is defined by the signed radius P (Fig. 9.7).

The sign of the radius value is given according to the size of the traversing angle as given below.

Less than or equal to 180°     P+
Greater than 180°              P–

No radii may be programmed, when the distance between the circle end point and circle start point is less than 10 μm, i.e. a complete circle must be programmed using the interpolation parameters $I$, $J$ or $K$.

### Example of Circular Interpolation by Specifying the Radius (Fig. 9.10)

The circle centre point is determined by the signed radius.

N5   G03 G90 X60. Y15. P15. LF   The tool moves from point $P_1$ to point $P_2$
N10  G02 X45. Y30. P15. LF       The tool moves from point $P_2$ to point $P_1$.

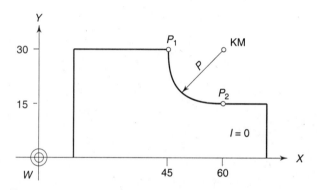

**Fig. 9.10**  Circular interpolation by specifying the radius

## 9.8  TOOL COMPENSATION

Tool compensation is a very useful feature in a control system and ensures that programming is independent of tool dimensions. The control system contains a memory in which both tool length and tool radius compensations are stored.

### 9.8.1  Tool Length Compensation

This is the distance between the gauge plane on the spindle nose and the tip of the cutter. It is essential to establish the length of each tool and store that in the system memory against the corresponding tool number before starting the actual cutting.

### 9.8.2  Tool Radius Compensation

During a turning or milling operation, an equidistant path of the tool tip centre or cutter centre has to be calculated and the dimensions have to be given to this tool path. This feature permits programming to the drawing dimensions of the workpiece and allows the calculation of the equidistant tool path to the control system. The actual tool radius is not specified in the program at all. This compensation can be given to the left of the workpiece (G41) or to the right of the workpiece (G42) as shown in Fig. 9.11. G40 cancels G41 or G42.

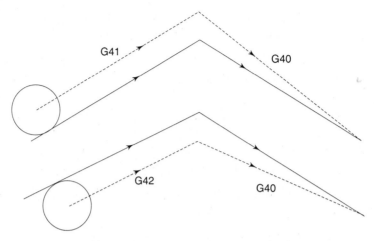

**Fig. 9.11** Tool radius compensation

## 9.9 SUBROUTINES (MACROS) (L)

When a component has repetitive pattern machining at different places, subroutine programming can be used to reduce the effort of writing a detailed program. The program for repetitive machining can be stored in the memory as a separate program and can be called in the main program whenever needed. When the last block in the subroutine is executed, control will automatically return to the main program.

Different controls allow nesting of subprograms up to different levels depending on the software capability. Usually nesting will be possible up to three levels (Fig. 9.12).

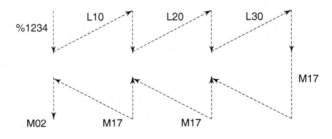

**Fig. 9.12** Subroutine nesting

### Example (Fig. 9.13)

```
%20
N5      G90     G71     G17*
N10     G0      X20     Y20*
N15     Z5  *
N20     L20 *
```

```
N25    G0     X60    Y20    *
N30    L20    *
N35    G0     X60    Y60    *
N40    L20    *
N45    G0     X20    Y60    *
N50    L20    *
N55    G0     Z100   *
N60    M02    *
(* End of block)
L20 *
N5     G91    *
N10    G1     Z-2    F200   *
N15    X20    *
N20    Y20    *
N25    X-20   *
N30    Y-20   *
N35    Z5     *
N40    G90    *
N45    M17    *
```

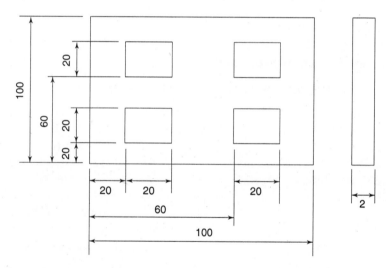

**Fig. 9.13** Example of subroutine

## 9.10 CANNED CYCLES (G81-G89)

A canned cycle (fixed cycle) defines (in accordance with DIN 66025) a series of machining sequences for drilling, boring, tapping, etc.

The canned cycles G81 to G89 are stored as subroutines L81 to L89.

The user may deviate from a standard fixed cycle and redefine it to suit his specific machine or tooling requirements. The parameters R00 to R11 (Sub-section 9.10.1) are used by subroutines to define the variable values necessary to correctly execute a fixed cycle (e.g., reference plane coordinates, the hole depth, feed rate and dwell time). Prior to a subroutine call, all necessary parameters must be defined in the main program.

A fixed cycle call is initiated with G80 to G89. G81 to G89 are fixed cycles that are cancelled with G80. A boring cycle can be called with L81 to L89, however, L81 to L89 are not modal. L81–L89 is performed only once in the block in which it is programmed. At the end of a fixed cycle the tool is re-positioned at the starting point.

### 9.10.1 R-Parameters Used in Cycles L81–L89

R00   Dwell time at the start point (deburr hole)
R01   First depth advance (incremental) entered without sign
R02   Reference plane or retract plane (absolute)
R03   Final depth
R04   Dwell time at hole bottom (break chips)
R05   Depth advance modifier entered without sign
R06   Reverse spindle rotation direction
R07   Return to the original spindle rotation direction used in the calling program (after R06 or M05)
R10   Retract plane
R11   Boring axis (axis numbers $X = 1$; $Y = 2$; $Z = 3$)

### 9.10.2 Using Canned Cycles in Programs

**CALL-UP G81 (DRILLING, BORING, CENTRING, BORING AXIS Z)**
**(Fig. 9.14)**

```
N8101   G90   S48   F460   LF              — Spindle ON
N8102   G00   D01   Z500   LF              — Activate tool offset
N8103   X100  Y150  LF                     — First drill position
N8104   G81   R02 360 R03 250 R11 3 LF     — Call cycle
N8105   X250  Y300  LF                     — Second drill position and
                                             automatic G81 call

N8110   G80   Z500  LF                     — Cancelling G81 and returning
                                             to starting plane
```

**CALL-UP WITH L81**

```
N8101   G90   S48   F460              LF
N8102   G00   D01   Z500              LF
N8103   X100  Y150                    LF
N8104   L81   R02 360 R03 250 R11 3   LF — Call up drilling cycle, first hole
N8105   X250  Y300                    LF
N8106   L81   R02 . . .               LF — Call-up drilling cycle, second hole
N8107   Z500                          LF
```

As opposed to the call-up with G81, here the drilling cycle must be called up a new at every new drill position.

## CALL-UP G82 (DRILLING, COUNTER SINKING) (Fig. 9.15)

```
N8201   ...   M03   F460                              LF
N8202   G00   D01   Z500                              LF
N8203   X100  Y150                                    LF
N8204   G82   R02 360  R03 250  R04 1  R11  3 LF
N8205   X250  Y300                                    LF
N8206   G80   Z500                                    LF
```

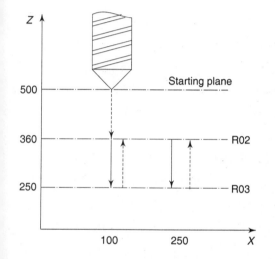

**Fig. 9.14** Call-up G81-drilling, boring, centering, boring axis Z

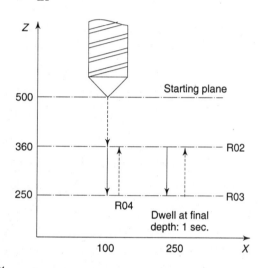

**Fig. 9.15** Call-up G82-drilling, countersinking

## CALL-UP G83 (DEEP HOLE DRILLING) (Fig. 9.16)

| | | |
|---|---|---|
| First drilling depth | 50 mm | R01 50 |
| Reference plane = retract plane | 146 mm | R02 146 |
| Final drilling depth | 5 mm | R03 5 |
| Dwell at starting point | 5 s | R00 5 |
| Dwell at final depth | 1 s | R04 1 |
| Degression value | 20 mm | R05 20 |
| Drilling axis (Z) | 3 | R11 3 |

```
N8301   ...   S48   M03   F460                        LF
N8302   G00   D01   Z500                              LF
N8303   X100  Y150                                    LF
```

```
N8304   G83   R01  50  R02  146  R03  5  R00  5
              R04  1  R05  20  R11  3              LF
N8305   X250  Y300                                 LF
N8306   G80   Z500                                 LF
```

At the rapid traverse advance with respect to the new drilling depth, a safety distance of 1 mm is kept (on account of the chips still remaining in the hole). With the inch system (G70) the safety distance must be changed accordingly.

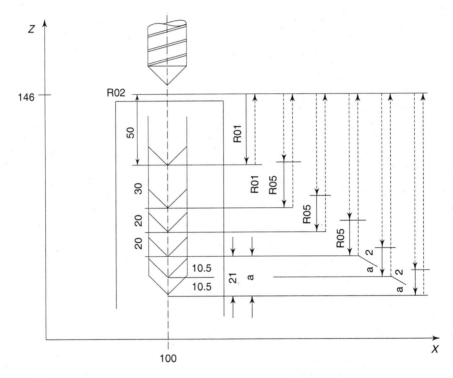

**Fig. 9.16** Call-up G83—deephole drilling

## CALL-UP G84 (Tapping Cycle) (Fig. 9.17)

```
N8401   ...   S48   M03   F460                     LF
N8402   G00   D01   Z500                           LF
N8403   X100  Y150                                 LF
N8404   G84   R02  360  R03  340  R06  04
              R07  03  R11  3                      LF
N8405   X250  Y300                                 LF
N8406   G80   Z500                                 LF
```

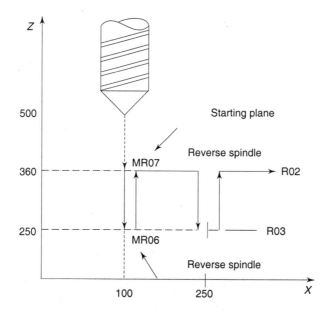

**Fig. 9.17** Call-up G84–tapping cycle

## CALL-UP G85 (BORING 1) (Fig. 9.18)

```
N8501   ...    S48   M03   F460                      LF
N8502   G00    D01   Z500                            LF
N8503   X100   Y150                                  LF
N8504   G85    R02 360 R03 250 R10 380 R11 3        LF
N8505   X250   Y300                                  LF
N8506   G80    Z500                                  LF
```

## CALL-UP G86 (BORING 2) (Fig. 9.19)

```
N8601   ...    S48   M03   F460                      LF
N8602   G00    D01   Z500                            LF
N8603   X100   Y150                                  LF
N8604   G86    R02 360 R03 250 R07 03
        R10 380 R11 3                                LF
N8605   X250   Y300                                  LF
N8606   G80    Z500                                  LF
```

## CALL-UP G87 (BORING 3) (Fig. 9.20)

```
N8701   ...    S48   M03   F460                      LF
N8702   G00    D01   Z500                            LF
N8703   X100   Y150                                  LF
```

**358** Mechatronics

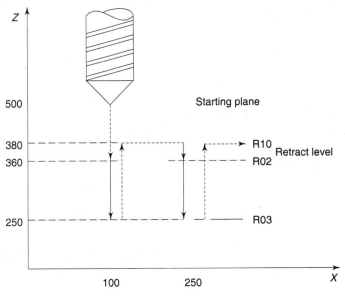

**Fig. 9.18** Call-up G85–boring 1

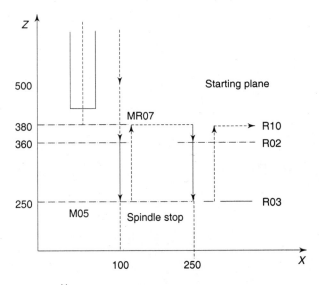

**Fig. 9.19** Call-up G86–boring 2

```
N8704    G87    R02 360 R03 250 R07 03 R11 3    LF
N8705    X250   Y300                              LF
N8706    G80    Z500                              LF
```

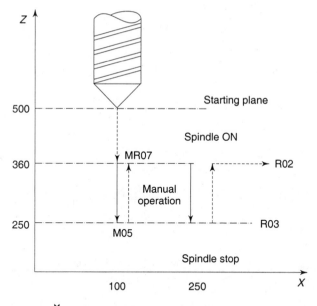

**Fig. 9.20** Call-up G87–boring 3

## CALL-UP G88 (BORING 4) (Fig. 9.21)

```
N8801   ...    S48    M03    F460              LF
N8802   G00    D01    Z500                     LF
N8803   X100   Y150                            LF
N8804   G88    R02    360    R03  250  R04  1
               R07  03  R11  3                 LF
N8805   X250   Y300                            LF
N8806   G80    Z500                            LF
```

## CALL-UP G89 (BORING 5) (Fig. 9.22)

```
N8901   ...    S48    M03    F460              LF
N8902   G00    D01    Z500                     LF
N8903   X100   Y150                            LF
N8904   G89    R02  360  R03  250  R04  1  R1103 LF
N8905   X250   Y300                            LF
N8906   G80    Z500                            LF
```

## 360 Mechatronics

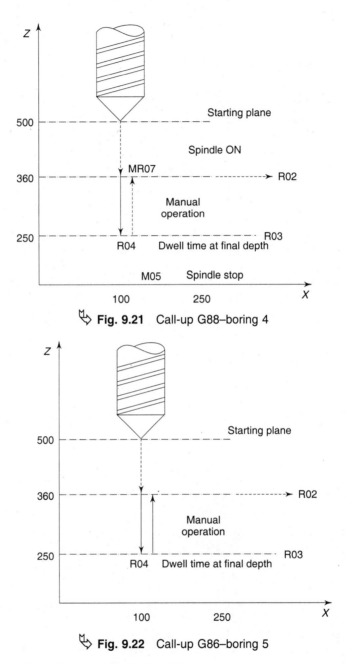

**Fig. 9.21** Call-up G88–boring 4

**Fig. 9.22** Call-up G86–boring 5

### Calling Boring Cycles in a Subroutine (Table 9.1)

If boring cycles are called in a subroutine, the following procedure is necessary:

%1
.
.

R02 360 R03 250 R00 81 R11 3   LF — Supply boring cycle parameters
L0101                            LF — Boring positions
M30*                             LF
L0101   (Boring positions)
GR00 X1 Y1                       LF — First boring location
X2 Y2                            LF — Second boring location
X5 Y5                            LF — Third boring location
X10                              LF — Fourth boring location
G80 M17                          LF — Deselect boring cycle and end of subroutine

TABLE 9.1

*Calling boring cycles in a subroutine*

| No. | Canned cycle Subroutine | Traverse rate into the part after positioning to the reference plane | At hole bottom Dwell | At hole bottom Spindle | Retract to the reference plane | Application example |
|---|---|---|---|---|---|---|
| 0. | L8000 | — | — | — | — | Cancels L81-L89 |
| 1. | L8100 | In feed | — | — | In rapid traverse | Drilling, Centering |
| 2. | L8200 | In feed | Yes | — | In rapid traverse | Drilling, Counter sinking |
| 3. | L8300 | In interrupted feed | — | — | In rapid traverse | Deep hole drilling, Chip breaking |
| 4. | L8400 | Feed per revolution | — | Rev. | In feed | Tapping |
| 5. | L8500 | In feed | — | — | In rapid traverse | Boring 1 |
| 6. | L8600 | In feed | — | Stop | In rapid traverse | Boring 2 |
| 7. | L8700 | In feed | — | Stop | Manual retraction | Boring 3 |
| 8. | L8800 | In feed | Yes | Stop | Manual retraction | Boring 4 |
| 9. | L8900 | In feed | Yes | — | In feed | Boring 5 |

## 9.11 MIRROR IMAGE

Two components which are identical and have the property that one is a mirror reflection of the other, can be manufactured by programming only for one component and by using the mirror imaging feature. This reduces the effort required for programming and reduces the amount of space occupied in the system memory. The contours are always mirrored about the coordinate axes. In mirror imaging the following operations occur:

- Sign reversal of the programmed coordinates (e.g. +X to –X).
- Interpolation direction (e.g. G02 to G03 or G03 to G02).
- Single axis mirrored in cutter radius compensation (CRC) plane (e.g. G41 to G42 or G42 to G41)
- Dual axis mirrored in CRC plane (e.g. G41 to G41; G42 to G42; G02 to G02; G03 to G03)

The following values are not mirrored
- Length offset dimensions
- Zero offset

## Example (Fig. 9.23)

```
%  1234  *
N5   G17 G71 G90 G41 *
N10  T1 L90 *
N15  G0 X0 Y0 S100 M03 D01 *
N20  Z5 *
N25  L123 *
N30  M21 *              (X-axis mirroring)
N35  G04 F2 *           (Dwell on account of PLC delay time)
N40  @31 *              (Emptying buffer)
N45  L123 *
N50  M20 *              (Cancel mirroring)
N55  M02 *
L123 *
N1 G1 Z-2 F56 *
N2 G1 Y20 *
N3 X15 Y35 *
N4 X35 *
N5 G03 X50 Y50 I0 J15 *
N6 G1 X80 *
```

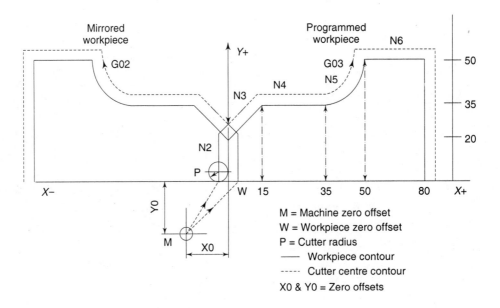

**Fig. 9.23** Mirror imaging

```
N7   Y0
N8   G0  Z5  *
N9   X0  Y0  *
N10  M17 *
```

## 9.12 PARAMETRIC PROGRAMMING (USER MACROS) AND R-PARAMETERS

Modern CNC systems offer the manual programmer increased computing power. Parametric programming allows the programmer to use canned cycles to save time. The cycles are uniquely written for the programmer and/or the machine combination, providing more flexibility. Parametric programming also allows the programmer to design his own canned cycles.

Parameters with address R are used in a program to represent the numeric value of an address, and can be assigned to all functions except N. They can be defined either in part programs or in subroutines. The number of R-parameters provided depends on the CNC system version used. System 3 (Hinumerik) gives R-parameters from R00 to R99, and a maximum of ten parameters can be programmed in one block.

**TABLE 9.2**

*Parameter linking*

| Arithmetic function | Programmed arithmetic operation |
|---|---|
| Definition | R1 = 100 |
| Assignment | R1 = R2 |
| Negation | R1 = – R2 |
| Addition | R1 = R2 + R3 |
| Subtraction | R1 = R2 – R3 |
| Multiplication | R1 = R2 · R3 |
| Division | R1 = R2/R3 |

*Note*: The result of arithmetic operation is written in the first parameter of a link, its initial value is thus overwritten and lost on linking. The values of the second and/or third parameters are retained.

**Example for Calculating R-Parameters**

```
%1234
N30  R01  10.78  R02  95.34  R03  -555.1  *
N35  X20.78 - R01 *                              X = 10
N40  Y44.9 + R02 *                               Y = 140.24
N45  Z10.1 - R03 *                               Z = 565.2
N50  X @10 R01 (SQUARE ROOT)                     X = 3.28
N55  Y @15 R01 (SINE)                            Y = 0.187
```

## 9.13 G96 S... CONSTANT CUTTING SPEED AND G97 CONSTANT SPEED

### 9.13.1 G96 S... Constant Cutting Speed

Depending on the programmed cutting speed, the controller determines the appropriate spindle speed on the latest turning diameter.

    N5 ... G96 S... LF    Constant cutting speed in (m/min)

The correlation between turning diameter, spindle speed and feed rates ensures an optimum matching of the program to the machine, the workpiece material, and the tool.

The zero point for the $X$-axis must be the turning axis. This is ensured by the reference point approach.

When calculating the spindle speed for constant cutting speed, the following values are taken into account:

- Actual value
- Tool length compensation
- Zero offset in $X$-direction
- —Settable Zero offset G54, G55, G56, G57 (Z0)
  —Additive Z0 G59
  —External additive compensation

The workpiece must not be shifted out of the turning axis by a zero offset in the $X$-axis. The latter can be used to shift the tool holder. G96 S... is therefore always referred to the workpiece zero.

### 9.13.2 G97 Constant Speed

The constant cutting speed is frozen with function G97. The last actual speed value is used as the constant speed.

G97 is used to avoid undesirable changes in speed in the event of intermediate blocks in the $X$-direction without machining. Additionally, an undesired change in speed caused by the PLC cycle time (S-word acknowledgement) can be avoided by G97.

## 9.14 MACHINING CYCLES

### 9.14.1 L95 Stock Removal Cycle

    R20    Subroutine number under which the contour is stored
    R21    Start point of the contour in $X$ (absolute)
    R22    Start point of the contour in $Z$ (absolute)
    R24    Finishing cut depth in $X$ (incremental)
    R25    Finishing cut depth in $Z$ (incremental)
    R26    Roughing depth in $X$ or $Z$ (incremental)

R27   Tool nose radius compensation (G41, G42)
R29   Type determination for roughing and finishing

The parameters to be entered are shown in the Fig. 9.24.

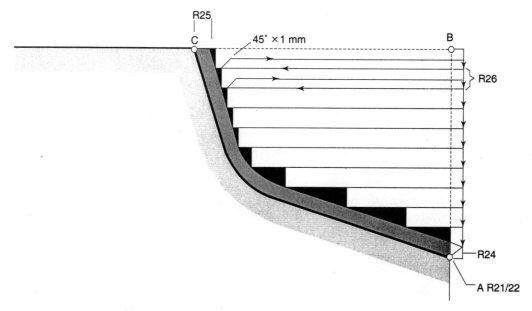

**Fig. 9.24**   Machining cycle—L95 stock removal cycle

## 9.14.2  L97 Thread Cutting Cycle

This cycle is used for longitudinal cutting of outside threads, inside threads and taper threads. The feed of the tool is automatic and decreases quadratically, the cut depth remains constant.

Before calling up cycle L97, the following R parameters must be assigned a value:

R20   Thread pitch
R21   Start point of thread in $X$ (absolute)
R22   Start point of thread in $Z$ (absolute)
R23   Number of compound feeds
R24   Thread depth (incremental); sign required to define inside or outside thread, + inside thread/– outside thread
R25   Finishing depth
R26   Approach path
R27   Run out path
R28   Number of roughing cuts
R29   Infeed angle
R31   End point of thread in $X$ (absolute)
R32   End point of thread in $Z$ (absolute)

The values must be assigned in two blocks, since a maximum of ten R-parameters per block can be assigned.

The individual parameter values are represented in Figs 9.25–9.30.

### R20 Thread Pitch

The parameter represents the value of the thread pitch. It is always written as a paraxial value without sign.

Minimum 0.001 mm
Maximum 40 mm

### R21 and R22 Start Point of Thread (Fig. 9.25)

The parameters R21 and R22 represent the original starting points of the thread (A). The start point of the thread cycle is at point B which lies at the parameter R26 (approach path) before the thread output point. The start point B lies on the X-axis, 1 mm over the parameter value R21. This raised plane is generated automatically in the controller. The thread cycle can be called up from any cutting position, the infeed to point B is effected at rapid traverse rate.

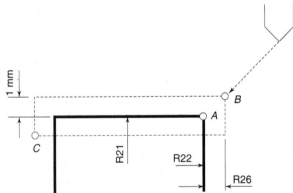

**Fig. 9.25** Thread cutting cycle—R21 and R22 start point of thread

### R23 Idle Passes

Any number of idle passes can be selected. They are entered using parameter R23, e.g. 3. Idle passes are entered as R23 3.

### R24 Thread Depth

The thread depth is entered using parameter R24. The sign determines the infeed direction, i.e., whether it is an outside or inside thread (+ inside thread, − outside thread) (Fig. 9.26).

### R25 Finishing Cut Depth

R25 gives the finishing cut depth (Fig. 9.27). When a finish cut depth is programmed, this depth is subtracted from the thread depth and the remaining value divided into

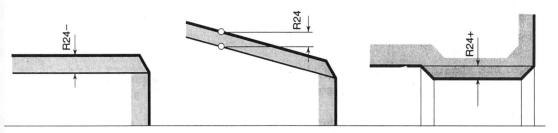

&#x21AA; **Fig. 9.26** Thread cutting cycle-thread depth

roughing cuts. After the roughing cuts have been completed, first a finishing cut is made and then the number of idle passes run as programmed under R23.

    R24 = thread depth
    R25 = finishing cut depth
    S = roughing depth, automatically calculated and divided into roughing cuts.

### R26 Approach Path
### R27 Run Out Path

The approach and run out paths are entered without signs.

&#x21AA; **Fig. 9.27** Thread cutting cycle—R25 finishing cut depth

The parameters represent paraxial incremental values. In the case of taper threads, the control calculates the approach and run out path distances in relation to the taper and determines the corner points $B$ and $C$ (Fig. 9.28).

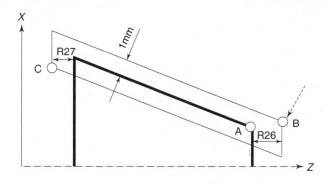

&#x21AA; **Fig. 9.28** Thread cutting cycle—R27 run out path

### R28 Number of Roughing Cuts

The parameter value determines the number of thread roughing cuts. The control automatically calculates the individual infeed depth. This ensures that the cut pressure from the first to the last roughing cut is the same.

### R29 Infeed Angle for Longitudinal or Transverse Thread

The infeed of the tool is at any desired angle possible, for longitudinal or transverse threads (in the case of taper threads the inclined position cannot be used). The angle input is made without sign (Fig. 9.29).

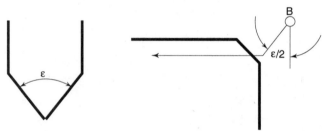

**Fig. 9.29** Thread cutting cycle—R29 infeed angle for longitudinal or transverse thread

In a cycle the angle is used according to the machining direction.

### R31 and R32 Thread End Point

The parameters R31 and R32 represent the original end points of the thread $A$ (Fig. 9.30).

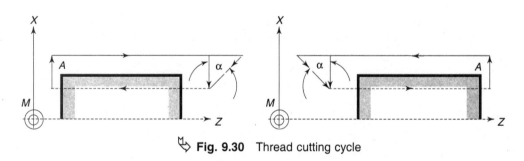

**Fig. 9.30** Thread cutting cycle

## 9.14.3 L903 Plunge Cutting Cycle (Fig. 9.31)

The plunge cutting cycle enables the machining of external and internal grooves. A value must be assigned to the following parameters before cycle L903 is called.

- R21  Outer diameter or inner diameter
- R22  Starting point in Z
- R23  Control parameter
- R24  Finishing allowance in $X$

R25   Finishing allowance in Z
R26   Infeed in X
R27   Plunge cutting width
R28   Dwell time at depth
R29   Angle
R30   Radius or chamfer at plunge cut base
R31   Plunge cut diameter
R32   Radius or chamfer at plunge cut edge

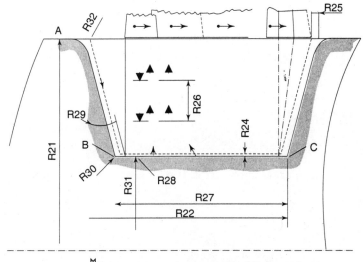

**Fig. 9.31** L903—plunge cutting cycle

Before the plunge cutting cycle is called in a machining program, the tool offset of the first tool must be selected. The tool offset for the second tool of the plunge cutting tool must be stored in the tool offset memory under the next higher offset number. If the tool offset for the first tool is T01, the tool offset for the second tool must be T02.

### R21/R22 Diameter (Outer or Inner)/Starting Point (Fig. 9.32)

Parameters R21 and R22 determine the starting point. The controller automatically approaches the point programmed by R21 and R22. With an outer groove, the Z-direction is first traversed, and with an inner groove the X-direction is first traversed. In the X-direction a safety clearance of 1 mm is maintained.

### R23 Control Parameter

This control parameter determines the type of groove: outer or inner, starting right or left.

    R23   1   Outer/left
    R23   2   Outer/right
    R23   3   Inner/left
    R23   4   Inner/right

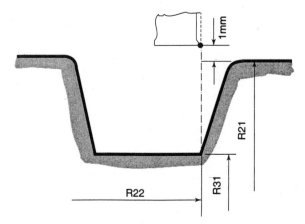

▷ **Fig. 9.32** R21/R22 diameter (outer or inner) and starting point

## R24/25 Finishing Allowance

The finishing allowance entered can be of different magnitude for the X and Z directions. The same plunge cutting tool is used for machining. The finishing allowance chosen should be large enough to ensure that the chamfers at the groove base are not damaged during plunge cutting.

## R26 Infeed in X (Fig. 9.33)

The maximum incremental infeed path is determined with the infeed (R26). For chip breaking, the tool is retracted by 1 mm.

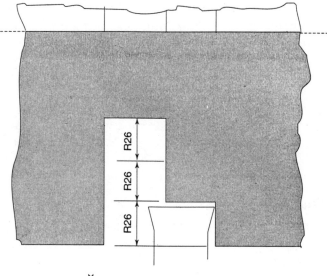

▷ **Fig. 9.33** R26 infeed X

## R27 Groove Width

If the groove is wider than the plunge cutting tool, the infeed in the Z-direction is subdivided into sections of the same size. The maximum infeed in Z depends on the tool width. It is 95% of the tool width. This ensures an overlap in cutting.

## R28 Dwell Time at Depth

The chosen dwell time must be large enough to ensure that at least one spindle revolution takes place.

## R29 Angle

The flank angle R29 may be 0° to 89°.

## R30/R32 Radius or Chamfer

Radii or chamfers at the groove base and/or groove edge can be inserted via parameters R30/R32.
   + sign    radius
   – sign    chamfer

## R31 Diameter of Groove

## Machining Sequence

**1. Radial Grooves (Fig. 9.34)** When a groove requiring one or more cuts perpendicular to the axis of rotation is involved, before withdrawal from the groove, a clearance of 1 mm is allowed from the second cut in the Z-direction.

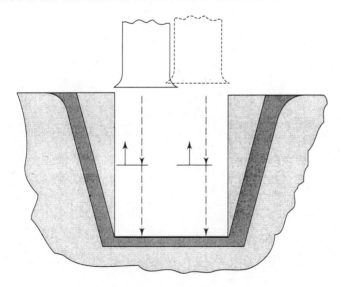

**Fig. 9.34** Machining of radial grooves

**2. Machining Flanks (Fig. 9.35)** During machining of flanks, where an angle has been programmed with R29. The infeed in the Z-direction takes place in several steps if the tool width is less than the flank width.

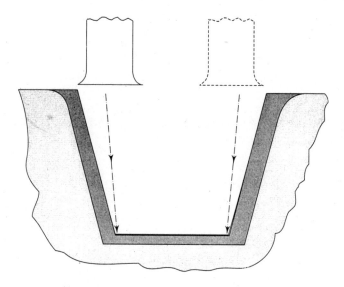

**Fig. 9.35** Machining of flanks

**3. Finishing (Fig. 9.36)** Machining with the finishing allowance is done parallel to the contour, down to the mid-point of the groove.

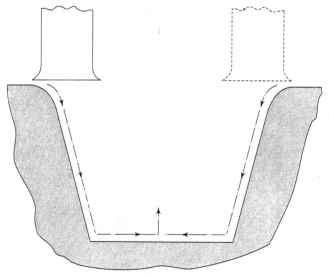

**Fig. 9.36** Machining with the finishing allowance

*Example (Fig. 9.37)*

```
       .
       .
N5        R21 60. R22 100. R23 2 R24 1. R25 0.5
          R26 10. R27 20. R28 1 R29 10. R30-1. LF
```

N10        R31 40. R32 2. L90301 LF

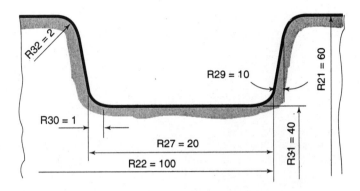

**Fig. 9.37** Machining with the finishing allowance

## 9.15 PROGRAMMING EXAMPLE FOR MACHINING CENTRE (FIG. 9.38)

### Operations

1. Bore one hole diameter 60H7
2. Drill two holes diameter 10
3. Drill and tap one hole M12
4. Drill and tap four holes M10
5. Chamfer bore $1 \times 45°$

### Tools Used

| | | |
|---|---|---|
| T1 | – | Core drill diameter 55.0 |
| T2 | – | High speed steel centre drill diameter 6.3 |
| T3 | – | Carbide tipped taper shank drill diameter 10.0 |
| T4 | – | Carbide tipped taper shank drill diameter 10.25 |
| T5 | – | Carbide tipped taper shank drill diameter 8.5 |
| T6 | – | Semi-finish boring bar diameter 59.5 |
| T7 | – | Finish boring bar diameter 60H7 |
| T8 | – | Machine tap M12 |
| T9 | – | Machine tap M10 |
| T10 | – | Chamfering end mill diameter 40.0 |

### Zero Offsets

1. X0, Y0 corresponds to the point marked
2. Z0 corresponds to the top of the workpiece from the table. (Workpiece thickness is 20 mm)

**374** Mechatronics

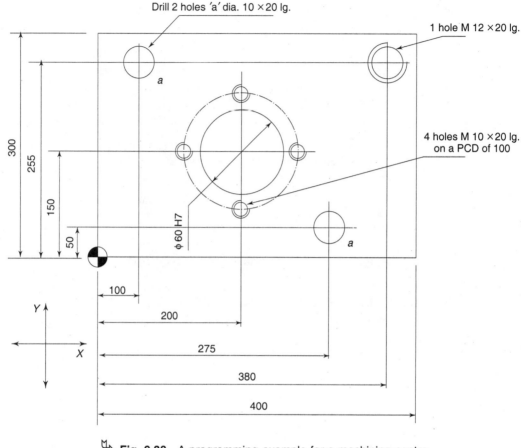

**Fig. 9.38** A programming example for a machining centre

```
% 1
N5  G17 G71 G90 *
N10 G94 G54 *
N15 T1 L90 (CORE DRILL DIAMETER 55.0) *
N20 S425 F60 M03 X100 Y200 D01 *
N25 G81 R02 5 R03-25 R11 03 *
N30 G80 *
N35 T2 L90 (CENTRE DRILL DIAMETER 6.3) *
N40 S1000 F100 M03 D02 X100 Y255 *
N45 G82 R02 5 R03-7 R04 1 R11 03 *
N50 X380 *
N55 X275 Y50 *
N60 G80 *
N65 T3 L90 (CARBIDE TIPPED TAPER SHANK DRILL DIAMETER 10.0) *
N70 S950 F90 M03 D03 X100 Y255 *
```

```
N75  G81 R02 5 R03- 25 R11 03 *
N80  X275 Y50 *
N85  G80 *
N90  T4 L90 (CARBIDE TIPPED TAPER SHANK DRILL DIAMETER 10.25) *
N95  S950 F90 M03 D04 X380 Y255 *
N100 G81 R02 5 R03-25 R11 03 *
N105 G80 *
N110 T5 L90 (CARBIDE TIPPED TAPER SHANK DRILL DIAMETER 8.5) *
N115 S1100 F110 M03 D05 *
N120 L100 R55 81 R02 5 R03-25 R11 03 *
N122 G80
N125 T6 L90 (SEMI FINISH BORING BAR DIAMETER 59.5) *
N130 S450 F45 M03 D06 X200 Y150 *
N135 G81 R02 5 R03-25 R11 03 *
N140 G80 *
N145 T7 L90 (FINISH BORING BAR DIAMETER 60H7) *
N150 S650 F65 M03 D07 X200 Y150 *
N155 G86 R02 5 R03-25 R07 03 R10 10 R11 03 *
N160 G80 *
N165 T8 L90 (MACHINE TAP M12) *
N170 S238 F416 M03 D08 X380 Y255 *
N175 G84 R02 5 R03-25 R06 04 R07 03 R11 03 *
N180 G80 *
N185 T9 L90 (MACHINE TAP M10) *
N190 S300 F450 M03 D09 *
N195 L100 R55 84 R02 5 R03-25 R06 04 R07 03 R11 03 *
N197 G80
N200 T10 L90 (CHAMFER END MILL DIAMETER 40) *
N205 S600 F700 M03 D10 X200 Y150 *
N210 Z-1 *
N215 G01 X190 Y150 *
N220 G02 X190 Y150 I10 J0 *
N225 G01 X200 Y150 *
N230 T0 L90 *
N235 M30 *
```

**L100 00**

```
N5  G10 X200 Y150 P50 A0 *
N10 G0 GR55 *
N15 G10 A90 *
N20 G10 A180 *
N25 G10 A270 *
N30 G80 *
N35 M17 *
```

## 9.16 PROGRAMMING EXAMPLE FOR TURNING CENTRE (FIG. 9.39)

### Operations
1. Turning for diameters
2. Grooving — 5 mm width
3. Threading — M30 × 2.5

### Tools Used
1. Turning tool
2. Grooving tool — 2 mm width (insert type)
3. Threading tool — 2.5 mm pitch (insert type)

```
% 6
N5   G0 X500 Z200 *
N10  T0101 M46
N15  G92 S2000
N20  G96 S200 M04 *
N25  R20=06 R21=0 R22=5 R24=0.5 R25=0.1 R26=3 *
N30  R27=0 R29=41 *
N35  L95 F0.2 M08 *
N40  L98 *
N45  T0808 M46 *
N50  G97 S600 M04 *
N65  R21=30 R22=-50 R23=2 R24=0.2 R25=0.05 R26=5 *
```

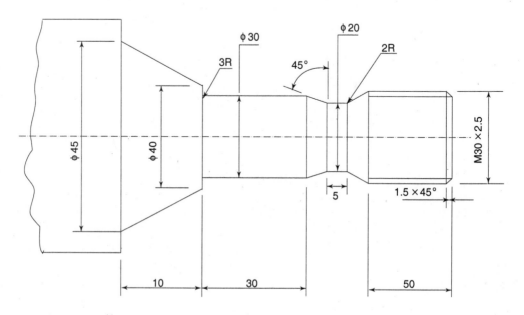

**Fig. 9.39** A programming example for a turning centre

```
N70   R27=5  R28=1  R29=45 R30=2 R31=20 R32=0*
N75   L903  F0.1  M08  *
N80   L98  *
N85   T0606  M46  *
N90   G95  S200  M03*
N110  R20=2.5  R21=30  R22=0  R23=3  R24=-1.58  R25=0.05  R26=5  *
N115  R27=2  R28=20  R29=30  R31=30  R32=-50  *
N120  L97  M08  *
N125  L98  *
N130  M30  *
```

**L06**  *
```
N5   G01  Z0  *
N10  X30  B-1.5  *
N15  Z-95  B3  *
N20  X40  *
N25  X45  Z-105  *
N30  X50  *
N35  M17  *
```

**L98**
```
N05  G0  D0  X500  Z100  M09*
N10  M17
```

CHAPTER **10**

# Testing of Machine Tools

10.1 **INTRODUCTION**

Due to the increasing demands for greater accuracy, higher productivity and improved reliability of machine tools, the manufacturing industry at large, is required to rely less on the skill of machine operators and place more emphasis on the performance of the machine tool. This has made it imperative for machine tool builders to continuously study and improve upon the testing and inspection of machine tools produced. Of late, participation of the users themselves in testing machine tools at the manufacturers, end is increasing.

Since it is essential to ensure that the basic requirements of accuracy, productivity and reliability are met on a machine tool, certain tests have been devised to assess these performances. Based on the long years of experience of machine tool manufacturers, users and the research organisations, various recommendations have been made for the performance evaluation of machine tools. They are National Aerospace Standards (NAS, USA), University of Manchester Institute of Science and Technology (UMIST, UK), etc.

Performance evaluation is done on every prototype before the machine is subjected to field trials. This assessment testing is done on a prototype of the machine tool to evaluate and approve both the design objectives and the machine tool specification. In case of machines in series production, machine acceptance testing is carried out according to a test chart which includes tests for both geometrical accuracy and working accuracy. In case the test results are not satisfactory as per the design objective, extensive investigation and analysis through measurement is carried out to analyse the problem and evolve remedial measures. This type of analytical testing also forms a part of the machine tool performance evaluation involving elaborate instrumentation and a high level of mathematical analysis.

The quality of a machine tool depends mainly on various factors such as the accuracy of different sub-units, their alignment with respect to one another and the functional interrelationship. Hence, the sub-assemblies of a machine tool are tested separately on a test rig before the final assembly. Some of the important sub-assemblies tested are headstocks, tool turrets, pallet changers, index tables, automatic tool changers, etc.

The machine tool testing is generally done in the following six areas:
1. Verification of technical specifications
2. Verification of functional aspects
3. Verification during idle running
4. Verification of the accuracy of the machine tool and the workpiece accuracy
5. Verification of the metal removal capability
6. Other tests

## 10.2 VERIFICATION OF TECHNICAL SPECIFICATIONS

The main technical specifications of a machine tool reveal the general information about the machine tool and the control system, such as the machine's workholding capacity, spindle speed ranges, axes traverse rates and traverse limits. The specification also includes details of the spindle, tool carrier, tailstock, index table, pallet changer, ratings of different motors, the overall size of the machine, and the weight of the machine. The specification of a machine provides the information on the positioning accuracy, working accuracy, etc. These specifications are intended to help the machine tool user in the selection of a machine tool for the required application. Hence, all the parameters mentioned in the technical specification are verified. A typical specification sheet for a machining centre is given in Annexure 10.1.

## 10.3 VERIFICATION OF FUNCTIONAL ASPECTS

A functional check of a machine tool is carried out before taking up a machine tool for further testing. In this regard, first a visual inspection of the machine tool, its subassemblies, elements and technical documents is done to ensure completeness. This is followed by a preliminary testing of all the functions of the machine and the various operating elements in accordance with specifications.

Further, verification of the electric, hydraulic and pneumatic systems is carried out in conformity with

(a) International Electrotechnical Commission (IEC) standard IEC-204
(b) Joint Industrial Council (JIC) hydraulic standards for industrial equipment and general purpose machine tools, H-1-1973
(c) International Standards Organisation (ISO) standards for Pneumatic Systems ISO-4414

The computer numerical control (CNC) system is also verified in accordance with the CNC system manual for the satisfactory functioning of its different features.

Verification of effectiveness of the chip and coolant management is also critical especially on modern machine tools. Their verification is generally based on specific checklists prepared for each type of machine. Specific checklists, depending on the machine, can also be evolved on the basis of functional requirements.

## 10.4 VERIFICATION DURING IDLE RUNNING

### 10.4.1 Idle Torque

Measurement of the idle torque is carried out both for the main spindle and the feed axes.

The drag torque, required just to rotate the main spindle in the idle condition is measured by using a spring balance along with suitable fixtures. In case of a main spindle drive, the drag torque is also measured for different speed ranges, if provided. The drag torque of the spindle should be minimum, since the idle power loss on the spindle should not exceed 30% of the rated power. The drag torque on the main spindle is an indication of the correctness of the headstock assembly, viz. the spindle bearing preload, alignment of the front and the rear bearing housings, the perpendicularity of the bearing seating surfaces, and the belt tension.

For the slides, the torque required to move the axes under no load condition should be less, such that sufficient torque capacity is available to carry out the expected machining operations.

Torque wrenches are used to measure the drag torque of the axes and measurements are done at different positions of the slide.

On CNC machines, the no-load torque required for the axis feed drive is measured by measuring the no-load armature current drawn by the axis motor at different feed rates including that during rapid traverse. The measurement of no-load torque of axis feed motion ensures correctness of the assembly of the ballscrew end bearings, the nut and slide, and the alignment between the ballscrew and the guideway.

### 10.4.2 Idle Power

The power for a machine tool main spindle is normally supplied by an electric motor. The power fed to the motor is partly lost in the motor itself and partly lost in the transmission system of the machine and only the remaining power is available for cutting process. The idle power consumed should be minimum to avoid wastage of power and heating up of machine elements, especially the main spindle unit. Excessive heat generation affects the component accuracy and results in malfunctioning of the machine and the electrical system.

The main spindle idle power measurement is carried out after ensuring the correctness of the headstock assembly by measurement of drag torque. Idle power is measured with a wattmeter connected to the input terminals of the motor. The machine is run at various speeds up to the maximum speed and power input to the motor is noted.

The power loss at any speed is expressed as percentage of the rated motor power, using the expression

$$N = \frac{N_i}{N_m} \times 100$$

where   $N$ = percentage power loss
   $N_i$ = measured input power under idle run
   $N_m$ = rated output power of the motor.

The spindle power losses determine the power available for rough machining. In general, power losses are restricted to 30% of the rated output of the motor.

### 10.4.3 Vibration

Forced vibration in machinery is caused by an imbalance in the rotating elements such as pulley, flywheel, gear and spindle. Interference between gears, errors in mounting of bearings, irregularities of belts, and imbalance in the electric motor also causes vibration.

Forced vibration produces harmful effects like noise, faster wear of machine elements, and inferior quality of the machined surface.

Forced vibration in a machine tool is measured under idle running condition of the machine, and at all operating speeds, with the slides positioned at mid-traverse. The measurement is carried out with a vibration pick-up such as an accelerometer, velocity pick-up, etc. in conjunction with a signal conditioner and a vibration display. The vibration measurement is carried out in the direction which is most critical that could affect the performance. The position and direction of measurement for different machine tools are listed in Table 10.1. The velocity of vibration (rms value) is the limiting criterion, which is normally in the frequency range of 10 to 1000 Hz, for turning centres and machining centres. For vibration levels below 10 Hz, the amplitude of vibration (peak to peak) is the limiting criterion. In case of grinding machines, where a large amplitude of the motion of headstock or slide can affect surface finish, the maximum displacement limiting standards apply. The recommended limits for permissible velocities of absolute vibration for different machine tools are given in Table 10.2.

In case where a detailed investigation of some forced vibrations is to be done, it is done by using fast fourier transform analyser. In this the amplitude of vibration is recorded against the frequency. The value is correlated with the rotational frequency of various elements such as shafts, gear tooth, etc. for further analysis.

It is a normal practice to check the electric motor independently for forced vibration before mounting it on the machine. This is done by suspending the motor by ropes. During measurement, a full length key of half height is introduced into the keyway on the motor shaft. The recommended maximum velocity of vibration for electric motors based on ISO 2373 is given in Table 10.3.

The equipment normally used for the measurement of vibrations are:

1. Electrodynamic vibration pick-up with integrating/differentiation type amplifier
2. Accelerometers with integrating type charge amplifier
3. Oscilloscopes
4. Fast fourier type spectrum analyser

Figure 10.1 gives the velocity and amplitude relations of absolute vibrations.

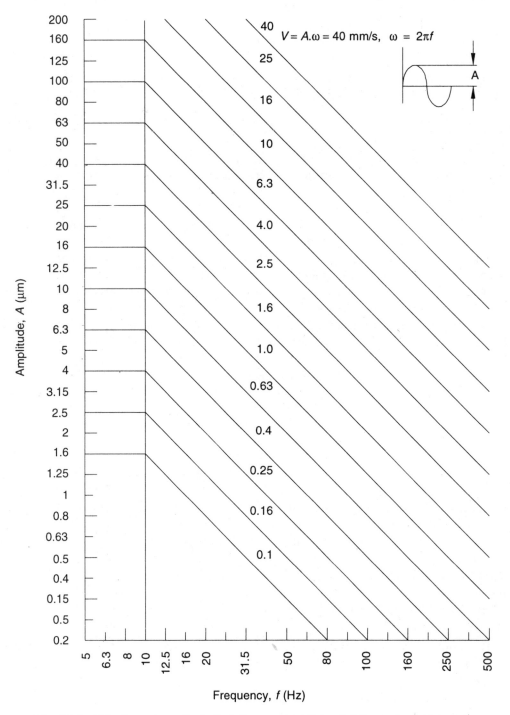

**Fig. 10.1** Velocity and amplitude limitations on absolute vibrations of machine tools

## TABLE 10.1

**Position and direction of measurement of forced vibration**

| Machine | Pick-up Positions | Direction of Pick-up | Remarks |
|---|---|---|---|
| Lathe | Headstock near front and rear bearings | In a plane containing the spindle axis and tool perpendicular to spindle axis | Without chuck |
| Cylindrical grinding machine | Wheel head front bearing Work head front bearing | Horizontal and perpendicular to spindle axis | Without workpiece and with grinding wheel of maximum size mounted |
| Surface grinding machine | Wheel head front bearing | Vertical and horizontal and perpendicular to the spindle axis | With balanced grinding wheel mounted |
| Boring machine | Spindle head near front bearing | Vertical and horizontal and perpendicular to the spindle axis | Without boring bar and face plate |
| Vertical milling machine/ machining centre | Spindle head near front bearing | Longitudinal and transverse direction in horizontal and vertical direction | Without cutter |
| Horizontal milling machine/ machining centre | Spindle head near front bearing | Horizontal and vertical direction perpendicular to the spindle axis and along axial direction | Without arbor |
| Drilling machine | On drill head positioned in the middle of the slide near lower bearing | Perpendicular to the plane passing through the spindle axis and column | Without taper sleeves or chuck |

## TABLE 10.2

**Recommended permissible velocities of absolute vibrations**

| Type of machine | | | Permissible velocity of absolute vibration $V_{max}$ (mm/s) for machine class | | |
|---|---|---|---|---|---|
| | | | I | II | III |
| Centre lathes | *$(D_{max}) \leq$ | 500 mm | 1 | 1.6 | 2.5 |
| Centre lathes | $(D_{max}) >$ | 500 mm | 1.6 | 2.5 | 4 |
| Cylindrical grinding machines | $(D_{max}) \leq$ | 300 mm | 0.16 | 0.25 | 0.4 |
| Cylindrical grinding machines | $(D_{max}) >$ | 300 mm | 0.25 | 0.4 | 0.4 |
| Internal grinding machines | $(D_{max}) \leq$ | 300 mm | 0.25 | 0.4 | 0.4 |
| Internal grinding machines | $(D_{max}) >$ | 300 mm | 0.25 | 0.4 | 0.63 |
| Centreless grinding machines | $(D_{max}) \leq$ | 80 mm | 0.25 | 0.4 | 0.63 |
| Centreless grinding machines | $(D_{max}) >$ | 80 mm | 0.4 | 0.63 | 1 |
| Horizontal boring machines | Spindle dia. up to | 100 mm | 0.4 | 0.63 | 1 |
| Vertical boring mills | Table dia. up to | 2000 mm | — | 0.4 | 0.63 |
| Knee type and table type milling machines | Table width $\leq$ | 400 mm | 0.63 | 1 | 1.6 |
| Knee type and table type milling machines | Table width $\geq$ | 400 mm | 1 | 1.6 | 2.5 |

* $D_{max}$ = Maximum workpiece diameter

**TABLE 10.3**

*Recommended permissible limits of vibration severity for electric motors (Ref: ISO:2373)*

| Quality grade | Speed (rev/min) | Maximum rms value of the vibration velocity (mm/s) for the shaft height H (mm) | | |
|---|---|---|---|---|
| | | $80 \leq H \leq 132$ | $132 \leq H \leq 225$ | $225 \leq H \leq 400$ |
| N (Normal) | 600 to 3,600 | 1.8 | 2.8 | 4.5 |
| R (Reduced) | 600 to 1,800 | 0.71 | 1.12 | 1.8 |
| | > 1,800 to 3,600 | 1.12 | 1.8 | 2.8 |
| S (Special) | 600 to 1,800 | 0.45 | 0.71 | 1.12 |
| | > 1,800 to 3,600 | 0.71 | 1.12 | 1.8 |

### 10.4.4 Noise

Noise in machine tools occurs due to errors in gears, low quality and faulty bearings, centre distance errors, etc. Excessive noise in machine tools has a hazardous effect on the health and physical comfort of the operator and those working in the vicinity. Hence, as a statutory requirement, stipulations regarding the maximum noise level on the machine tools have to be complied with. Noise is measured in terms of the sound pressure level expressed in decibels (dB) by comparing the actual pressure and basic absolute pressure of 20 micropascals in logarithmic scale. A condenser microphone and sound level meter are used for the measurement of noise in decibels. Specifications of a sound level meter for general use is given in Bureau of Indian Standards' (previously Indian Standards Institution) IS:3932. The sound level meter consists of frequency weighted networks A, B, C, and also a linear network for use.

For the measurement of noise, the frequency weighting network A is adopted, which has frequency gain characteristics closely simulating of a normal human ear. The sound pressure level is designated as dB(A) when weighting network A is used for measurement. Noise is measured at all the operating speeds of the machine tool under no load with microphone oriented towards the machine. In CNC machines, the main spindle speed is infinitely variable and hence noise measurement is carried out at a few selected speeds covering the entire speed range from the minimum to the maximum spindle speed. The measurement is carried out at the operator's normal working position, at a height of 1.5 m from floor level, and also at various points around the machine at a distance of 1 m away from the closest machine surface. Measurements on CNC machines are carried out with and without the axes motion in order to evaluate the noise emanating due to the slide motion. The generally recommended maximum limit of the sound pressure level is 85 dB(A). IS 4758 explains the methods of noise measurement on machine tools. No specific national/international standards are available recommending the maximum allowable sound pressure level. However, industrial standards dealing with noise measurement procedures are available.

The noise level measures depend on environment conditions. The noise measured gets affected because of the reflection of noise emitted by the machine from surrounding barriers like a structural wall and other machines located near the machine under consideration. An ideal environment for noise measurement is an anechoic chamber. The machine is to be placed in this chamber for carrying out measurement. The walls of anechoic chamber are designed to absorb the noise generated by the machine with minimum reflection. The chamber is also isolated from the outside noise. In the absence of such a chamber, it is preferable to locate the machine in an environment where the sound pressure level falls by 6 dB(A) for every doubling of the distance from the source. This is with a view to minimise the effect of reflected noise on the measured value.

Noise measurement also gets affected by background noise, i.e. the noise emitted by sources surrounding the machine. To eliminate the effect of background noise, measurement is made when the background noise is at least 10 dB(A) lower than the noise level measured. If difference is less than 10 dB(A), the following correction is applied on the measured sound pressure level (SPL).

| Difference in SPL with and without running the machine dB(A) | Value to be subtracted from the measured SPL dB(A) |
|---|---|
| 6–8 | 1 |
| 9–10 | 0.5 |

*Note*: When the difference in SPL is less than 6 dB(A), it is not recommended to carry out the measurement.

For a detailed investigation of the noise, a record of frequency versus sound pressure level is made in a way similar to the one made for forced vibration analysis. This helps to identify the sources of noise.

### Sound Power Level for Machine Tools

The international practice of specifying noise levels for machine tools, which generate a continuous, steady state of noise, is in terms of the sound power level. This is similar to quantifying luminescence of an electric bulb.

In case of sound pressure level, the values measured are affected by various factors such as the distance at which measurement is taken and the environmental factor of acoustic reverberation characteristic of the room or hall in which the machine is tested.

The sound power level is used for machine tools to overcome any such ambiguity. Details of the procedure for measurement and evaluation of the sound power level are given in Indian Machine Tool Manufacturers' Association (IMTMA) standard *IMTMASIS: 2-1987*. Also german national standard DIN:45635 gives the procedure for the measurement of both sound pressure and sound power levels.

### 10.4.5 Thermal Effects

The rise in temperature of machine tool elements has a significant effect on the accuracy and performance of the machine. Temperature rise leads to thermal deformation of the different machine elements, causing misalignment between machine sub-units, thus affecting the accuracy of the workpiece. Excessive temperature rise adversely affects the properties of the lubricant, leading to improper lubrication.

The major heat source in a machine tool are the main spindle bearings. This is due to the viscous friction torque (drag torque) on the main spindle assembly. The other sources are: hydraulic elements like pumps, motors, throttle valves, and electrical elements such as motors, clutches and transformers.

After carrying out a running-in procedure for the main spindle front and rear bearings, the main spindle bearing temperature rise test is conducted by running the spindle at

- Two-thirds of the maximum spindle speed till the temperature gets stabilised.
- Maximum spindle speed for 30 minutes, immediately after running at two-thirds of the maximum speed.

The spindle is run at about two-thirds of the maximum speed, and the temperature of the bearings supporting the spindle is measured at frequent intervals of time (at 5 min. intervals initially, and at 15 min. intervals after half an hour of running, depending on the rate of temperature rise). The measurement is carried out as close to the bearings as possible on the housing. Figure 10.2 shows a typical set-up for measuring the main spindle front and rear bearing temperatures. The spindle is run continuously till the temperature rise gets stabilised or until the temperature rise shows a rapid increasing trend, leading to the likely seizure of spindle bearings. In the later case, spindle running should be discontinued and the cause for rapid temperature rise should be investigated. The temperature rise is deemed to be established when the rate of rise over a period of

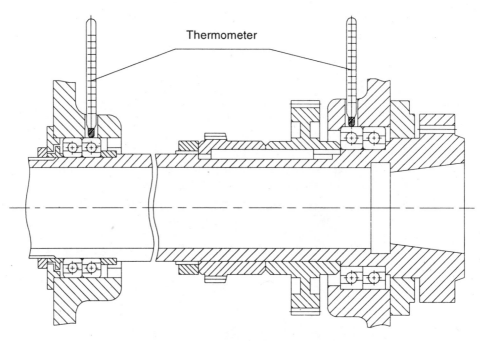

**Fig. 10.2** Set-up for main spindle thermal stabilisation test

15 min. is less than 5% of the total rise (above ambient) at that instant. The ambient temperature is also noted at regular intervals. The maximum temperature attained by the spindle bearing lubricating oil, if any, and the oil in hydraulic system are also measured.

Figure 10.3 shows a typical time-temperature characteristic. During this test, the spindle shift due to thermal growth is also checked using a test mandrel and dial indicators. The set-up for measurement of the main spindle shift and a typical thermal shift characteristic is shown in Figs 10.4 and 10.5 respectively. For a lathe, the dial indicators $A$ and $B$ show the shift in secondary plane and the dial indicators $C$ and $D$ shows the shift in primary plane. Figure 10.4 is useful to find out whether the spindle is moving up or down, uniformly or is having angular tilt for time interval $t_1$ to $t_2$ or $t_1$ to $t_n$. Any large shift of the spindle in the principal direction is always accompanied by high spindle bearing temperature. This test result indicates the quality of spindle assembly.

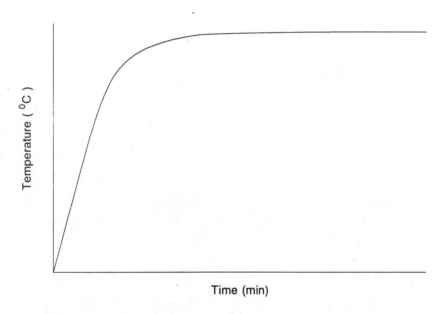

**Fig. 10.3** Typical time-temperature characteristic for spindle bearing temperature

## Computer Aided Analysis of Thermal Effects

Thermal effects on machine tools can also be investigated using a computer aided analysis.

Thermal analysis is done for sub-assemblies containing heat generating elements, viz. headstock with spindle-bearing system. The analysis is done to estimate the temperature rise in machine structure due to the heat generated during idle running of the main spindle.

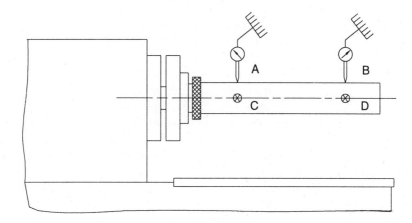

A, B, C, D : Dial indicator positions

**Fig. 10.4** Test set-up for the measurement of main spindle shift with rise in temperature

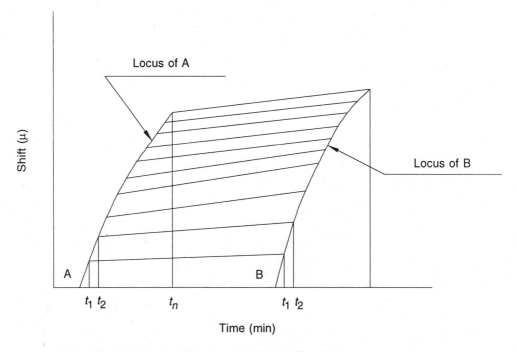

A, B : Dial indicator positions as depicted in Fig. 10.4

**Fig. 10.5** Typical plot of main spindle thermal shift

Finite element modelling of headstock, spindle-bearing system, etc. is carried out by identifying the heat sources and by structuring the model using a three-dimensional system for solids, plates, beams, etc. to arrive at the thermal boundary condition.

The resulting thermal deformation is also calculated to study the effect of the temperature rise on spindle shift in the principal and secondary planes.

The analysis for thermal deformation can also be extended to machine elements and sub-assemblies. This analysis helps to implement the necessary modifications to improve machine tool performance.

The results of the analysis are always compared with test results. Figure 10.6 shows the analytical prediction of isotherms on a headstock. Table 10.4 gives the corresponding

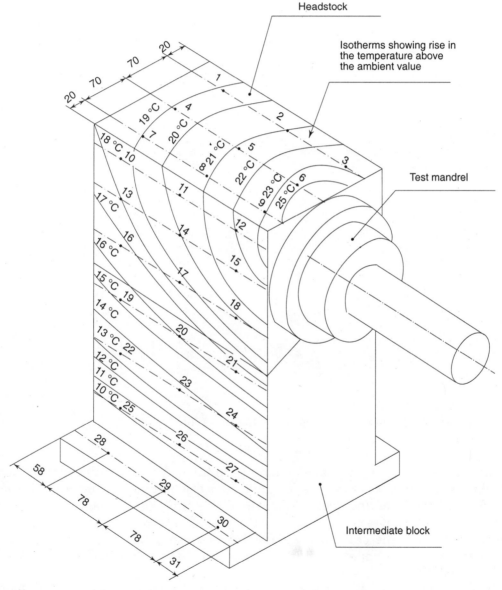

**Fig. 10.6** Plot of analytical prediction of isotherms on the headstock of a turning centre

temperature rise (above ambient) arrived at analytically and also measured experimentally.

TABLE 10.4

*Temperature rise (analytical Vs experimental) on locations identified on headstock in Fig. 10.6*

| Point of Measurement (see Fig. 10.6) | Temperature rise above ambient (°C) | |
|---|---|---|
| | Analytical | Experimental |
| Front Bearing | 26.0 | 30.3 |
| Rear Bearing | 20.6 | 24.7 |
| 4 | 19.5 | 19.0 |
| 5 | 21.3 | 21.1 |
| 6 | 24.2 | 22.5 |
| 7 | 18.9 | 18.7 |
| 8 | 20.8 | 20.5 |
| 9 | 23.1 | 22.5 |
| 13 | 18.3 | 17.8 |
| 14 | 20.0 | 20.0 |
| 15 | 21.8 | 23.0 |
| 16 | 16.8 | 16.8 |
| 17 | 18.7 | 18.0 |
| 18 | 19.8 | 22.0 |
| 19 | 15.3 | 15.8 |
| 20 | 16.0 | 18.3 |
| 21 | 16.7 | 16.8 |
| 22 | 13.3 | 9.7 |
| 23 | 13.7 | 10.7 |
| 24 | 14.0 | 10.8 |
| 25 | 10.1 | 7.8 |
| 26 | 9.7 | 8.0 |
| 27 | 9.6 | 8.3 |

### 10.4.6 Analysis of Dynamic Behaviour of Machine Tool

*Axis Drives*

The dynamic behaviour of a continuous path control fitted to a machine tool influences the machine's working accuracy. The dynamic response of a contouring control is characterised by acceleration and deceleration times or paths. Fast response of a machine tool not only leads to short non-cutting times but also leads to increased accuracy in the contouring mode. On the other hand it is often necessary, particularly, in the case of large moving masses to limit acceleration and deceleration values so as to avoid excessive inertial forces.

The dynamic behaviour of a continuous path or contouring control is determined by the functional modules responsible for generating the command variable and its actual value (refer to Fig. 10.7). Command value generation is effected by interpolation in a corresponding computing unit called the interpolator. The actual value is generated by the servo-system and the mechanical transmission system. The servo-system is based on position feedback control with infinitely variable feed drives. These drives are also provided with a velocity feedback loop using a tacho generator for improving the control behaviour and reducing the sensitivity to disturbances.

As already mentioned, the most common method of generating the actual position value is based on position feedback with continuously acting drives. The behaviour of the position feedback loop depends largely on the loop gain. This is the ratio of the actual speed to servo-error in the steady state. This ratio is normally constant over the active speed range.

If the loop gain is too low, excessive contour deviations will occur. On the other hand, if it is too high, destabilisation of controller will occur. However, the loop gain is set as high as possible.

Present day CNC systems automatically measure and display the servo-error and loop gain. The dynamic behaviour of the drive is characterised by its time constant and its damping factor. The time constant is the time elapsed between the step input and when the output reaches the required level. The damping factor will control the output overshoot or undershoot. The lower the time constant, the better is the dynamic behaviour. The step response of a drive gives both the time constant and the damping factor. A properly tuned drive should have a damping factor between 0.5 and 0.6, which corresponds to a minimum time constant with an overshoot of about 3–5%.

The characteristics of drive are measured by monitoring output velocity to a step input in the drive open-loop condition. The drive is tuned for the required time constant and overshoot or undershoot, input and output are recorded for future reference. The typical open loop-velocity step response characteristics are given in Fig. 10.8, indicating the definitions of the variables evaluated.

Axis drives should be tuned such that, in open-loop, the acceleration time to attain a rapid traverse rate from a standstill position should be within 100 ms and the velocity overshoot and undershoot are eliminated. While achieving this, it should be noted that the axis drive motor armature current should be limited within rated limit in the drive. The axis dynamic contouring capability can be measured using the Renishaw magnetic quick check ball bar system.

It consists of a ball-ended bar-length transducer which is held magnetically and kinematically between the machine spindle and a specially designed magnetic base clamped to the machine bed. The machine is programmed to move in a circle and the ball bar transducer detects the deviations which indicate the inaccuracies in the machine tool path (refer Fig. 10.9).

Data on deviation may be captured statically or dynamically, depending on the type of analysis required. An instant graph of the contouring ability can be plotted. Evaluation of circularity, repeatability, local squareness error, and linear scale error are also possible. With the help of data provided in the users manual, the user may diagnose error plots to obtain information about backlash compensation, servo gains, axis alignment, stick-slip effects, controller errors and the leadscrew vibration.

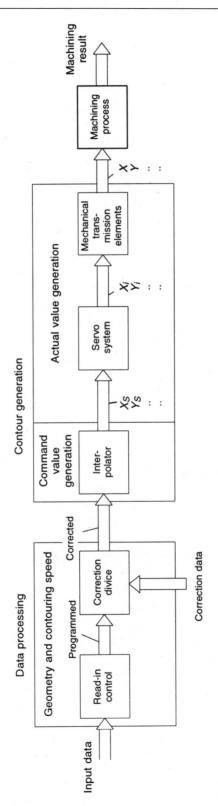

**Fig. 10.7** Functional modules of a contouring control for CNC machines

Definitions of response variables

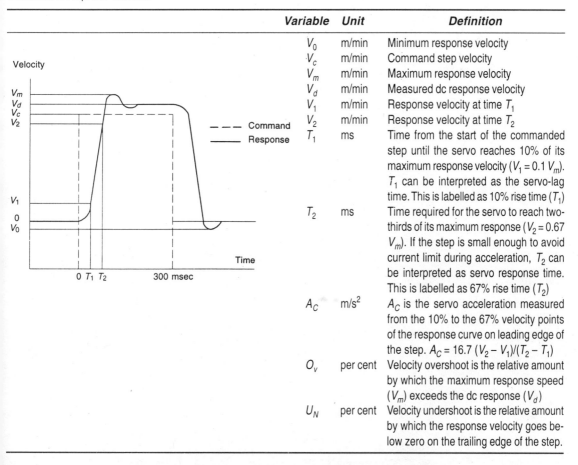

| Variable | Unit | Definition |
|---|---|---|
| $V_0$ | m/min | Minimum response velocity |
| $V_c$ | m/min | Command step velocity |
| $V_m$ | m/min | Maximum response velocity |
| $V_d$ | m/min | Measured dc response velocity |
| $V_1$ | m/min | Response velocity at time $T_1$ |
| $V_2$ | m/min | Response velocity at time $T_2$ |
| $T_1$ | ms | Time from the start of the commanded step until the servo reaches 10% of its maximum response velocity ($V_1 = 0.1\ V_m$). $T_1$ can be interpreted as the servo-lag time. This is labelled as 10% rise time ($T_1$) |
| $T_2$ | ms | Time required for the servo to reach two-thirds of its maximum response ($V_2 = 0.67\ V_m$). If the step is small enough to avoid current limit during acceleration, $T_2$ can be interpreted as servo response time. This is labelled as 67% rise time ($T_2$) |
| $A_C$ | m/s² | $A_C$ is the servo acceleration measured from the 10% to the 67% velocity points of the response curve on leading edge of the step. $A_C = 16.7\ (V_2 - V_1)/(T_2 - T_1)$ |
| $O_v$ | per cent | Velocity overshoot is the relative amount by which the maximum response speed ($V_m$) exceeds the dc response ($V_d$) |
| $U_N$ | per cent | Velocity undershoot is the relative amount by which the response velocity goes below zero on the trailing edge of the step. |

**Fig. 10.8** Typical plot of an open-loop velocity step response

The Renishaw quick check system can be used on CNC lathes and on any machine tool capable of performing circular interpolation—horizontally or vertically. Although, until now, only linear measurements were made, the increasing demands on the machine tool accuracy have made the circularity check an essential procedure.

When a machine performs a complete circular interpolation, its position, velocity and acceleration are constantly changing in two or more axes. Therefore, its capability to execute a perfect circle (contouring ability) can be influenced by many possible sources of error. Hence, the circularity plot can highlight the true dynamic capability of the machine.

### Main Spindle Drive

While analysing the dynamics involved in the main spindle drive performance, it has to be ensured that the spindle motor torque capacity matches with the spindle inertial

**394** Mechatronics

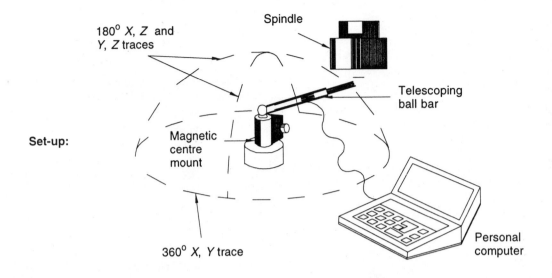

**Set-up:**

**Results:**

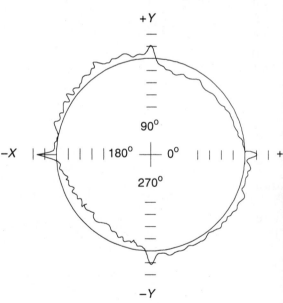

Renishaw ball bar system

Dynamic ISO 231-1
    Circularity   :  1.33 μ
    Maximum   :  73.2 μ
    Minimum    :  59.9 μ

Vertical machining centre :

| | | |
|---|---|---|
| L | : | 150.000 mm |
| R | : | 150.000 mm |
| f/r | : | 3000.000 mm/min ACW |
| s/rate | : | 75–80 values/sec |
| Data start | : | 180.000° |
| Data end | : | 180.000° |
| Machine start | : | 0.000° |
| Machine end | : | 0.000° |

| | | |
|---|---|---|
| ACW | : | Anticlockwise |
| f/r | : | Feed rate |
| L | : | Left |
| R | : | Right |
| s/rate | : | Scanning rate |

↳ **Fig. 10.9** Contouring envelope of quick check ball bar system along with a typical profile error plot

effects. This will facilitate in optimising the spindle drive characteristics. Matching this with feed drive acceleration is required during tapping operations in machining centres. Further, the non-productive spindle acceleration and deceleration time can be reduced. The test for spindle acceleration and deceleration is conducted by simultaneous monitoring of the spindle velocity (tachometer output) and the spindle command and by recording them on storage oscilloscopes or ultraviolet recorders. Other analysis can be done as is done in the case of axis drive characteristics.

## 10.5 VERIFICATION OF MACHINE TOOL ACCURACY AND WORKPIECE ACCURACY

### 10.5.1 Geometrical Accuracy Test

The geometrical accuracy tests are carried out on any machine tool basically to check the accuracy of the machine axes movement, main spindle accuracy, etc. The concept of geometrical accuracy of a machine was first introduced by Prof. G. Schlesinger of Germany. After an extensive study, he conceived a series of alignment tests to be carried out on machine tools. These tests have finally been formulated as test charts at the national and international level, viz. ISO, DIN, Verein Deutscher Ingenieure/Deutsche Gesellschaft fuer Qualitaet (VDI/DGQ), etc. In addition, standards at the level of the machine tool manufacturers' association, like IMTMA, have also been formulated. All these test charts deal only with the verification of accuracy and do not apply to the testing of the machine for vibration, noise and other functional characteristics such as feeds and speeds.

As the result of the geometrical accuracy test is an indication of the accuracy of the movement or rotation of the various machine elements, the test should be conducted only after the completion of the final assembly of the machine, so that the overall effects of the various other elements on the geometrical accuracy can be discerned.

The geometrical accuracy tests are done as per test charts which indicate the required test to be conducted along with the type of equipment to be used and the limiting values of the test results recommended. A code for machine tool testing has been formulated by BIS (*IS:2063-1962*) corresponding to a similar international recommendation, ISO-R-230.

The major inspection equipment used for conducting the geometrical accuracy tests are:

(a) Dial gauges/level type of indicators
(b) Level
(c) Auto-collimator
(d) Laser measuring systems
(e) Test mandrels
(f) Squares
(g) Straight edges

In case of CNC turning centres, the test chart includes aspects to be verified for machine elements and sub-assemblies, viz. bed, carriage, headstock spindle, tailstock, tool turret with and without rotary tool feature, etc. Tests are also included for measurement of straightness of carriage movement, spindle run out, spindle indexing accuracy, parallelism of spindle axis to carriage movement, parallelism of axis of centres with the movement of carriage or turret, turret indexing repeatability and coaxiality of turret bore with the spindle axis, squareness of transverse movement of turret with spindle axis, axis positioning accuracy and repeatability.

Similarly, in the case of machining centres, test charts specify the verification of bed, carriage, spindle, table/pallet, etc. Measurement of flatness of table/pallet top surface, parallelism of table/pallet surface to axis movement, squareness amongst axes movements, spindle taper run out, parallelism of spindle axis to axes movements, table indexing accuracy, run out of pallet face, axis positioning accuracy and repeatability of pallet positioning are some of the important tests carried out. Information contained in a representative test chart for geometrical accuracies of machining centres and turning centres are given in Annexure 10.2.

## 10.5.2 Positioning Accuracy and Repeatability Test

The positioning accuracy and repeatability of each machine tool axis is one of the most important parameters to be evaluated for ensuring satisfactory operation of CNC machines. The main parameters measured are positioning accuracy, repeatability and lost motion. The positioning accuracy is normally checked using a laser measuring system and is determined in accordance with the methods detailed in ISO standard 230-2. However, many machine tool builders still follow positioning accuracy evaluation procedures different from the ISO standard, viz. VDI/DGQ guidelines, National Machine Tool Builders' Association (NMTBA, USA) guidelines or Japanese Institute for Standards' (JIS) guidelines. It is important therefore, that all four methods be understood. Table 10.5 gives a comparison of different machine tool positioning accuracy evaluation methods.

Figure 10.10 shows a schematic layout of a typical laser measurement system configured for carrying out the axis positioning accuracy and repeatability measurement of a horizontal machining centre. Figures 10.11 and 10.12 show the optical arrangement of the laser measurement set up for measuring linear positioning errors in $X$ and $Y$ axes of a horizontal machining centre.

The assessment involves a statistical test which measures the actual displacement when an axis is moved to a programmed target position. The errors between the targeted and actual distances are usually recorded from the laser interferometer display unit.

A set of errors may be linear, which is made up of progressive and symmetrical errors, ignoring the effect of repeatability. Cyclic errors can be considered as any error which repeats itself at least once along the axis. A progressive error, as its name indicates, is the error component that changes its value with distance and is regarded as non-cyclic.

Generally, the problems of progressive errors are only associated with machines that use the ballscrew as part of the positioning system. This is because the machine's controller will only be monitoring the axis servo rotation and any pitch error of the ballscrew will show up as a positioning error on the machine. Most machine tool controls offer a software compensation facility where the progressive pitch errors are entered as a com-

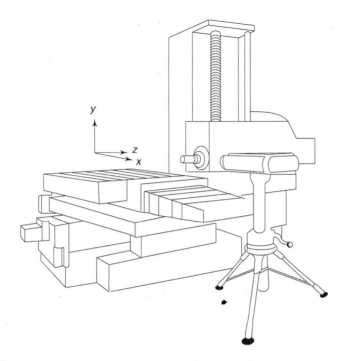

**Fig. 10.10** Machining centre with laser head and tripped set-up

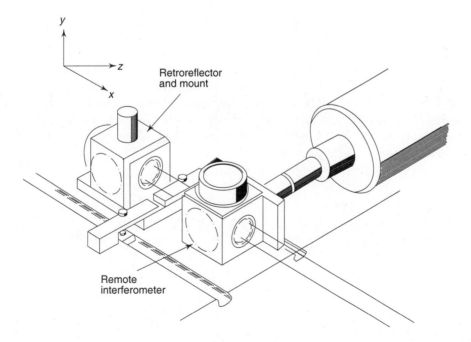

**Fig. 10.11** Measurement of linear positioning error in the X-axis

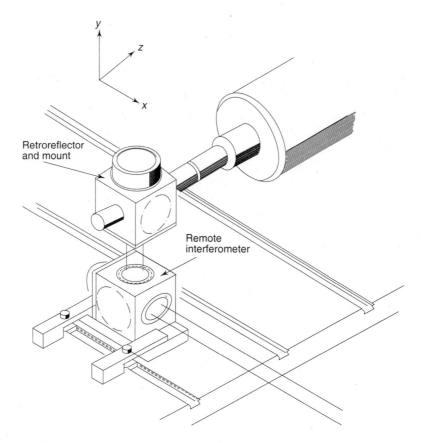

**Fig. 10.12** Measurement of linear positioning error in the Y-axis

pensation parameter. However, the cyclic error component of the axis positioning inaccuracy and axis repeatability error can only be reduced or the accuracy improved by using quality machine elements and by following the right assembly procedures besides following other standard precautions.

## TABLE 10.5

### Comparison of different machine tool positioning accuracy evaluation standards

| Parameter | ISO Standard | VDI Guidelines | NMTBA Definition | JIS Standard |
|---|---|---|---|---|
| Target position | Position to which the moving part (axis) is programmed | Same as ISO | Same as ISO | Same as ISO |
| Actual position | Measured position reached by the moving part | Same as ISO | Same as ISO | Same as ISO |
| Number of target positions required | Five per metre up to 2 m. More for longer | Depends on length of axis. Minimum of five | Unspecified; example shown is 20 | Depends on length. Every 50 mm up to 1000 mm; then every 100 mm |
| Number of runs to the target positions | Minimum of five for each direction | Ten per metre for each direction | Minimum of seven | One run in each direction for positional accuracy and seven for repeatability |
| Unidirectional/ bi-directional | Unidirectional involves a series of measurements in which the approach to target is always from same direction. Bi-directional refers to movement in both directions. Bi-directional recommended | Same as ISO. Recommends bi-directional | Same as ISO. Recommends unidirectional | Same as ISO. Recommends bi-directional |
| Positional deviation | Difference between actual position reached and target | Different from ISO. Maximum difference of the mean values of the actual positions versus the individual target positions along an axis | Agrees with ISO definition, however, noted as deviation *from target* | Term not noted since only one run is made to a target: two for bi-directional |
| Mean positional deviation | Algebraic mean of the positional deviations at a target position | Agrees with ISO, however, noted as *mean value*. | Agrees with ISO; noted as *mean* | Not considered |
| Reversal error | Value of the difference between the mean positional deviations at a position for the two directions of approach | Same as ISO | Same as ISO. Referred to as *lost motion* | Not considered |
| Standard deviation | $\sqrt{\frac{1}{(n-1)}\sum_{i=1}^{n}(X_{ij}-\overline{X}_j)^2}$<br>where,<br>$n$ = number of runs<br>$i$ = any one run<br>$X_{ij}$ = positional deviation for any one run<br>$\overline{X}$ = mean positional deviation | Same as ISO | Same as ISO | Not considered |

(Contd)

Table 10.5 (*Contd*)

| | | | | |
|---|---|---|---|---|
| Positional accuracy | Maximum difference between extreme values of $X + 3\sigma$ and $X - 3\sigma$ regardless of the position or direction of motion. Applies to unidirectional as well as bi-directional. Due to reversal error, the spread will be greater and the positional accuracy less for bi-directional | No specific term for *accuracy*, however the term *positional uncertainty* as described is comparable to the ISO definition for positional accuracy, although the calculations differ | Comparable to ISO. Referred to as *accuracy* | Differs considerably from any one of the other three. Positioning accuracy is measured as the largest variation of any actual position from a target position |
| Repeatability (unidirectional or bi-directional) | Spread at the target position having the largest spread | Comparable to the ISO standard | Comparable to the ISO standard | Differs considerably from the other three, and is expressed as a value based on dividing the value read at the target position of maximum spread by two |

In the case of evaluation of a rotational accuracy, an optical polygon in conjunction with an auto-collimator is the most common method of measuring the rotary indexing accuracy. A schematic set-up for the table indexing accuracy test is given in Fig. 10.13. The precision index tables such as those produced by Messrs Moore Special Tool Co Ltd., USA and Messrs A.C. Davis Ltd., USA enable the indexing accuracy of high precision axes to be evaluated. A summary of the test condition and test program for conducting the positioning accuracy test according to the ISO 230-2 standard is given below.

(a) The temperature of the environment of the machine and the measurement set-up should be controlled to 20 ± 0.5 °C. If the measuring instrument incorporates compensation for environmental factors such as air pressure, humidity, air temperature and machine temperature, these should be used to yield results corrected to 20 °C. The location of the sensors used to compensate the above factors should be selected appropriately.

(b) All necessary levelling operations, geometric alignment tests and functional checks are to be completed satisfactorily before starting the axis positioning accuracy and repeatability tests.

(c) The test should be preceded by an appropriate warm up operation of the machine.

(d) The machine should be programmed to position the slide at a series of target positions on the axis.

(e) The machine should be programmed to move between the target positions at a rapid feed rate.

(f) The target positions should be selected such that they represent an adequate sampling of any position errors with a fairly uniform spacing of the target positions.

(g) Each target position should be reached five times in each direction. The test cycle should be as shown in Fig. 10.14.

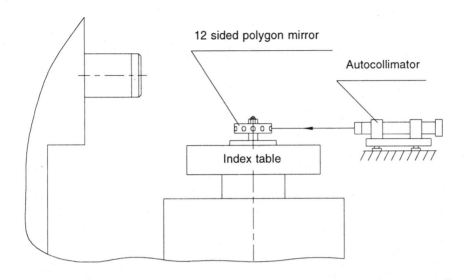

**Fig. 10.13** Schematic set-up for table indexing accuracy test

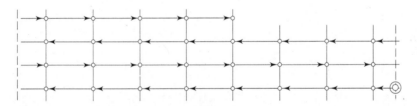

**Fig. 10.14** Linear test cycle for axis positioning accuracy and repeatability measurement

(h) In case of a rotary axes calibration, tests should be made at the principal positions 0°, 90°, 180°, 270° and if a measuring instrument with a continuous scale of adequate accuracy is available, other target angular positions can also be taken up for evaluation.

A sample positioning accuracy measurement data and test results are given in Table 10.6(a) and (b). Both the tables should be read together as follows.
(a) Item 1 gives the target positions ($P_j$) along with the approach directions.
(b) Positioning error, i.e. error of the position with reference to the commanded position is given in item 2 for all the five target positions. The mean of five unidirectional positional deviation ($\overline{X}$) is given in item 3 and item 4 gives the corresponding unidirectional standard deviation ($S_j$).

## TABLE 10.6(a)

### Sample positioning accuracy measurement data and test results

| | | | 1 | | 2 | | 3 | |
|---|---|---|---|---|---|---|---|---|
| 1. | Number of target positions max. $j = m$ | | | | | | | |
| | Target position, mm, $P_j$ | | 0 | | 150.71 | | 300.85 | |
| | Approach direction | | ↑ | ↓ | ↑ | ↓ | ↑ | ↓ |
| 2. | Positional deviation, $X_{ij}$ (μm), max $i = n$ | $i = 1$ | −2.6 | 2.3 | 0.2 | 5.4 | −5.0 | −1.5 |
| | | $i = 2$ | −1.6 | 1.1 | 0.7 | 4.6 | −5.6 | −1.0 |
| | | $i = 3$ | −2.7 | 3.2 | 0.4 | 5.7 | −4.8 | −0.8 |
| | | $i = 4$ | −1.8 | 2.5 | 1.8 | 5.4 | −2.7 | −2.6 |
| | | $i = 5$ | −0.9 | 1.5 | 1.5 | 5.9 | −2.8 | −0.8 |
| 3. | Mean unidirectional positional deviation $\overline{X}_j = \frac{1}{n}\sum_{i=1}^{n} X_{ij}$ | | −1.92 | 2.12 | 0.92 | 5.4 | −4.18 | −1.34 |
| 4. | Standard deviation $S_j = \sqrt{\frac{1}{(n-1)}\sum_{i=n}^{n}(X_{ij} - \overline{X}_j)^2}$ | | 0.75 | 0.83 | 0.70 | 0.50 | 1.34 | 0.76 |
| 5. | $3S_j$ | | 2.24 | 2.49 | 2.09 | 1.49 | 4.01 | 2.28 |
| 6. | $6S_j$ | | 4.48 | 4.98 | 4.18 | 2.98 | 8.02 | 4.56 |
| 7. | $\overline{X}_j + 3S_j$ | | 0.32 | 4.61 | 3.01 | 6.89 | −0.16 | 0.94 |
| 8. | $\overline{X}_j − 3S_j$ | | −4.16 | −0.36 | −1.17 | 3.92 | −8.18 | −3.62 |
| 9. | $B_j = \overline{X}_j\!\uparrow - \overline{X}_j\!\downarrow$ | | −4.04 | | −4.48 | | −2.84 | |
| 10. | $3S_j\!\uparrow + 3S_j\!\downarrow + \|B_j\|$ | | 8.77 | | 8.06 | | 9.13 | |
| 11. | Greater value of item 6↑; 6↓ and 10 | | 8.77 | | 8.06 | | 9.13 | |

Accuracy $A = (\overline{X}_j + 3\,S_j)$ MAXIMUM − $(\overline{X}_j − 3S_j)$ MINIMUM =

| Repeatability of positioning | Unidir | $R\!\uparrow = 6\,S_j\!\uparrow$ MAXIMUM (AT $j = 3$) = |
|---|---|---|
| | | $R\!\uparrow = 6\,S_j\!\downarrow$ MAXIMUM (AT $j = 3$) = |
| | Bi–dir. | MAXIMUM OF ITEM 11 (AT $j = 3$) = |
| Mean reversal value | | $\overline{B} = \frac{1}{m}\sum_{j=1}^{m} B_j =$ |

## TABLE 10.6(b)

### Sample positioning accuracy measurement data and test results

| | | | 4 | | 5 | | 6 | |
|---|---|---|---|---|---|---|---|---|
| 1. | Number of target positions max. $j = m$ | | | | | | | |
| | Target position, mm, $P_j$ | | 450.33 | | 601.11 | | 750.31 | |
| | Approach direction | | ↑ | ↓ | ↑ | ↓ | ↑ | ↓ |
| 2. | Positional deviation, $X_{ij}$ (μm), max $i = n$ | $i = 1$ | 0.6 | 2.4 | −8.4 | −9.0 | −4.9 | −5.9 |
| | | $i = 2$ | 1.6 | 2.1 | −7.4 | −9.4 | −3.7 | −5.9 |
| | | $i = 3$ | 0.5 | 3.6 | −7.8 | −8.1 | −4.7 | −4.0 |
| | | $i = 4$ | 2.3 | 2.2 | −6.1 | −8.3 | −3.4 | −4.3 |
| | | $i = 5$ | 2.3 | 2.9 | −6.9 | −7.8 | −2.3 | −2.9 |
| 3. | Mean unidirectional positional deviation $\bar{X}_j = \frac{1}{n}\sum_{i=1}^{n} X_{ij}$ | | 1.46 | 2.64 | −7.32 | −8.52 | −3.80 | −4.60 |
| 4. | Standard deviation $S_j = \sqrt{\frac{1}{(n-1)}\sum_{i=n}^{n}(X_{ij} - \bar{X}_j)^2}$ | | 0.88 | 0.62 | 0.88 | 0.66 | 1.05 | 1.30 |
| 5. | $3S_j$ | | 2.64 | 1.86 | 2.63 | 1.98 | 3.16 | 3.89 |
| 6. | $6S_j$ | | 5.28 | 3.72 | 5.26 | 3.96 | 6.32 | 7.78 |
| 7. | $\bar{X}_j + 3S_j$ | | 4.10 | 4.50 | −4.69 | −6.54 | −0.64 | −0.71 |
| 8. | $\bar{X}_j − 3S_j$ | | −1.18 | 0.78 | −9.95 | −10.50 | −6.96 | −8.49 |
| 9. | $B_j = \bar{X}_j\uparrow - \bar{X}_j\downarrow$ | | −1.18 | | 1.20 | | 0.80 | |
| 10. | $3S_j\uparrow + 3S_j\downarrow + |B_j|$ | | 5.68 | | 5.81 | | 7.85 | |
| 11. | Greater value of item $6\uparrow; 6\downarrow$ and 10 | | 5.68 | | 5.81 | | 7.85 | |
| | Accuracy | | | | 17.39 | | | |
| | Repeatability of positioning | Unidir | | | 8.02 7.78 | | | |
| | | Bi-dir. | | | 9.13 | | | |
| | Mean reversal value | | | | − 1.76 | | | |

(c) Reversal value ($B_j$) for each of the target positions is calculated using the mean unidirectional positional deviation, arrived at item 3 and is given in item 9.

(d) Item 11 determines the maximum value of 6 $S_j$, in positive direction, 6 $S_j$ in negative direction (item 6) and item 10 for each target position.

(e) Positioning accuracy, defined as $A = (X_j + 3S_j)_{max} - (X_j - 3S_j)_{min}$ is calculated accordingly.
(f) Unidirectional repeatability in positive (↑) and negative (↓) direction is obtained by determining the maximum of $6S_j$ (item 6) for both directions of motion respectively.
(g) Bi-directional repeatability is indicated by the maximum of the values listed in item 11.
(h) Mean reversal value ($\overline{B}$) is obtained by averaging the reversal value at each target position (item 9).
(i) Calculated values for accuracy, repeatability and mean reversal values for the sample test data are given in Table 10.6(b).

*Positioning error* as explained above is the error between the desired position of tool or workpiece and the actual position when the tool or workpiece is moved from a reference point to any other point in a specified axis. This error is caused by a combination of the following errors: axis geometric errors such as, pitch yaw, roll or lack of straightness in the specified axis or in its perpendicular axis, errors in parallelism between the specified axis and the axis of built-in measuring system, and the Abbe offset errors due to distance between the specified axis and the axis of built-in measuring system.

*Planar error*, defined as the error between any two points in a specified plane or area set up by two perpendicular axes, arises due to the positioning errors of both the axes.

Similarly, *volumetric error*, defined as the error between any two points in a specified working volume, arises due to the respective axes positioning errors for a system of three mutually perpendicular axes.

Since, machine tool axis positioning accuracy determines the final accuracy of a machined workpiece, the overall planar error as applicable for a turning centre (x-z plane) or volumetric error as applicable for a machining centre (x-y-z axes) should be determined. In this regard, measurement of individual machine tool axis geometrical parameters such as pitch, yaw, orthogonality and straightness errors in each of the two or three axes has to be carried out. The above measured error values are synthesised along with the error values of the axis positioning accuracy measurement detailed earlier, in order to find an overall planar error or volumetric error between any two points in the working plane or volume of the machine tool.

This method of analysis will enable a machine tool builder to pin-point the parameter which contributes most to the planar or volumetric error. The necessary measures can then be taken if the planar or volumetric accuracy is outside acceptable limits.

Figure 10.15 shows a laser measurement set-up for carrying out the measurement of yaw error in the y-axis of a horizontal machining centre. Squareness error and straightness can be measured either by using a high quality granite square or by using a laser measurement system. Figure 10.16 shows a laser measurement set-up for measuring the vertical straightness error in the x-axis of a horizontal machining centre. Figure 10.17 shows the set-up for measurement of horizontal straightness error in the y-axis and squareness error between the y- and x-axis of the horizontal machining centre.

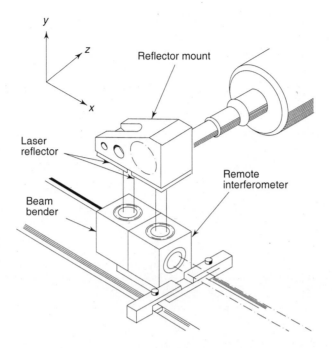

**Fig. 10.15** Measurement of yaw error in the *y*-axis

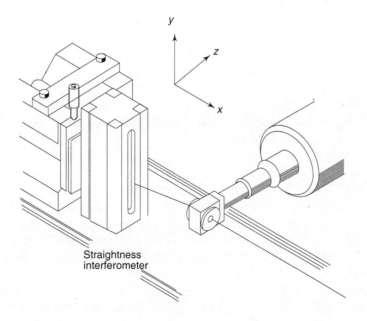

**Fig. 10.16** Measurement of vertical straightness error in the *x*-axis

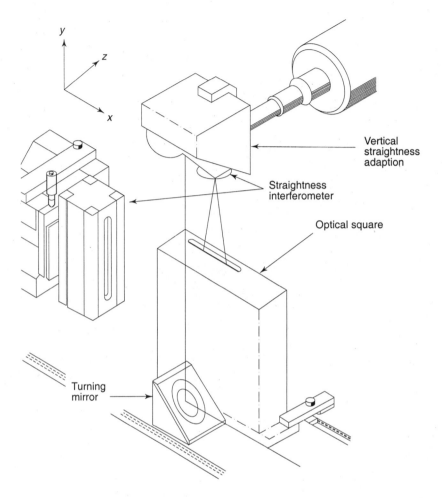

**Fig. 10.17** Measurement of horizontal straightness error in the Y-axis and the squareness error between the y- and x-axis

### 10.5.3 Working Accuracy Tests

As the geometrical accuracy tests are conducted at very slow speeds and feeds or under static condition, the effects of dynamic factors, clearances in parts, rigidity, weight of workpieces, etc. are not taken into account. To take care of this, tests are included for a few representative practical working accuracy tests to determine the capability of the machine to produce accurate components.

In case of a CNC turning centre, the component accuracy parameters verified during working accuracy tests include circularity, cylindricity, flatness, profile accuracy and threading accuracy. Similarly for a machining centre, component accuracy parameters like roundness and cylindricity of bores, flatness, straightness, squareness and parallelism of surfaces and coordinate boring accuracies are evaluated.

For each of the above tests, the permissible values are given in the relevant test chart. The working accuracy tests are conducted in a controlled manner with regard to the work and tool materials and dimension, cutting condition, machine thermal stabilisation, etc.

Figures 10.18 and 10.19 give the typical working accuracy test pieces suitable for a CNC turning centre and a machining centre respectively. Profile accuracy turning test piece shown in Fig. 10.18 is a test piece held between the chuck and the centre. A similar profile turning test piece is used also for chuckers. During this test, accuracy parameters like profile turning accuracy, slow and steep taper interpolation, and cylindricity are evaluated. As per the IMTMA test chart, the turned profile should be contained within 0.030 to 0.045 mm for different machine sizes.

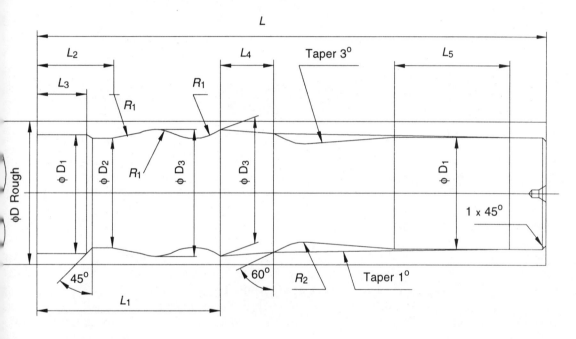

D : Diameter (mm)
L : Length (mm)
R : Radius (mm)

**Fig. 10.18** Profile accuracy test piece for CNC turning centres

A composite milling accuracy test piece is machined on a machining centre (see Fig. 10.19) to evaluate accuracy parameters such as circular contouring accuracy, i.e. roundness, linear contouring accuracy for canted square machining, flatness, parallelism and squareness for milling surface and angularity in ramp cut. Information contained in a representative test chart for practical tests of machining centres and turning centres are given in Annexure 10.3.

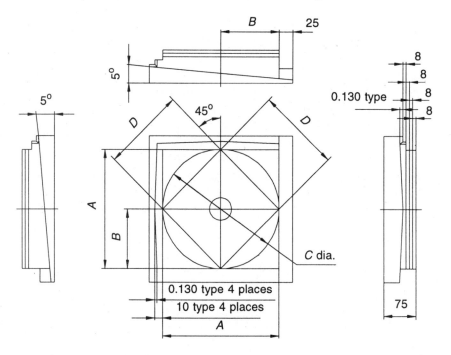

A, B, C, D : Test piece dimension (mm)

**Fig. 10.19** Test piece for composite milling test for machining centres

### 10.5.4 Process Capability Tests

Working accuracy tests establish the capability of the machine with regard to machining of a few test pieces under controlled machining conditions. This test does not assure the machine's capability on a continued production. The process capability test is carried out to determine by measurement of consecutive pieces produced in a production trial run whether the machine would be capable of producing the specified tolerances in a continuous production run. The basic objectives of the process capability test is to determine both the inherent and systematic variations due to the machine or process influencing factors.

The test results are analysed statistically to obtain the product dimension variation which results from variability in performance of the machine itself when used on representative production parts.

Regarding methods of test, VDI/DGQ 3441 standard *Statistical testing of the operational positioning accuracy of machine tools—Basis* is followed. There are two choices for the test procedure, viz. standard test and short test.

In the *standard test*, the random sampling covers the whole of running time between two tool changes starting with a sharpened tool. This takes care of the tool wear and the other systematic machine variations. For the standard test, in case of single spindle machines, ten individual random samples are generally taken for every five parts.

If a quick, low cost investigation into the overall picture of the operational scatter is required, the *short tests* may be sufficient, e.g. when machining with a single spindle machine with sharpened tools, as far as possible, 50 parts are run off one after another. This test is satisfactory for machines which are not tied to a particular component.

Compared with the standard test, the short test offers quicker information, a relatively simple procedure and little expense.

In either of the tests, the dimensional variations of machine parts are brought out by different causes. The principal causes are:

- Operational scatter of the machine
- Adjustment and measurement uncertainty of the machine
- Trend due to tool wear, thermal aspects, etc.

As a guideline, in the case of chip forming machining where tool wear only causes a small dimensional change, an operational scatter of up to 80% of the workpiece tolerance may be allowed, for instance, when drilling. In the case of machining processes with trend, the operational scatter should not exceed 60% of the workpiece tolerance. To give a clearer appreciation of the VDI/DGQ test, a sample test record used for determining the operational scatter of a machine by machining workpieces is shown in Tables 10.7 and 10.8.

The test workpiece is a turned part. The diameter is machined to a nominal size of 61.0 mm. The machining tolerance is 20 µm. Table 10.7 contains all the necessary data concerning the machine itself, the test workpieces, the machining condition and information about the measuring gauges used to check the dimensions.

The measured diameter values are recorded in the order of machining in item 1 of Table 10.8 of the test record. There are $n = 5$ successive individual values/ for one random sample $j = 1$ to $m$ (recorded in one column); there being $m = 10$ random samples in all. The average values ($\overline{X}_j$), for each random sample line, are plotted on the $\overline{X}$ chart. Trend amongst the samples $T_N = 15$ µm is derived from $t$ the $\overline{X}$ chart for 50 machined parts (item 10 in Table 10.8). The aggregate of the mean values of $\overline{X}$, i.e. $\overline{\overline{X}} = 8.7$ µm is obtained from the ($\overline{X}$) values (item 6). Standard deviation, $S_R = 2.8$ µm (item 8) is determined by the average range $\overline{R}$ and includes trend effect and measurement uncertainty. Relative to the tolerances, if these factors are too high their effect must be eliminated theoretically. In the example, both corrections were made for the sake of completeness.

After apportioning total trend value to each of the measured values, the revised data is tabulated (item 12 of Table 10.8). Based on the above, the mean corrected range $\overline{R}_{corr}$ works out to be 5.6 µm (Item 13 in Table 10.8). The standard deviation $S^*_R$ adjusted for the effect of trend is calculated as shown in item 14 of Table 10.8. This measurement still contains the influence of measurement uncertainty of the measuring procedure. To overcome this, the standard deviation of measurement uncertainty, $S_{\text{Requipt}}$ (Item 15 in Table 10.8) was calculated and found to be 0.9 µm using methods stated in the VDI/DGQ 3441 standard. Using $S_{\text{Requipt}}$ and $S_R$, the overall standard deviation $S_R$ becomes 2.2 µm (Item 16 of Table 10.8) and the operational scatter of the machine $A_s = 6 S_R$, then becomes 13.2 µm (Item 17 in Table 10.8). Since component tolerance is 20 µm, the operational scatter of the machine is 13.2 µm, the process capability index (PCI) for the machine under specific operational conditions laid down in Table 10.7 is:

TABLE 10.7

*Determination of the operational accuracy of a lathe (as per VDI/DGQ:3441)—Test conditions*

| Machine : | Type : |
|---|---|
| Manufacturer : | Year of manufacture : |
| Swing over bed : 300 mm | Last overhaul : |
| Clamping : Hydraulic chuck : 250 $\phi$ | Foundation : |
| Positioning device : CNC | |
| Testing place : | |
| Remarks : | |

**Test workpiece :**

| Work material | : C45 |
|---|---|
| Dimensions | : |
| Diameter | : 63 mm |
| Length | : 35 mm |
| Gauge length | : 16 mm |
| Standard reference dimension | : 61.000 mm |
| Tolerance | : +20 µm |
| Remarks | : |
| Cutting conditions | |

Working drawing (dimensions: 35, 16, $\phi 63$, $\phi 61.000$, 2)

Standard values to VDI/DGQ 3441
  rpm          : 560
  Depth of cut : 0.5 mm
Remarks :
Tooling           : Type of carbide  : P10
Cutting geometry  : Approach angle   = 50°
                    Orthogonal rake  = 8°
                    Clearance angle  = 6°

Deviating values
  Cutting speed : 106 m/min
  Feed          : 0.1 mm/rev

Radius of nose = 0.5 mm
Back rake      = 9°

Short designation :
Measuring gauge :
(Designation)
Scale reading      : 1 µm                Location with :                Date :
Measurement uncertainty : (of VDI/DGQ: 3441 Section 6.4)
                   ≤ 0.2 tolerance        ☐     (1)
        4 $S_{Requipt}$  ≥ 0.2 tolerance  ☐     (2)                     Checked :
In case (2) 4 $S_{Requipt}$ must be considered in relation to the actual machine scatter (Item 16 in the test record)

## TABLE 10.8

**Determination of the operational accuracy of a CNC lathe (as per VDI/DGQ:3441)**
**—test record evaluation**

| | Tabular Values in μm as Deviation from the Set Dimension | | | | | | | | | | | mm |
|---|---|---|---|---|---|---|---|---|---|---|---|---|
| 1. | Clamping position | | I | II | III | IV | V | | | | | |
| | Random sample $j$ (max. $j = m$) | | 1 | 2 | 3 | 4 | 5 | 6 | 7 | 8 | 9 | 10 |
| | Test piece $i =$ (max. $i = n$) | 1 | 1 | 2 | 4 | 5 | 6 | 7 | 9 | 11 | 10 | 14 |
| | | 2 | 0 | 2 | 3 | 6 | 5 | 6 | 8 | 10 | 11 | 13 |
| | | 3 | 1 | 3 | 5 | 9 | 6 | 10 | 12 | 13 | 13 | 16 |
| | | 4 | 2 | 2 | 6 | 6 | 7 | 9 | 11 | 14 | 15 | 19 |
| | | 5 | 4 | 6 | 8 | 10 | 11 | 13 | 16 | 19 | 18 | 22 |
| 2. | $\Sigma X_i$ | | 8 | 15 | 26 | 36 | 35 | 45 | 56 | 67 | 67 | 82 |
| 3. | $\bar{X}_j = \dfrac{\Sigma X_i}{n}$ | | 1.6 | 3 | 5.2 | 7.2 | 7 | 9 | 11.2 | 13.4 | 13.4 | 16.4 | 87.4 |
| 4. | $R_j = (X_i)_{max} - (X_i)_{min}$ | | 4 | 4 | 5 | 5 | 6 | 7 | 8 | 9 | 8 | 9 | 65 |

5.

$\bar{X}$ chart

Trend (as far as necessary)

Trend = 15 μm

**Analysis**

6. Total mean value $\quad \bar{\bar{X}} = \dfrac{\Sigma \bar{X}_j}{m} = \dfrac{87.4}{10} = 8.7$ μm

7. Range mean value $\quad \bar{R} = \dfrac{\Sigma R_j}{m} = \dfrac{65}{10} = 6.5$ μm

8. Standard deviation $\quad S_R = \dfrac{\bar{R}}{d_n} = \dfrac{6.5}{2.326} = 2.8$ μm
   ($d_n = 2.326$ for $n = 5$)

9. Number of pieces machined $\quad N = 50$

10. TN (from $\bar{X}$ chart) $\quad = 15$ μm

11. $T_S$ (trend per piece) $\quad = \dfrac{T_N}{N-1} = \dfrac{15}{49} = 0.31$ μm

*(Contd)*

Table 10.8 (Contd)

| 12. | Allowance for trend (VDI/DGQ:3441, section 6.5.2.2) | | | | | | | | | | | | |
|---|---|---|---|---|---|---|---|---|---|---|---|---|---|
| | Random sample, $j$ | | 1 | 2 | 3 | 4 | 5 | 6 | 7 | 8 | 9 | 10 | |
| | Test piece $i =$ | 1 | 1 | 0.45 | 0.9 | 0.35 | −0.2 | −0.75 | −0.3 | 0.15 | −2.4 | 0.05 | |
| | | 2 | −0.31 | 0.14 | −0.41 | 1.04 | −1.51 | −2.06 | −1.61 | −1.16 | −1.71 | −1.26 | |
| | | 3 | 0.38 | 0.83 | 1.28 | 3.73 | −0.82 | 7.53 | 2.08 | 1.53 | −0.02 | 1.43 | |
| | | 4 | 1.07 | −0.43 | 1.97 | 0.42 | −0.73 | 0.32 | 0.77 | 2.22 | 1.67 | 4.12 | |
| | | 5 | 2.76 | 3.21 | 3.66 | 4.11 | 3.56 | 4.01 | 5.40 | 6.91 | 4.36 | 6.81 | $\Sigma R_{jc}$ |
| | $R_j$ corr | | 3.07 | 3.69 | 4.07 | 3.76 | 5.07 | 6.07 | 7.07 | 5.07 | 6.76 | 8.07 | 55.7 |

13. $\overline{R}_{corr} = \dfrac{\Sigma R_{jc}}{m} = \dfrac{55.7}{10} = 5.6 \, \mu m$

14. $S_R^* = \overline{R}_{corr} = \dfrac{5.6}{2.326} = 2.4 \, \mu m$

15. $S_{Requipt} = 0.9 \, \mu m$ (Regard to measurement uncertainty (as necessary in VDI/DGQ:3441 section 6.4)

16. $S_R = \sqrt{S_R^{*2} - S_{Requipt}^*} = 2.2 \, \mu m$

17. Operational scatter $A_S = 6$, $S_R = 13.2 \, \mu m$ ($S_R$ taken from Items 8, 14 or 16 as appropriate)

Date:       Signature:

$$\text{PCI} = \dfrac{\text{Tolerance}}{\text{Operational scatter}} = \dfrac{20}{13.2} = 1.52$$

As mentioned earlier, the normal process capability involves both machine related and process related parametric variation. One can study the actual process only if the machine parameters are considered and is designated as $C_m$ or $C_{mk}$. Process capability study involving both machine and process parameters are designated as $C_p$ or $C_{pk}$. Table 10.9 lists the process parameters which influence the machine performance. In machines where adjustment may be possible both for position setting and cutting parameter selection, and is targeted for manufacturing component of specific nature having both upper and lower tolerance limits, index $C_m$ or $C_p$ is used to denote the machine or process capability without influence of disposition of the mean dimension. In this case, machine capability index $C_m$ or process capability index $C_p$ is given by:

$$C_m \text{ or } C_p = \dfrac{T}{6 S_R} \quad \text{[Refer Fig. 10.20(a)]}$$

Where, $T$ = component tolerance limit
         $S_R$ = standard deviation.

In this case it has been assumed that middle value of the tolerance band and production dimension mean are identical and any deviation that occurs during production can be adjusted easily and hence need not be considered. Referring to Tables 10.7 and 10.8, the PCI, either $C_m$ or $C_p$ is found to be 1.52.

## TABLE 10.9

**Machine and process parameters influencing the machine performance**

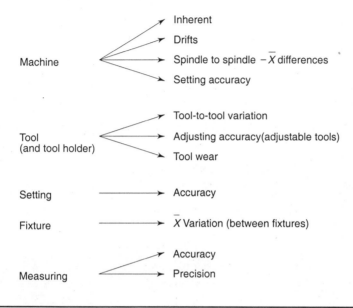

Any machine which has its production dimension mean not easily settable or changeable will have shifting of the production dimension mean with respect to the required tolerance band during changeovers of tools, jobs, etc.

In such a case, one has to consider the effect of these deviations obtained during process on production dimension mean. This effect is shown in Fig. 10.20 wherein a value $D$ is specified which is the minimum of the dimension between tolerance limits and the production dimension mean.

Now, the overall machine process capability is defined by

$$C_{mk} \text{ or } C_{pk} = \frac{D}{3 S_R} \quad \text{[refer Fig. 10.20(b)]}$$

Where, $D$ = minimum distance between a tolerance limit and the average of the produced parts

$S_R$ = standard deviation.

Referring to the example given in Tables 10.7 and 10.8, $D_{\min}$ is 8.7 micron. Hence,

$$C_{mk} \text{ or } C_{pk} = \frac{8.7}{3 \times 2.2} = 1.32$$

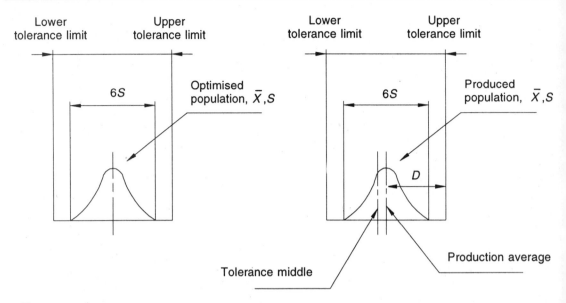

**Fig. 10.20** Distribution of machined component dimension vis-a-vis schematic representation of $C_m$, $C_p$ and $C_{mk}$, $C_{pk}$

The effect of variability of the production mean on process capability $C_m/C_p$ and $C_{mk}/C_{pk}$ is given in Figs 10.21 and 10.22. It can be seen that though there is a shift in production mean to the left by 1.3 micron, $C_m$, $C_p$ will be identified as 1.52, however, $C_{mk}/C_{pk}$ is taking that into consideration and gives a process capability of 1.32.

Normally the acceptable limits for machine capability index $C_m$, $C_{mk}$ and process capability index $C_p$, $C_{pk}$ are 1.66 and 1.33 respectively.

For processes which are centred with respect to the nominal of the specification $C_m/C_p$ and $C_{mk}/C_{pk}$, both will be equal (see Fig. 10.22) and so evaluation of $C_m/C_p$ will suffice. For instance, in the case of turning centres and machining centres, all the CNC programs are aimed at the nominal of the dimensional specifications, whereas in the conventional work centres like gear hobbing and grinding, there are chances that the specifications are not held at the nominal due to differences in tools and job setting. In case of the former, $C_m$ or $C_p$ would be preferable for evaluating the machine or process variability and in the latter case evaluation of $C_{mk}$ or $C_{pk}$ would be preferable. Generally, for machines and processes which are not adjustable to the required accuracy during production, machine or process evaluation is done through the capability index $C_{mk}$ or $C_{pk}$.

The important factors necessary to be considered while evaluating $C_{mk}$ and $C_{pk}$ are as follows.

$C_{mk}$ related

(a) Workpiece mounting spindle rigidity and repeatability
(b) Machine positional repeatability

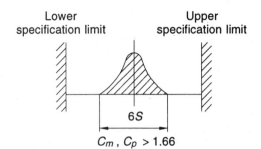
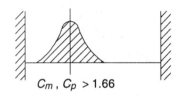
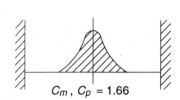

D : Minimum distance (mm)

**Fig. 10.21** Schematic representation of evaluation of capability indices $C_m$ and $C_p$

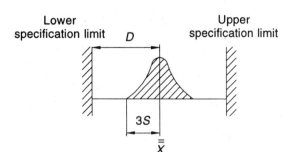
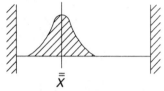
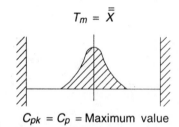

D : Minimum distance (mm)

**Fig. 10.22** Schematic representation of evaluation of capability indices $C_{mk}$ and $C_{pk}$

(c) Machine form errors
(d) Spindle thermal drifts.

$C_{pk}$ related

(a) Tool wear
(b) Tool-to-tool variation
(c) Workpiece clamping deformation
(d) Variation in incoming material stock and type
(e) Measurement error

### 10.5.5 Spindle Running Accuracy

The variation in position of the rotational axis during one revolution in dynamic condition is called the spindle running accuracy. This variation directly influences the circularity error of the machined workpieces. Hence, the spindle running accuracy test is carried out on such machines where the circularity requirement on the workpiece is stringent. The method of carrying out the test is explained below.

The set-up for measurement of the spindle running accuracy along with a schematic oscilloscope plot is shown in Fig. 10.23. A high accuracy steel or ceramic ball, with a circularity of the order of 0.1 μm, is mounted with an eccentricity of about 10 μm on the spindle to be tested. Two displacement pick-ups (e.g., inductive type) are mounted perpendicular to each other on a rigid fixture clamped to the spindle head casting. The stylus of the pick-up rests against a thin leaf spring which, in turn, contacts the ball with a slight pressure. Springs prevent damage to the pick-up stylus during spindle rotation. The output signals from the pick-up are fed to the horizontal and vertical plates of an oscilloscope (sensitivity of the oscilloscope is set equal for both the plates).

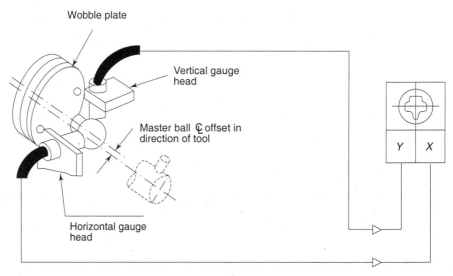

**Fig. 10.23** Set-up for the measurement of spindle running accuracy along with a typical oscilloscope plot

When the spindle rotates, the eccentricity of the ball results in a displacement of the pick-up styli, and a perfect circle with its radius proportional to the eccentricity will be traced on the screen of the oscilloscope, if the spindle runs accurately. Any shift in the spindle axis, during running, is superimposed on the circle and is a measure of the spindle running accuracy.

## 10.6 METAL REMOVAL CAPABILITY TEST

The cutting performance of a machine tool can be assessed from the result of actual cutting tests or by inference from the results of vibration measurements (indirect tests) on the machine tool structure. Both approaches have their special merits. However, irrespective of whether one chooses the indirect type of test (exciter test) or the direct test (cutting test), the results have to be interpreted in terms of some accepted standard test conditions. Otherwise' assessment of the performance of one machine or the relative performances of similar machines cannot be made. The reason for this is that all cutting conditions influence chatter and unless one has precise knowledge of the interrelationship between factors such as workpiece material and size, tool geometry, cutting speed, feed, and tool and workpiece overhang comparisons of machine performance would be impossible. It would be extremely helpful, therefore, from the viewpoints of both users and manufacturers, if some standard conditions could be adopted for testing purposes on different sizes and types of universal machines. This will not only facilitate in specifying the performance of the machine tool but also reducing the cost of testing.

The power utilisation of a machine is related to the metal removal rate and for roughing operations this is some measure of the potential productivity. The metal removal rate generally depends upon three main factors, viz. cutting speed, feed and depth of cut. In a machine tool, for a given speed and feed, the depth of cut is limited by the onset of chatter.

In most of the machine tools the installed power of the spindle will indicate the machine's possible maximum metal removal capability; hence full power utilisation test will be conducted.

However, a machine's chatter stability will vary with not only speed and feed, but also with the orientation and workpiece or tool mounting condition such as workpiece material, overhang of workpiece and tool. In such a case it may not be able to machine workpieces at full power capability and is expected to work at a limited power capability.

In specific type of cutting such as

(a) plunge turning with form tools
(b) slab milling with form cutters or set of cutters
(c) boring with extended boring spindle

it is difficult to analyse the cutting process both from the chatter conditions and cutting power utilisation. For these cases, it is more meaningful to express the limiting conditions in terms of the limit width of cut, $b_{lim}$. This parameter is directly related to the dynamic compliance of the machine and is predominantly governed by structural design of the machine, and hence the limit width of cut test is important. From this the metal cutting performance of the machines are evaluated by

(i) Full power test
(ii) Limit power test
(iii) Limit chip width test

The performance of a machine tool depends amongst other things, on the workpiece material, condition of clamping, its homogeneity, cutting tool material and geometry, cutting speed, feed and depth of cut. Hence, in order to arrive at a comparable basis for performance evaluation amongst different machines, a need for selecting a general test suitable for the above test criteria under conditions applicable for different types of machine tools, viz. CNC turning centre, machining centre, etc. was evolved after deliberations at Production Engineering Research Association (PERA, UK), Machine Tool Industry Research Association (MTIRA, UK), University of Manchester Institute of Science and Technology (UMIST, UK) and Institution of Production Engineering Research (CIRP, France). Figure 10.24 details the factors to be considered while specifying standard cutting conditions and test procedure for machine tools.

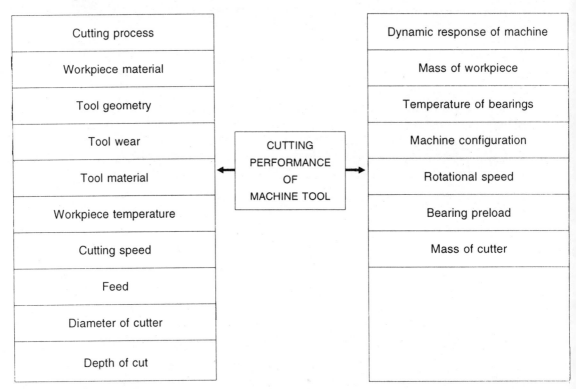

**Fig. 10.24** Factors to be considered when specifying standard cutting conditions and test procedures

## 10.6.1 Cutting Conditions for CNC Turning Centres

In turning centres the mass of workpiece, the spindle rigidity in radial direction, the clamping of the workpiece and the workpiece overhang are the most important factors in the system. It is, therefore, important to select the size of the test workpiece with due consideration to these factors.

The two basic clamping configurations are those with the workpiece held in chuck with overhang and longer workpieces held between the chuck and the centre.

There are five test set-ups for the cutting tests and these are outlined below (refer to Table 10.10).

**TABLE 10.10**

*Cutting conditions and acceptable values for metal removal capability tests on CNC turning centres*

| Sl. No. | Sketch | Test | Cutting Conditions ||||||| Acceptable Values ||||
|---|---|---|---|---|---|---|---|---|---|---|---|---|
| | | | $S$ | $V$ | $d_w$ | $\theta$ | $L$ | Tool | $P_{max}$ | $P_{lim}$ | Class | $b_{lim}$ |
| 1. | 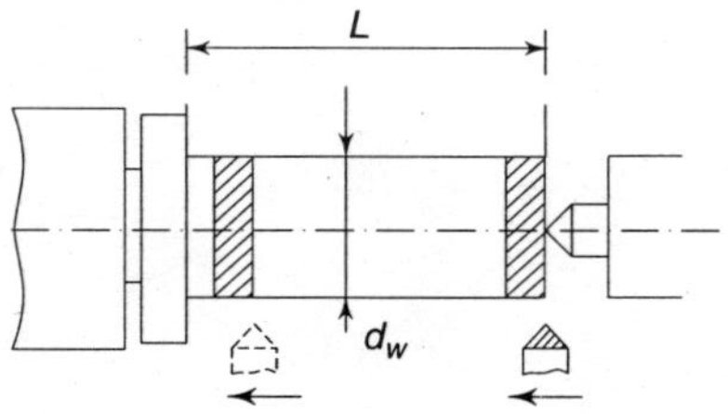 | $A_1$ | $\dfrac{D_{max}}{1000}$ mm/rev | $V_{30}$ | 0.2 $D_{max}$ to 0.25 $D_{max}$ | 45° | 0.7 $L_{max}$ < 6 $d_w$ | CC | ✓ | — | — | — |
| 2. | 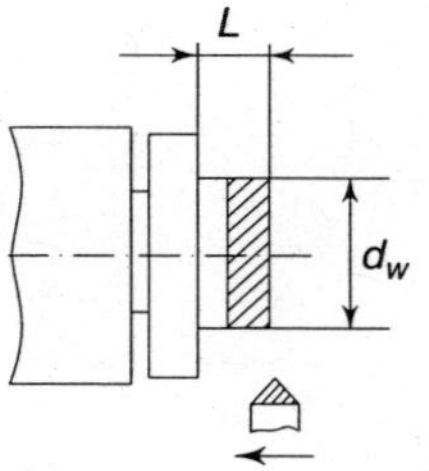 | $A_1$ | $\dfrac{D_{max}}{1000}$ mm/rev | $V_{30}$ | 0.3 $D_{max}$ to 0.4 $D_{max}$ | 45° | 0.15 $D_{max}$ to 0.17 $D_{max}$ | CC | ✓ | — | — | — |
| 3. | 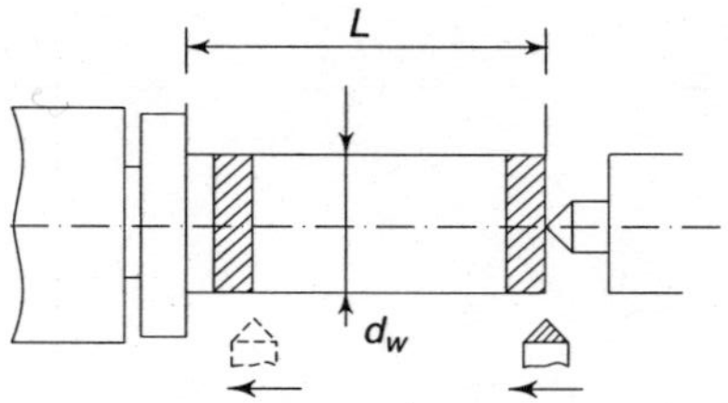 | $A_2$ | $\dfrac{D_{max}}{1000}$ mm/rev | $V_{30}$ | 0.3 $D_{max}$ | 45° | 0.8 $L_{max}$ to 0.9 $L_{max}$ < 6 $d_w$ | CC | — | $\geq 0.7\ P_{max}$ | — | — |

*(Contd)*

Table 10.10 (Contd)

| Sl. No. | Sketch | Test | Cutting Conditions | | | | | | Acceptable Values | | | |
|---|---|---|---|---|---|---|---|---|---|---|---|---|
| | | | $S$ | $V$ | $d_w$ | $\theta$ | $L$ | Tool | $P_{max}$ | $P_{lim}$ | Class | $b_{lim}$ |
| 4. | | B | 0.1/ 0.2 mm/rev | 100 m/min | 0.2 $D_{max}$ to 0.25 $D_{max}$ | 0° | 0.7 $L_{max} <$ 6 $d_w$ | CC | — | — | O > A > B > C < | 0.03 $D_{max}$ 0.02 $D_{max}$ 0.01 $D_{max}$ < 0.01 $D_{max}$ |
| | | | 0.1/ 0.2 mm/rev | 20 m/min | 0.2 $D_{max}$ to 0.25 $D_{max}$ | 0° | 0.7 $L_{max} <$ 6 $d_w$ | HSS | — | — | O > A > B > C < | 0.03 $D_{max}$ 0.02 $D_{max}$ 0.01 $D_{max}$ < 0.01 $D_{max}$ |
| 5. | | B | 0.1/ 0.2 mm/rev | 100 m/min | 0.2 $D_{max}$ | 0° | 0.3 $D_{max}$ | CC | — | — | O > A > B > C < | 0.03 $D_{max}$ 0.02 $D_{max}$ 0.01 $D_{max}$ < 0.01 $D_{max}$ |
| | | | 0.1/ 0.2 mm/rev | 20 m/min | 0.2 $D_{max}$ | 0° | 0.3 $D_{max}$ | HSS | — | — | O > A > B > C < | 0.03 $D_{max}$ 0.02 $D_{max}$ 0.01 $D_{max}$ < 0.01 $D_{max}$ |

1. *Full power test ($A_1$)*: The workpiece is held between the chuck and the centre, and the test is carried out at the chuck and the tailstock ends of the workpiece. The size of the workpiece is such that its diameter ($d_n$) is 0.2 to 0.25 of the maximum swing over the bed ($D_{max}$). The length of the workpiece is 0.7 of the maximum length between chuck and tailstock ($L_{max}$), but this is restricted to six times the workpiece diameter for long beds in order to eliminate the problems associated with slender workpieces. The objective of the test is to utilise full power of the machine ($P_{max}$) without chatter.
2. *Full power test ($A_1$)*: This test is carried out with the workpiece clamped in the chuck only and with an average overhang. The workpiece diameter ($d_w$) is specified as 0.3 to 0.4 of the maximum swing over the bed ($D_{max}$) and an overhang ($L$) of 0.15 to 0.17 of $D_{max}$.
3. *Limited power test ($A_2$)*: In this test a heavy workpiece is held between the chuck and the centre. The diameter $d_w$ is specified as 0.3 $D_{max}$ and length as 0.8 to 0.9 $L_{max}$ or less than 6 $d_w$. The test is carried out at both the tailstock and headstock end of the workpiece and owing to the large mass involved the machine is not expected to use the full power but use atleast limited power $P_{lim}$ which is 0.7 of the maximum power. The specified feed of 0.001 $D_{max}$/rev is the usual feed for heavier roughing operations for middle sized lathes with an average installed power. However, in high powered lathes higher feeds are usual and is in the order of 0.002 $D_{max}$. Therefore, if with the basic recommended feed the required power is not reached without chatter, the limit widths of the cut for $S = 0.001\ D_{max/rev}$ are recorded and the test is repeated with $S = 0.002\ D_{max}$/rev, where finally the $A_1$ or $A_2$ criterion is applied. The cutting speed ($V_{30}$) for test $A_1$ and $A_2$ is decided by the 30 minutes tool life criterion.
4. *Limit width of chip test (B)*: The fourth test is a plunge cut operation with varying tool width and an approach angle of $\theta = 0°$. Here the criterion is to check the limit width of the tool ($b_{lim}$) with which machining can be performed without tool chatter using both $S = 0.1$ mm/rev and 0.2 mm/rev. The workpiece set-up is the same as test (1) and the test is carried out at the headstock and tailstock end with both carbide and high speed steel tools.
5. This is also a *B* test with the workpiece held in the chuck only with an overhang of $L = 0.2\ D_{max}$ and a diameter of $d_w = 0.2\ D_{max}$/rev. The test is carried out with both carbide and high speed steel tools at the headstock.

The smallest value of $b_{lim}$ for the *B* tests proposed in tests 4 and 5 above can be used to classify the chatter performance of the machine. There are four classes proposed and the specified values are based upon previous experience. The classes are specified for both high speed steel and carbide tools and are summarised along with the above tests in Table 10.10.

## 10.6.2 Cutting Conditions for CNC Machining Centres

The cutting tests for machining centres include face milling and boring operations. The cutting test conditions and a schematic representation of the tool-workpiece for horizontal machining centres are given in Table 10.11. A guideline for the selection of the face

**TABLE 10.11**

*Cutting conditions and acceptable values for metal removal capability tests for horizontal machining centres*

| Test | Hh | Hp | V | S | Remarks |
|---|---|---|---|---|---|
| $A_1$ | $0.6\ H_{max}$ | $0.3\ H_{max}$ | 150 m/min | 0.25 mm/tooth | $P_{max}$ for orientations (a) (b) (c) (d) (e) (f) (Refer to Fig. 10.25) |
| $A_2$ | $H_{max} - D_c$ | $0.6\ H_{max}$ | 150 m/min | 0.25 mm/tooth | $P'_{lim} \geq 0.75\ P_{max}$ for orientations (a) (b) (c) (d) (e) (f) (Refer to Fig. 10.25) |

milling cutter is given in Table 10.12. This guideline has been specified on the basis of the spindle diameter and the availability of standard cutters.

### TABLE 10.12
**Selection of face milling cutter**

| Boring Spindle Diameter | Cutter Diameter, $D_c$ | Remarks |
|---|---|---|
| 63 mm | 200 mm | Cutter clamped directly to spindle |
| 70, 75 mm | 250 mm | sleeve flange or via standard adaptor |
| 90, 100 mm | 315 mm | to face plate |

When the face milling cutter is mounted in the most rigid configuration, i.e. clamped directly onto the spindle, then stability generally depends upon the directional orientation and the height of the cutter above the top surface of the table. In case of a horizontal machining centre, full power ($P_{max}$) utilisation is required with the spindle up to a height of 0.6 of $H_{max}$ (see Table 10.11), where $H_{max}$ is the maximum vertical traverse of the horizontal machining centre. Since frame milling is often used in practice, all the basic orientations shown in (a) to (f) of Fig. 10.25 must be considered and also the possi-

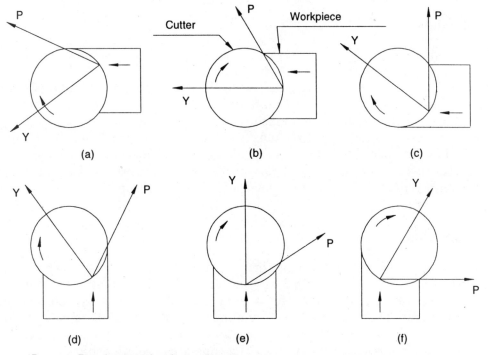

P : Resultant cutting force direction
Y : Direction of tool displacement

**Fig. 10.25** Different cutter-workpiece orientations in milling

bility of improving the stability by changes in speed if the specified power cannot be achieved. If full power is satisfied only for some orientations, then the depth of cut $A_{\lim}$ should be recorded for the others.

The limit power test ($A_2$) for a horizontal machining centre is specified when the cutter is at the top of the column, i.e. where the machine is in the most compliant configuration. For this case the utilisable power can be greater than 0.75 of the nominal power.

In the case of a vertical machining centre, the cutting test conditions and a schematic representation of the tool-workpiece are given in Table 10.13. The selection of cutter is to be done in accordance with Table 10.12. Tests should be conducted for both the full power test ($A_1$) and the limit power test ($A_2$) as mentioned in Table 10.13. Tool workpiece orientation will remain the same as mentioned in (a) to (f) of Fig. 10.25.

The boring test parameters are mentioned in Table 10.14 and is carried out with the boring tool mounted on the boring bar clamped to the spindle. The limit cutting condition in boring is a function of the boring bar length and for the test, two sizes of the boring bar length are specified, viz. one with the extension of 3 $d_s$ and the other 5 $d_s$, where $d_s$ is the boring bar diameter. Limit conditions are not classified but for the larger extension a minimum of 2 mm depth of cut is required.

## 10.7 OTHER TESTS

### 10.7.1 Rigidity

Machine rigidity values are necessary while investigating the machine's capacity to consume full rated power without any chatter. In this regard, both rigidity of machine tool spindles and that of axes feed drive are to be checked. Theoretical value of static stiffness of machine tool structures is estimated by a computer aided design (CAD) analysis. Static rigidity measurement on the machine tool gives practical results. Rigidity measurements are carried out in the directions of relevance to a specific machine. Sufficiency of rigidity thus obtained should be reviewed along with the machine performance achieved during metal cutting and accuracy tests with respect to the expected value. A statically rigid machine reduces the resultant form error of the component and improves the workpiece accuracy.

### 10.7.2 Reliability

Reliability is an important aspect of the proper performance of a machine. It is the ability of the machine and the CNC system to operate consistently and satisfactorily, under specified working conditions, for a given time. In some cases, the machine may continue to operate but generate major or minor errors such as dimensional variations and in impaired surface finish. The generation of these errors may be attributed to process element failures characterised by the time-order relationships.

Tests are formulated to check the functional reliability of the machine and the effects of a short-term working cycle on machine performance. The tests are of short-time and informative in content. They may be referred to as *short term reliability acceptance test*. The tests are carried out in two parts with: (a) full load and (b) no load.

## TABLE 10.13

**Cutting conditions and acceptable values for metal removal capability tests for vertical machining centre**

| Test | S | V | B | H | Remarks |
|------|---|---|---|---|---------|
| $A_1$ | 0.3 mm/tooth | 150 m/min | 0.75 of $D_c$ | 0.25 to 0.3 $H_{max}$ | Full power in one or more in configurations (a) (b) (c) (Refer to Fig. 10.25) |
| $A_2$ | 0.3 mm/tooth | 150 m/min | 0.75 of $D_c$ | 0.6 of $H_{max}$ | $P_{lim} \geq 0.75\, P_{max}$ One or more in configurations (a) (b) (c) (Refer to Fig. 10.25) |

$H_{max}$ = Maximum vertical traverse
$L_{max}$ = Maximum longitudinal traverse

## TABLE 10.14
### Cutting conditions for boring tests to be carried out in machining centres

| Test | L | S | V | Tool | Remarks |
|---|---|---|---|---|---|
| B | $3d_s$ | 0.1 mm/rev | 100 m/min | CC | — |
| B | $5d_s$ | 0.1 mm/rev | 100 m/min | CC | $b_{lim} \geq 2$ mm |

The full load reliability test is a rough machining test, subjecting the spindle drive system to full power and the guideways to high bearing pressure, combined with high working feed velocities. The test is carried out continuously for one to two hours, depending upon the machine, during which period no failures should take place. During the test, the temperature rise of the spindle bearings, lubricating oil and hydraulic oil, electric motors, etc. are measured. These should be within the normally permissible limits as specified. During the test, there may be a deterioration in the machine's accuracy. Therefore, it is recommended to repeat the geometrical and working accuracy tests, and ensure that the accuracies do not get altered after the full load reliability test.

The no load reliability test is formulated mainly for computerised numerically controlled machines. Under this test, the machine and the control system under no load should cycle automatically under the CNC system for a minimum period of eight hours. During the test run, no failure of any kind is to occur. In case a failure or a malfunction occurs, the test cycle is started again, after rectifying the fault, and the test is repeated for eight hours. The program used for the cycling test must position the machine over almost the entire working range and include all machine functions such as constant cutting speed, speed changeover, speed range changeover, oriented stop, canned cycle and tool change to qualify the entire complexity of the machine's function. The machine is programmed to return to the starting point at the end of each cycle, and upon completion of the cycle, the cycle is repeated with a time dwell of about 60 seconds.

At the time of starting the test, all the linear axes positions at their starting points are registered against dial indicators. The set points are selected at one end of each axis travel, such that the indicators do not hinder the machine movements during the test. If the machine is provided with a home position, then the home position points may be selected as set points. At frequent intervals (say 30 minutes), the slide positions are noted, and the deviations during the eight hour test is expected to be within the tolerances specified by the manufacturer. The test duration can be either 8 hours, 16 hours or 24 hours depending upon the intended use of the machine at the user's end.

## 10.8 SAFETY ASPECTS

Safety is an important aspect; it needs to be considered at the machine design stage itself. Various types of safety features are designed to protect the machine, the operator and the environment. The safety device provided must be simple and should serve its purpose to the fullest extent, without itself being the source of an accident.

The safety features usually provided on the CNC machine are checked for their proper functioning. The safety aspects are stressed more in the case of NC machines which are highly automated, with the least manual involvement. Many safety and interlock devices (both hardware and software) are essentially built into these machines. Some of the safety features usually to be provided on CNC machines are listed below:

- Overtravel of all slides is to be limited by safety switches, stops and software limits.
- Emergency stop is to be provided for all modes of operations.
- Collisions of slide with another slide or a fixed member, e.g. collisions of the saddle with the tailstock in a lathe, is to be avoided by providing hardware or software limits.

- Lubrication failure of the spindle bearing system is to be indicated by a lamp, and all functions have to come to a stop after a preset time.
- When feed hold is ON, all slide motions are to be inoperative.
- When workpiece is not clamped, the spindle should not be run. For example, when the chuck is open and the tailstock is withdrawn, the spindle should not be run in a lathe.
- When the spindle is running, the workpiece should not be unclamped. For example, chuck jaws and tailstock should be inoperative when the spindle is in the running condition.
- Low hydraulic or pneumatic system pressure and chuck clamping pressure should lead to the automatic stoppage of the machine.
- Software limit switches are to be provided to restrict the total slide travel, maximum feed velocity and maximum spindle speed.
- While the turret is indexing, all the slides are to be at rest.
- Improper turret clamping should be displayed on the CRT screen and all the slides are to be inoperative in that condition.
- Interlock is to be provided between the spindle brake and the power draw bar to prevent spindle rotation when cutters are being loaded and unloaded on the spindle.
- Improper tool clamping in the spindle nose should inhibit spindle rotation and the error is to be displayed on the CRT screen.
- Main motor is to be protected against overload.
- Spindle speed override and feed rate override are not to be effective in the thread cutting mode.
- Sensing of high motor temperature and overload should lead to the stopping of the machine after a preset time.
- The workpiece should not get unclamped in the event of a power failure or an emergency stop.
- Number of manual lubrication points to be kept as few as possible.
- Load meter should be provided to indicate the load on the spindle drive motor, so that the load on the motor can be maintained within safe limits, by reducing the feed rate or depth of cut to avoid any interruption during machining due to the stopping of the machine on account of overloading.
- Drives are to be provided with overload safety; the machine should come to a stop when overload occurs.
- When any of the machine functions, like spindle rotation, slide movement, turret indexing, etc. come to a stop without being programmed to stop, the machine should come to a stop and is operative only in manual mode till the fault is rectified.
- Machine guards should guard operator against chip and coolant. Suitable interlocks are to be provided to check the position of guards before the machine is operated in auto mode.
- Effective spindle braking system and an interlock against improper release of tool to be provided.
- Work area enclosure should ensure safe machine operation, machine setting including setting of tools and jigs, safety of vision and vibration.

## ANNEXURE

### Typical specification sheet for a machining centre*

**Standard Accessories**

- Chip collection arrangement
- Complete work area enclosure
- Twin pallet changer
- Heat exchanger for electricals
- Continuous rotary table
- Electrical equipment suitable for 3 phase 415 ± 10V, 50 Hz ± 3% AC supply
- Set of instruction manuals
- Foundation bolt kits
- Machine light
- Hand tools

**Specifications***

| | | |
|---|---|---|
| **Table** | | |
| Pallet table surface | mm. | 320 × 320 |
| Maximum permissible load | kg. | 250 |
| Table T-slots (No. × width × pitch) | | 3 × 12 × 100 |
| Indexing positions | No. | 3,60,000 (Continuous) |
| **Axes Travel and Ranges** | | |
| Longitudinal | mm. | 420 |
| Cross | mm. | 350 |
| Vertical | mm. | 350 |
| Distance from spindle face to centre line of table (min.) | mm. | 100 |
| (max.) | mm. | 450 |
| **Spindle** | | |
| Spindle taper | | BT - 40/ISO 40 |
| Spindle diameter | mm. | 70 |
| Spindle speed | rpm. | 40-4000 (60-6000 Opt.) |
| Spindle Power | kW. | 5.5 Continuous |
| | | 7.5 Intermittent (30 Min. rating) |
| **Feed Rates** | | |
| Rapid traverse rate (all axes) | mm./min | 20000 |
| Feed rate | mm./min | 1-6000 |
| **Automatic Tool Changer** | | |
| No. of tools | | 24 (40 optional) |
| Tool selection | | Random/Bidirectional |
| Max. tool dia—all pockets full | mm. | 80 |
| —alt. pocket empty | mm. | 125 |
| Max. tool length | mm. | 250 |
| Max. tool weight | kg. | 8 |
| Tool change time | sec. | 3.5 |
| Chip to chip time | sec. | 7.0 |
| **Automatic Pallet Changer** | | |
| No. of pallets | No. | 2 |
| Pallet change time | sec. | 12 |
| **Accuracies (As per JIS Standard)** | | |
| With encoder feed back | | |
| Positioning accuracy, all linear axes | mm. | ± 0.005 : over full |
| Repeatability | mm. | ± 0.003 : traverse |
| **CNC System** | | Hinumerik 3100 m |
| **Power Requirement** | KVA | 25 |
| **Floor Space Requirement** | | |
| (L × W × H) | mm. | 3200 × 2500 × 2400 |
| **Machine Weight** | kg. | 5500 |

\* Illustrations and specifications are not binding in detail and are subject to modification without notice. For further details, contact us.
*Note*: Standard and special accessories/equipment listed are provisional; final list is as per the quotation.
*Source*: IMTMAS

## ANNEXURE 10.2

### Representative test chart for geometrical accuracies of machining centres and turning centres

| Test No. 11 | Parallelism between the spindle axis and Z axis in the XZ and YZ planes | | | | | |
|---|---|---|---|---|---|---|

| | | | | | Deviations | |
|---|---|---|---|---|---|---|
| Figure | Objects | Measuring Instruments | Reference to IS:2063-1962 &/or instructions for testing | | Permissible | Actual |
| | Verification of the parallelism of the spindle axis to the movement of:<br>A - the table saddle on the bed<br>B - the table on the bed<br>C - the column on the bed<br>D - the column on its saddle<br>E - the spindle head on its slide.<br>F - the spindle head on its slide<br>a in the horizontal plane<br>b in the vertical plane | Test mandrel and dial gauge | 5.3.2.2. (b)<br>X axis in centre of the travel, Y axis to be locked. | | (a) 0.02 for 300<br>(b) 0.02 for 300 | |

*Note*: Test shall be repeated for atleast FOUR positions of test mandrel in the spindle nose. The maximum value shall be the measured value.

## C. Headstock Spindle

| 1 | 2 | 3 | 4 | 5 | 6 | 7 | 8 | 9 |
|---|---|---|---|---|---|---|---|---|
| G4 | | (a) Measurement of periodic axial slip<br><br>(b) Measurement of camming of face plate mounting surface | Dial gauge and possibly a special device for application of force | 5.5.2, 5.5.2.1 (b) 5.5.2.2. and 5.3.3 Note: if force 'F' is required, it has to be specified by the manufacturer. | (a) 0.005<br>(b) 0.010* | 0.010<br>0.015* | 0.015<br>0.020* | 0.020<br>0.025* |
| G5 | | Measurement of the runout of spindle nose (chuck locating diameter) | Dial gauge | 5.5.1.2 and 5.5.2 Note: If force 'F' is required, it has to be specified by the manufacturer. | 0.005 | 0.010 | 0.015 | 0.020 |
| G6 | | Measurement of the runout of the spindle locating bore, for such machines which are provided with a locating bore for mounting work holding fixture. | Dial gauge | 5.5.1.2 | 0.005 | 0.010 | 0.015 | 0.020 |

\* Including periodic axial slip

**432** Mechatronics

## ANNEXURE 10.3

**Representative test chart for practical tests of machining centres and turning centres**

| Machine No. | | Test Chart for Horizontal Spindle Machining Centres | | Customer | |
|---|---|---|---|---|---|
| Type | | II. Practical Tests (All dimensions in mm) | | Inspector | |

| Test No. | Figure | Object | Measuring Instruments | Reference to IS:2063-1962 &/or instructions for testing | Deviations | |
|---|---|---|---|---|---|---|
| | | | | | Permissible | Actual |
| 1 | | Verification of<br>(a) Roundness of bores<br>(b) Cylindricity of bores | Bore gauge/talyrond | 3.1, 3.2, 2, 4.1, 4.2. 5.3. 4. 2 & 5.5. 1 (d)<br>Before commencing the tests make sure that the mounting surface of test piece is flat. | (a) 0.01<br>(b) 0.015 | |
| 2 | | Verification of<br>(a) Flatness of milled faces.<br>(b) Step difference at the end of cycle (machined by face milling cutter) | Straight edge & dial gauge<br>Or<br>Surface plate & dial gauge | (a) 5.2, 5.2.2 & 5.2.3<br>Before commencing the tests, make sure that the mounting surface of test piece is flat | (a) 0.015<br>(b) 0.015 | |
| 3 | | Vertification of<br>(a) Squareness of faces<br>(b) Parallelism of faces (machined by face milling cutter) | (a) Master square and dial gauge<br>(b) Dial gauge | (a) 5.4, 5.4.1.1 & 5.4.2.5<br>The faces are machined by indexing the table (pallet) by 90° for each face.<br>(b) 5.3.2, 5.3.2.2, 5.3.2.3 | (a) 0.02/300<br>(b) 0.03/300 | |

Testing of Machine Tools    433

**Test Chart For Computer Numerically Controlled Turning Centres**

Type  
Machine No.  
Order No.  
Date  
Customer  
Inspector

## II. Practical Tests
### All dimensions in Millmetres

| Sl. No. | Figure | Nature of Test | Cutting Condition | Checks to be applied | Measuring instruments | Reference test code | Permissible Deviations for Turning Diameters for Range | | | |
|---|---|---|---|---|---|---|---|---|---|---|
| | | | | | | | I Upto 160 | II 160-315 | III 315-630 | IV 630-1250 |
| 1 | 2 | 3 | 4 | 5 | 6 | 7 | 8 | 9 | 10 | 11 |
| P1 | | Turning a cylindrical test piece held in the standard work holding device with single point tools mounted on the turret. | Machining operation to be carried out for external and internal machining with both roughing and finishing cuts. | (a) Circularity Variation in radius close to the spindle nose and at the end of the test piece. (b) Consistency of machined diameters Measure variation between diameters $D_1-D_2$ and $d_1-d_2$ | Roundness measuring machine, micrometer | Clauses 3.1, 4.1 & 4.2 | (a) 0.005 (b) 0.01 | | 0.01 0.02 | |

| Dimensions for Range | Diameters | | | | |
|---|---|---|---|---|---|
| | $D_3$ | $D_4$ | $D_5$ | $D_6$ | |
| I & II | 100–120 | 50–75 | 125 | 100 | |
| III & IV | 150–175 | 60–85 | 155 | 140 | |

| Sl. No. | Figure | Nature of Test | Cutting Condition | Checks to be applied | Measuring instruments | Reference test code | I | II | III | IV |
|---|---|---|---|---|---|---|---|---|---|---|
| P2 | | Facing test piece held in the standard work holding device with a single point tool on one station of the facing turret. Test piece material, together with type and form of tool, feed, depth of cut and cutting speed is to be specified by the manufacturer consistent with 2.4 (Sheet 1) | | Flatness of faced surface | Straightedge and slip gauges | Clauses 3.1, 4.1 and 4.2 | 0.02/300ϕ | | 0.03/300ϕ Concave only | |

\* Test pieces can be made of free cutting steel or cast iron.  
@ $D_{min}$ = 0.5 Maximum turning diameter

CHAPTER **11**

# Industrial Design, Aesthetics and Ergonomics

### 11.1 INTRODUCTION

Fielden has defined design as "the use of scientific principles, technical information and imagination in the definition of a mechanical structure, machine or system to perform prespecified functions with the maximum economy and efficiency". In contrast, industrial design, is concerned with how to strike a balance between the external features and functional performance to create a coherent whole quality product. It also ensures that the product is safe and easy to produce, use, control and maintain. Thus, the industrial design function extends to interfacing with ergonomics and technology on the one hand and satisfying the customer on the other. The stress on industrial design will contribute to the achievement of a high quality product, which in turn, increases the competitiveness of a firm.

Figure 11.1 shows the main factors involved and the scope of industrial design effort in the design of an industrial product.

Aesthetics constitute a fundamental part of engineering design. Aesthetic quality in engineering design can be achieved through elegance of concept and elegance of realisation. When both are achieved, a completely satisfying product is obtained. Too often, importance is given to the functional aspects and not to the product presentation, resulting in a crude product.

The opportunities for achieving aesthetic quality in engineering design are considerable. For apt selection of visual qualities, one of the functions of aesthetics in machine tool design is to indicate the function and purpose. Simple and easy-to-perceive forms which recognise the function and purpose are the two factors in aesthetic design of any product. There is an almost infinite variety of materials, capable of being endowed with different shapes and characteristics arising from appropriate forms and from alternative manufacturing processes. There are also a variety of different surface treatments, finishes such as paint and metal. Thus there is a commercial value for aesthetics. An aesthetically designed product will always be more acceptable in the market, resulting in better saleability.

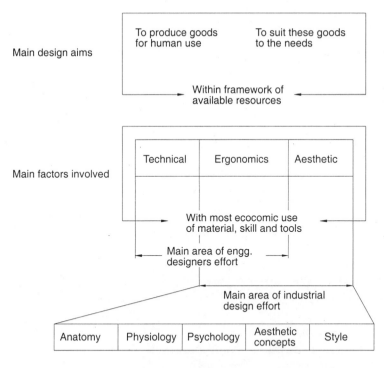

↳ **Fig. 11.1** Product design concept

## 11.2 ELEMENTS OF PRODUCT DESIGN

The three elements that define product design are product physiogonomy, product physiography and product anatomy; these give rise to aesthetics, ergonomics and anatomy to form the basis of design prognosis.

### 11.2.1 Product Physiogonomy—Aesthetics

Aesthetics is important because the sum total of perception of quality of the product is communicated through it. The physical characteristics of products can evoke different feelings in different people and individuals form judgement on their basis. The product communicates about itself through its physiogonomy. The moment one sets eyes on the product, feelings of like, dislike or neutrality emanate, depending on the physical presentation of the product. Therefore, physiogonomy (aesthetics) establishes a bond between the user and the product. In a competitive market flooded with different brands of similar products, the sales volume will depend on product physiogonomy. Industrial designers working in the area of physiogonomy deal with semantics, semoitics, perception, cognition and study of culture. Good product physiogonomy helps to,

- establish communication between the product and the user.
- improve the environment in which the product is used

- create satisfaction to the user
- increase productivity
- increase sales of the product

Now the question is, what actually makes a product aesthetically pleasing? There are certain elements that determine the aesthetics of a product and some principles that govern the manipulation of these elements. These principles of perception and cognition are mainly based on Gestalt school of psychology which deals with aspects like proximity, symmetry, harmony, continuity, rhythm, etc. In addition to these, material, form, proportion, colour, texture, graphics, finish, etc. also contribute to aesthetic improvement. While designing, the following aspects are considered.

- Aesthetics is not cosmetics, it is not beautifying a product.
- Appearance must be built into the product and not applied to it.
- Materials and processes are essential factors, to be considered while determining the form of a product.
- Form of a product must communicate the functions of the product.
- Colours influence the perception of a product in three ways:
    - Impression—visual, interest, brightness, and illuminance
    - Expression—emotionally warm, heavy, repulsive
    - Construction—symbolically, red indicates danger.
- Texture on the surface is important to project the quality.
- Material and manufacturing process influence the aesthetics of the product.

## Elements of Aesthetics

***Concept of Unity*** In product design, aesthetics attempts to obtain harmony of all parts fitted together in such proportions that nothing can be added, diminished or altered. This is the concept of unity and is illustrated in Fig. 11.2. A simpler interpretation of unity has three main aspects. Firstly, the product should appear complete, no part should appear missing or endowed with any superfluous elements. Items (a) to (c) in Fig. 11.2 depict this concept. Secondly, there should be a harmonious relationship between all the parts contributing to the product as a whole and thirdly, the emphasis should be placed on actually seeing the product as a whole and not regarding any one part of it to the exclusion of others. This is illustrated in item (d) in Fig. 11.2, wherein the eyes tend to see two squares one over the other, rather than a single rectangle. In item (e), the rectangle is a divided one but its overall quality is retained with a less dominant character. Arranging component parts to produce an apparent whole may be a highly sensitive process. In item (f) there is a tendency to see three equal strips joined together rather than the rectangle as a whole, but when width of the strip is increased in item (g), an impression of unity returns. Items (h) and (i) underline the sensitivity of judgement which may be needed to make individual parts in any product contribute to the whole. Finally, items (j) and (k) illustrate the elevation view of a simple container, and its designer has embodied a curved form to make it attractive. The approach to unity depends upon constantly keeping the whole subject in mind in order to obtain a successful relationship of its component parts.

## Industrial Design, Aesthetics and Ergonomics

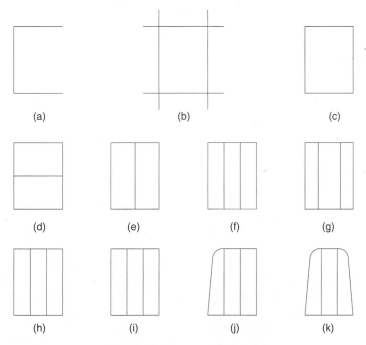

**Fig. 11.2** The concept of unity

***Concept of Order with Variety*** To achieve unity the component elements in a product should be related to the whole, i.e. each element should be related to the other. This relationship is referred to as order. In music, order is provided by a particular rhythm, but this rhythm by itself could be extremely monotonous. The music appeals because of the melody created by variations in rhythms. Variations played without any control create confusion. This need for order coupled with variety is required for various aesthetic experiences.

Figure 11.3 illustrates this concept. The square in item (b) has order but no variety since all sides are equal; the rectangle in item (c) may be preferred to item (b) because, while the shape is still highly ordered, some variety exists in lengths of the sides. A similar and equally elementary comparison may be made between a circle in item (d) and an elipse in item (e). The circle has no variety; its curvature is constant. But the elipse, though still an orderly shape, has a continuously changing curvature.

***Concept of Purpose*** The most successful approach to a clear statement of purpose is made by starting with the simplest sketch of a machine's main elements. Ideally, these elements would bear a character which describes their functions. What would be recognised and appreciated is the structural composition. If this can be expressed in a simple way, the concept of purpose is achieved.

***Style and Environment*** It is possible for different designers to embody all the basic concepts in products performing similar functions, and yet these products could possibly

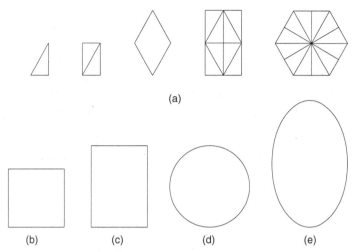

**Fig. 11.3** The concept of order

be distinguished from each other. Therefore an aesthetic response may well include recognition of a particular mode of expression. Architecture and products differ from works of art in that they cannot help being related with their environment. Relating the product to its environment could be regarded as another aspect of style.

***House Style*** Just as a person may be judged in part by his appearance, a manufacturer may be judged by the homogeneity among products from the firm. If the product appears out-of-date, badly finished, crude and ungainly, these characteristics may detract from the impression one wishes to create in the market. This impression will be built up, not only from the appearance of the product, but also from the appearance of catalogues, advertisements, stationery, exhibition, equipment, delivery vehicle and the premises. If good aesthetic standards can be maintained in these features as well, it is likely that a stimulating influence can be created on potential customers; an impression that the manufacturer really cares about the things he produces is formed. Further, if there are visual links among the products, trade literature will be regarded as operating in a well planned manner. This visual continuity will make the manufacturer more discernible in the markets; especially if one is operating in international markets. The first impression on potential customers may well be through the visual appeal of catalogues or displays in trade exhibitions. Subsequently, there will be the impressions made by products or through correspondence, instructions, and service manuals. Another approach to house style is providing satisfying service to customers and not merely selling the products.

### 11.2.2 Product Physiography—Ergonomics

Ergonomics is concerned with assurance of safety and well being of the operators at workplace while maintaining an optimum level of productivity. It deals with methods of designing the machine operations, work stations and work environment, so that they match human capabilities and limitations. By applying ergonomic principles and data effectively, it is possible to optimise the design of a product, job, workstation, training methods and the system's safety.

The main objective of ergonomics is to achieve internal compatibility between the human and machine components, and external compatibility between the system and the environment in which it operates.

A broad outline of systems design is as follows:

(a) Specify the inputs and outputs to the system.
(b) Specify a set of functions capable of transferring the inputs into the outputs.
(c) Specify which functions are to be allocated to the people and which to the machine.
(d) Specify necessary training procedures, job aids and man-machine designs.
(e) Specify any changes needed to ensure compatability between human components, machine components and environments.

Human sub-tasks should, as far as possible, be integrated into jobs that offer variety and satisfaction to each operator and provide scope for the exercise of initiatives that improve system performance. An operator should carry out emergency operations in case of machine failure, routine maintenance, and be responsible for adjusting the system to reduce stoppages.

The detection of incompatibilities between human capabilities and demands of the machine system requires participation of specialists in design. The compatibility is investigated in relation to objectives, cost and completion date, and the complete requirement of the system.

### 11.2.3 Ergonomics in Machine Tool Design

It is well known that machine tools are basic to production. Progress in production technology is governed by advances in machine tools and manufacturing techniques. The advances in machine tool manufacturing techniques take into account various aspects such as, augmenting higher productivity, reduction in labour cost, improving labour conditions, reducing physical labour, need for automation, provision of a clean environment, occupational safety, and all other factors connected with the man-machine interface. Man-machine system consideration is of great importance and value to modern industries. Industries are now realising the human limitations to processes and controls, and are engaged in designing systems based on ergonomic principles. Thus, ergonomics attempts to match successfully the machine to man, so that the whole system works effectively. Before the product is designed, it is essential to have a good knowledge of the working of the various operational elements. Since, in ergonomics the human operator is regarded as an integral part of the system, an understanding of the physical and mental ability of human beings constitutes an important part of the work. While designing a new machine, the operator's work zone should be given due importance so that while working on the machine, the operator can keep his natural position and operate all controls with ease. The design, therefore, has to be based on human anthropometric and biomechanic data. Easy identification of controls, low operating forces, logical grouping and pleasing colour schemes, find a major place in the development of modern machines. Man serves machines within a total system and plays a vital role in eliminating errors which make him an important constituent of the total system reliability. Finding procedures for reducing the man-machine error is a very important means of raising productivity. This is the essence of ergonomics.

Ergonomic principles, in addition to augmenting the functional aspects, can also influence the aesthetic aspects of the product. This is assured by selecting a proper shape and size for control elements, graphic displays and colour contrast for better readability and their optimum placement. Thus, ergonomics ensures a blend of functional aspect of the product with that of aesthetics. Ergonomics in short is a combination of science and art.

### 11.2.3 Ergonomics in Machine Tool Safety

Safety in machine tools and in working environment has gained much importance in recent years. This is because safety to the operators and capital intensive equipment have become a pre-requisite in production environments of the modern factories. In fact, health acts and product liability have now led the governments to enforce legislation on occupational safety. The recently laid down recommendations on acceptable shop noise levels and mandatory safety regulations have increased the liability of the machine tool builder to fall in line with the regulations. This is more so because the newer machine tools require higher speeds and more power to obtain increased output. The increase in cutting speeds have far reaching effects on the design of safety guards for machine tools.

Occupational Safety and Health Administration Act (OSHA), and the Health and Safety at Works Act of 1975 demand that the designers ensure that a machine is safe and is in no way hazardous to health. It also makes it mandatory for importers of machines to ensure that a satisfactory standard of protection is built into the machines. This factor is very important, especially in the case of high speed grinding machines. Since the kinetic energy of wheel increases at the square of the speed, the chances of wheel failure due to an operator's carelessness has to be positively eliminated. Grinding machine guards should be capable of absorbing the energy on braking and in containing the fragments.

In pressworking, constant refinements in man-machine systems have now assured utmost safety in operations. Barrier guards interlocking gate guards, tripping devices, swing out panels and light curtains are some of the notable fail-safe mechanisms developed by careful study of man-machine systems.

Colour schemes for indicating pinch-points or hazardous area, easy identification of controls and prominent caution or warning labels on the machine have now become a part of the machine tool design for maximum safety in operation.

### 11.2.4 Product Safety Audit

Since, safety level of an equipment during operation should be such as not to harm the operator or other installations in its vicinity, every piece of equipment should be evaluated in terms of established conventions as regards safety. This is the basis of safety audit. Safety devices should be built into the basic product and not left to the discretion of purchasers.

For evaluating the overall safety of a product, it should be subjected to very stringent tests which go far beyond the normally encountered accidents. Even some seemingly unreasonable limits should be set so that the evaluation proves that the product is absolutely safe. The results of such a safety audit should constitute the basis for designing safety features into the product. The product must be designed for forseeable use, not solely for intended use. Once the functional aspects of the product are designed, a subjective analysis should be undertaken to articulate the type of use and misuse the product can be subjected to in the hands of all who may come in contact with it.

The following procedure can be followed in the design for improving the reliability and safety of the product.

(a) List the requisite product functions and needs.
(b) Identify the environments within which the product will be used.
(c) Anthropometric and biomechanic data of the operator population.
(d) Postulate all possible hazards, including estimates of probability of occurrence and seriousness of resulting harm.
(e) List out alternative design features or production techniques, including warning instructions that can be expected to effectively eliminate the hazards.
(f) Evaluate the alternative design and incorporate the safety aspect in the final design.

These factors should be taken care of after the initial design incorporating the requirements of applicable standards has been formulated.

In designing and manufacturing machine tools, incorporating all aspects of safety is the sole responsibility of the machine tool builders. If their products do not provide adequate safety to the operator, the machine tool builders are liable even much beyond the warranty period. This dimension has gained importance with the introduction of the concept of product liability. Though this concept applies to many of the western countries, it is also necessary for Indian machine tool builders to meet this liability inasmuch since their products compete in the international market. To this end, the product liability concept has to be taken into account in the designs. Thus, there is a need for instituting certain procedures such as product safety design review or product safety audit, so that the whole concept of safety is institutionalised.

## 11.3 ERGONOMIC FACTORS FOR ADVANCED MANUFACTURING SYSTEMS

Industrial productivity must be improved and maintained to meet the challenges of the global market. This is possible through technological changes or progress.

Manufacturing firms around the world have invested heavily in advanced manufacturing technologies: automation, programmable controllers, robotics, machining cells, flexible manufacturing systems, machine vision, CAD/CAM, hierarchical computer control strategies, in addition to a wide range of manufacturing management technologies. These integrated systems are rarely viewed from the sense of integrating people and machines, while both advanced manufacturing and computer integrated manufacturing have become synonymous with total automation. The process involves an increase in integration of both functions and associated data, and also needs an enormous information exchange among the different stages in a functionally divided production process.

Since human beings have a limited information processing power, the monitoring and controlling capabilities of the overall production process are restricted. But the development of powerful computer based tools now allows human beings to control more complex combinations of manufacturing tasks. This results in the elimination of many of the previous divisions of work and integration of tasks at the workplace. A typical computer integrated manufacturing (CIM) model is given in Fig. 11.4.

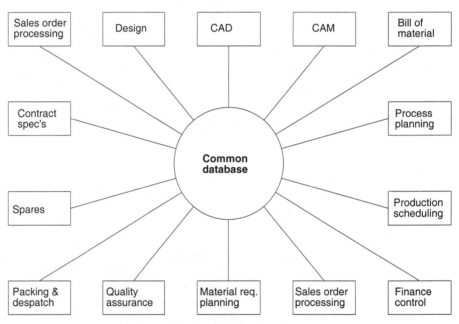

Fig. 11.4  CIM model

For establishing the degree of automation expressed by computer-integrated manufacturing, complex computing and control systems are required. Programming of these systems has to be tailored to suit each system in the form of flexible manufacturing system (FMS). The FMSs are not stand-alone pieces of equipment but consist of a number of sophisticated machines with well-integrated software and hardware modules. The various stations are interlinked by automated materials handling systems like rail guided vehicle, automated guided vehicle, etc. for transferring pallet mounted prismatic components. Figure 11.5 shows a layout of a typical flexible manufacturing system (FMS).

There should be a human-centred approach to the design of human machine systems focussing on how people are complemented by machines and vice versa. Once decision

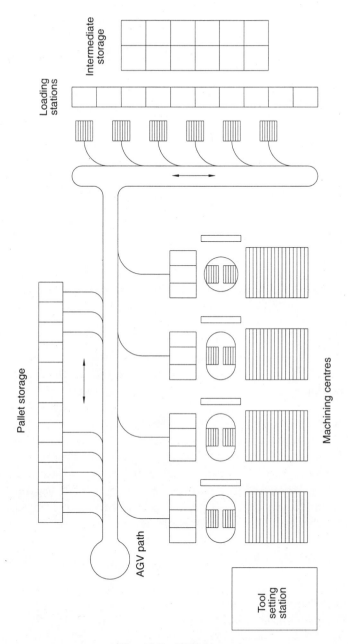

**Fig. 11.5** FMS layout

on allocation of function has been made, the various tasks need organising into job designs for the operators of the system and local organisation structures need designing to support their activity.

The idea of human-centred technology has gained prominance in the manufacturing technology only recently. The general qualities which form the basis of human-centred advanced manufacturing technology are:

(a) Human-centred technology accepts the present skill of the user and allows it to develop. Conventional technological design tends to incorporate this skill into the machine itself with the resultant de-skilling of the human.
(b) More degrees of freedom open to the users to shape their own working behaviour.
(c) Human-centred technology unites the planning, execution and monitoring components of work.
(d) Human-centred technology encourages social communication between the users.
(e) Human centred technology provides a healthy, safe and efficient work environment.

### 11.3.1 Machine Oriented Industrial Design

Earnest efforts are being made in the industry to leap into the future with state-of-the-art manufacturing plants. The goal of all technology push has been to get rid of human element to a large extent. There are indicators which confirm the machine bias of the current drive to improve the competitiveness in the industry. Machines are devices to replicate simplified human tasks. Machine can only capture a portion of the human workers' actual functions—leaving a gap, which the human worker is called upon to fill. The human element is the most flexible part of a system, filling the gaps or deficiencies in machine design. The attempt to eliminate the gap by means other than human labour leads to adding cost and complexity to a system which is already complex. This will invite new management systems and also a set of new experts. This will only be compounded by the current attempt to rationalise office work with computer assisted design and engineering, computerised manufacturing, resource planning systems, expert systems and other forms of artificial intelligence. Traditional production and industrial relations system are not suited to the changed conditions of the product market. Today, workers are better educated, their living standards are high and so are their expectations. This makes it difficult to retain workers for long in one organisation.

### 11.3.2 Factory Without People

Human wisdom is not easily displaced by the computer integrated workerless factory of the future. Manufacturing processes are becoming increasingly complex and building a software model of these processes is a monumental task. It may only be possible to achieve flexible process to fixed computer algorithms. In a factory without people, the major share of both shop floor and office work will be frozen into hardware/software, computer algorithms and knowledge bases. In addition to the obvious function of providing goods and services, production also serves as a key educational function, i.e. to give people hands-on experience under the guidance of knowledgeable colleagues. If the division of labour reduces the skill and knowledge of the worker, the factory without people practically eliminates the training and education function of productive work. The attempt to fully automate production could have an impact on both product and process innovation. The creative process at stake here, requires stimulus from a wide variety of sources other than computer stimulations of the real world. Most advances in technology today are generated along the learning of an on-going production process and not in factories without people.

### 11.3.3 Ergonomic Problems in New Technology

For maximum benefits from the implementation of a new technology such as FMS, industrial robots, CAM, etc., an understanding of the ergonomic problems involved is necessary. For automated systems, the overall ergonomic problems to be considered are job design, work place design, environmental design, evaluation, equipment design evaluation, communication design evaluation, human reliability measurement documentation design, system safety design, training and consideration of socio-psychological factors. Figure 11.6 gives the scope of ergonomics for automated systems.

*Socio-psychological Factors*

The new technology can create a reduction in the number of operators, requires a change in skill and work locations. The employees who have to move to a less skilled job will feel the effects of new technology directly. They face loss of job status, job interest, and job challenge. New technology can reduce the number of operators. New skills and jobs will emerge as a consequence of CNC machines. Labour-saving techniques which resulted from the adoption of new technology have improved the living standards and also productivity.

*Job Status*

For introduction of new technology, any industry requires qualified workers with the relevant training and experience. These employees will obviously command higher prestige and pay. Consequently workers would welcome new technology, since it provides them with high calibre and well-paid jobs. But, the union leaders and politicians may have opposing views.

*Organisational Relationship*

New technology may minimise the interaction among workers by reducing their numbers and increasing the distance between their workstations. For some, such an environment may cause loneliness. One can anticipate worker alienation in such conditions. But introduction of new technology enhances the interaction between workers and management.

*Worker Attitude and Satisfaction*

Operators manning new technology installations often express satisfaction in the variety of work, especially while dealing with different kinds of machinery. Also, the manning of complex installations increases the sense of responsibility.

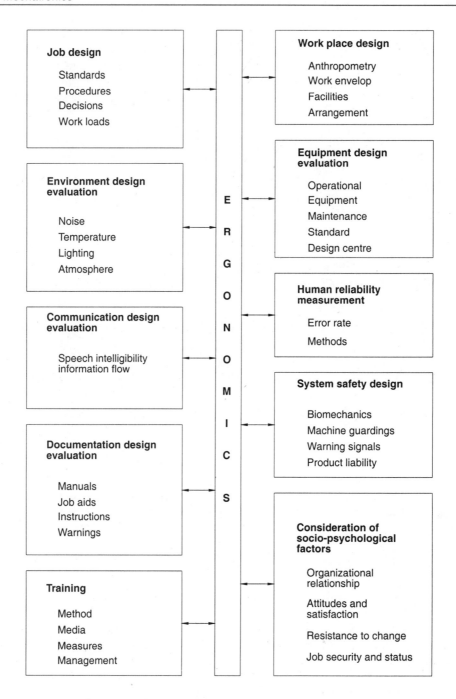

Fig. 11.6 Scope of ergonomics in an automated system

## System Safety Design

Industrial robots are installed to improve productivity and also the working life. Operators are prevented from working in hazardous and unhealthy working conditions, accidents and occupational diseases. Relieving operators from monotonous and repetitive work reduces psychological stress.

Where industrial robots improve working life, they also pose safety related problems for human beings and the other machines. Special workplace layout has to be prepared to ensure safety.

## Training

The new technology will require higher skills to program machine instructions, but once this is accomplished, production will be almost fully automatic and little labour will be required. Also, highly specialised, trained maintenance persons will be required. Hence, the responsibility of the production and maintenance engineers will undergo substantial change as a consequence of new technology.

## Workplace Design

The workplace layout should be compatible not only with system performance requirements but also with users. The user should be able to see the work clearly. Workplace dimensions should be compatible with anthropometric characteristics of the anticipated user. In designing the workplace, the physical relationship in terms of distances and other linear dimensions, often of major consequence in achieving production efficiency and physical and mental well being of operators is considered. For implementation of new technology, attention should be paid to overall workplace design.

In implementing new technology, the critical issue is how to make the transition from conventional manufacturing to advanced manufacturing so that the technology will be supported at all levels in an organisation. A wide range of ergonomic problems can be eliminated if the new technology is implemented, realising the importance of human elements in technology and managed in a better way. Management should give consideration to ergonomic factors also, along with technical and economical factors. With a concentrated management, while implementing new technology, it may be possible to minimise social disruption, increase worker job attitudes, satisfaction and job safety to enhance quality of working life.

CHAPTER **12**

# Introduction to Computers and CAD/CAM

## 12.1 INTRODUCTION TO COMPUTERS

The birth of computers is one of the revolutionary breakthroughs in scientific invention in the history of mankind. Today, the computer has become a part and parcel of every activity in business, design, information management, etc. It has become so integral a part of man's daily activity, that it is popularly known as personal computer or PC.

For any task a computer is used when man is ineffective in executing it with required efficiency and vice versa. This being so, it is useful to examine some of the individual characteristics of man and the computer in order to identify the processes that can be performed by each, and where one can aid the other. Table 12.1 compares the capabilities of man and the computer for various tasks. It can be seen that in most cases the two are complementary—for some tasks man is far superior to the computer, and in others the computer excels. It is therefore the marriage of these characteristics of each, which is so important in the application of computers. In short, the computer serves as an extension to the memory of man, enhances his analytical and logical power and relieves him from performing repetitive and routine tasks.

### 12.1.1 Elements of a Computer System

A computer is capable of performing certain fundamental operations internally:
- Representing data and instructions;
- Holding (storing) data and instructions;
- Moving data and instructions
- Interpreting and executing commands of instructions

For performing such operations the computer is equipped with hardware and software.

#### Hardware

Computer hardware includes equipment which can aid data preparation, input processing, secondary or auxiliary storage and output from the computer. The internal

## TABLE 12.1

### Characteristics of man and the computer

| Characteristics | Man | Computer |
|---|---|---|
| Method of logic and reasoning | Intuitive by experience, imagination & judgement | Systematic and stylised |
| Level of Intelligence | Learns rapidly, but sequentially, unreliable intelligence | Little learning capability but reliable level of intelligence |
| Method of information input | Large amount of input at one time by sight or hearing | Sequential and stylised input |
| Method of information output | Slow sequential output by speech or manual actions | Rapid stylised sequential output by the equivalent of manual actions |
| Organisation of information | informal and intuitive | Formal and detailed |
| Effort involved in engineering information | Small | Large |
| Storage of detailed information | Small capacity, highly time dependent | Large capacity, time dependent |
| Tolerance for repetitive and mundane work | Poor | Excellent |
| Ability to extract significant information | Good | Poor |
| Production of errors | Frequent | Rare |
| Tolerance for erroneous information | Good, intuitive correction of errors | Highly intolerant |
| Method of error detection | Intuitive | Systematic |
| Method of editing detection | Easy and instantaneous | Difficult and involved |
| Analysis capability | Good, intuitive analysis, poor numerical analysis | No intuitive analysis, good numerical analysis ability |

operations mentioned above are carried out by the central processing unit (CPU) of the computer. The CPU contains three sub units: the control unit, the processing or arithmetic unit and the storage unit. These units plus the input and output subsystems such as keyboards, scanners, printers and plotters form the hardware elements of the simple computer system.

## Software

Software comprises non-hardware aids namely, computer programs and computer routines which facilitate operation of the computer by the user on installation. The computer programs include programs for performing standard tasks such as storing data records, organising and maintaining files, translating programs written in a symbolic language and scheduling jobs through the computer. The term software can include user programs, but more commonly refers only to general programming and operating programs which are sourced through the hardware manufacturers or from independent software companies. Software is as vital to the effective use of a computer as is hardware.

## 12.1.2 Examples of Fields of Computer Application

A. *Scientific and engineering applications*
- Surveying
- Navigation
- Weather forecasting
- Computer-aided design and manufacturing
- Process control (e.g. in the chemical industry)
- Engineering drafting
- Software engineering
- Earthquake predictions
- Communication
- Defence
- Space sciences, etc.

B. *Business, commercial and administrative application*
- Preparation of pay roll
- Invoicing
- Financial accounting
- Word processing, document preparation
- Library management
- Large scale database management
- Large scale spreadsheet calculations
- Annual reports and balance sheet preparations
- Commercial graphics displays, presentations, animations, simulations for subsequent audio, video outputs in TV telecasts, movies, etc.

## 12.2 CAD/CAM SYSTEMS

Computer-aided design (CAD) involves the use of computer to assist in the designing of an individual part or a system, e.g. a machine tool. The design process usually involves computer graphics.

In computer-aided manufacturing (CAM), computer is used to assist in the manufacture of a part. CAM can be divided into two main classes:

1. On-line applications, namely the use of a computer to control manufacturing systems in real time, such as the CNC system of a machine tool.
2. Off-line applications, namely the use of a computer in production planning and non-real time assistance in the manufacturing of parts. Examples of off-line CAM are the preparation of a part program or display of the tool path in a machining simulation.

CAD/CAM is a unified software system in which the CAD portion is interfaced inside the computer with CAM system. The end result of a CAD/CAM system is usually a part program which can be listed out or directly fed into the control computer of a CNC machine. The functions of a typical CAD/CAM system are illustrated in Fig. 12.1.

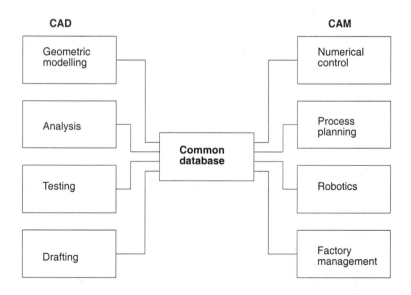

**Fig. 12.1** Functions of CAD/CAM

### 12.2.1 Computer-Aided Design

The CAD system is basically a design tool used to analyse various aspects of a designed product. CAD systems support the design process at all levels, namely the conceptual, preliminary and final stages of the design. It can also compute and analyse the various features of a product such as strength, stiffness and weight. Display of the designed object on the CRT is one of the valuable features of the CAD system. The computer graphics help the designer in studying the object by rotating it on the screen, separating it into segments, enlarging a specific portion of the object in order to observe it in detail and studying the motion of mechanisms with the aid of kinematic programs. Most CAD systems use interactive graphic systems which allow the user to interact directly with the computer in order to generate, manipulate and modify graphic displays. The end products of many CAD systems are drawings generated on a plotter interfaced with the computer. Modern CAD systems include the use of computers for analysis. The finite element method is one of the popular methods of analysing an object or a part represented by a model consisting of small elements, each of which has stress and deflection characteristics. Finite element analysis can output deflections or stress values directly on to the computer. It is also used for optimisation of the material content of any part or a section thereof.

CAD is also employed for industrial design applications by which the design concept of a product for improved aesthetics can be optimised. Solid modelling, animated movements and a variety of colour schemes available in the software enables application of the industrial design concepts.

## Elements of a CAD System

Typically, a stand-alone CAD system consists of a workstation including the following hardware components:

    Graphic terminal—monitor
    Input device—keyboard, mouse, scanner
    Output device—plotter, line printer
    Secondary storage media—tape, floppy disk

CAD software is a drawing tool for designing a product on a computer and consists of utilities to create, manipulate and draw the part using a computer. The following are the major functions of a CAD software.

- Drawing functions
- Editing functions
- Display functions
- Drawing aids
- Dimensional functions
- Plotting functions

### 12.2.2 Computer Aided Manufacturing

Computer-aided manufacturing (CAM) is a system which oversees the many aspects of manufacture by introducing a hierarchical computer structure to monitor and control the various phases of manufacturing process. A CAM system is an integration of CAD, engineering, manufacturing and sometimes includes the quality control functions. The CAM hardware elements include CNC machines, inspection equipment, digital computers and other related devices. CAM software is an interrelated mesh of computer programming systems that serve to monitor, process and ultimately control CAM hardware and the flow of manufacturing data. Because a CAM system oversees many aspects of the manufacturing process, database creation and information processing assumes greater importance; the CAM system relies upon both the real time and stored information. The level of complexity of a CAM database is directly proportional to the number of tasks required by the system. Figure 12.2 shows a simple CAM system, integrating the various phases of manufacturing. However, a large scale CAM system encompasses major areas related to the manufacturing process, i.e. production management and control, engineering analysis and design, and finance and marketing.

Thus, the main concept of a CAD/CAM system is the generation of a common database which is used for all the design and manufacturing activities. These include specification of the product, conceptual design, final design, drafting, manufacturing, and inspection. At each stage of this process data can be added, modified, used and distributed over networks of terminals. A single database provides a substantial reduction in human errors and a significant shortening of the time required from the introduction of a concept of a product to the manufacturing of the final product.

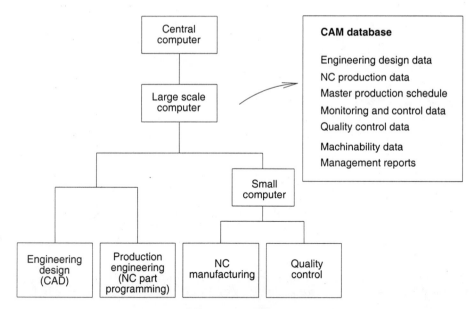

**Fig. 12.2** CAM system

# Further Reading

American Society for Metals, 'Metals Handbook', 9th ed., vol 1, *Properties and selection: Irons and steels*, 1978 and vol 4: *Heat treating*, 1981, ASM, Ohio, New York.
Bradley, B.A., et al, *Mechatronics: Electronics in products and processes*, Chapman & Hall, London, 1991.
W. Canning Co. Ltd., *Handbook of electroplating*, 19th ed., Canning Co. Ltd., Birmingham, 1960.
Central Machine Tool Institute, *Machine tool design handbook*, Tata McGraw-Hill, New Delhi, 1981.
Chambers, H.L. and C.A. Chacey, *Drafting and manual programming for numerical control*, Prentice-Hall, Englewood Cliffs, N.J., 1980.
Chute, G.M. and R.D. Chute, *Electronics in industry*, 5th ed., McGraw-Hill, New York, 1979.
Dubey, G.K., *Fundamentals of electrical drives*, Narosa Publishing House, India, 1995.
Gibbs, D., *CNC part programming: A practical guide*, Cassell, 1987.
Gibbs, D. and T.M. Crandell, *An introduction to CNC machining and programming*, Industrial Press. Inc., New York, 1991.
Gross, H., Ed., *Electrical feed drives for machine tools*, John Wiley & Sons, USA, 1983.
HMT Limited, *Production Technology*, Tata McGraw-Hill, New Delhi, 1981.
Hunt, V.D., *Mechatronics: Japan's newest threat*, Chapman & Hall, London, 1988.
Kamenichny, I., *Heat treatment handbook*, Mir Publishers, Moscow.
Malvino, A.P., *Transistor circuit approximations*, 3rd ed., Tata McGraw-Hill, New Delhi.
Millman, J. and Halkias, C.C., *Electronic devices and circuits*, McGraw-Hill, New York, 1982.
Otto, W. and K. Schaening, *International comparison of standard materials*, Deutsches Institut fuer Normung e.v. (DIN), Berlin, 1985.
Ramachandra, H.S., *Salt bath quality control and maintenance* (Unpublished).
Roberts, AD, and R.C. Prentice, *Programming for numerical control machines*, McGraw-Hill, New York, 1978.
Subrahmanyam, V., *Electric drives: Concepts and applications*, Tata McGraw-Hill, New Delhi, 1994.
Theraja, B.L., *A text book of electrical technology*, Niraj Construction & Development Co. (P) Ltd., India, 1987.
Schleisinger, G., *Testing of machine tools*, Pergamon Press, London, 1982.
UMIST, *Specification and tests for machine tools*, 2V, UMIST, Manchester, UK, 1970.

# Index

Acceptor  46, 48
Actuating mechanism  150
Advanced manufacturing system  20
Aesthetics  435
    Concept of house style  438
    Concept of order with variety  437
    Concept of purpose  437
    Concept of style and environment  437
    Concept of unity  436
Alternating current(ac)  33
Ampere  32
Amplifier  257
   Four quadrant  259
Annealing  111
Anode  47
Approach path  367
Assembly tool  236
Asymetrical face milling  93
Austentic transformation  111
Austentite  109
Autocollimator  395
Auxiliary function  348
Axes nomenclature  340

Backlash  152
Ball bar system  392
Ball bearing assembly  199
Ball bushing  141
Ballnut
   Assembly  203
   Failure  202
Ballscrew
   Accuracy  158
   Alignment  197
   Assembly  195

Bearing mounting assembly  199
   End bearing  156
   Failure  202
   Mounting  152, 199
   Pretensioning  157
   Stretching  157
   Thermal displacement  205
Base  58
   Cantilever  88
   Fixed  88
   Simply supported  88
Bearing
   Antifriction  172
   Ball  174
   Ceramic  177
   Failure  231
   Friction  83
   Hydrodynamic  172, 173
   Hydrostatic  172, 173
   Linear  142
   Maintenance  231
   Noise, Causes  227
   Preloading  232
   Roller  174
   Spindle  135, 169
Bearing vibration
   Causes  227
   Reduction  231
Bending moment  88
Bending of beam  88
Binary system  308
Block number  344
Block skip character  344
Braking
   Dynamic  264

# 456  Index

  Regenerative  264
Bushes
  Fluid pressurised  168
  Taperlock  168

CAD  452
CAD/CAM  450
CAM  452
CNC machine  6, 135, 338
  Advantages  6
CNC system  308, 338
  Modes of operation  315
Canned cycles  353
Capacitance  38, 39
Case hardening  112
Cathode  47
Cementite  109
Central processing unit  309
Chip formation  92
Chromium plating  116
Circular arc  152
Circular interpolation  348
Clearance angle  92
Close-loop response  277
Close-loop system  311
Coefficient of
  Friction  138
  Rolling resistance  83
  Sliding friction  81
  Static friction  81
Collector  58
Commutation, electronic  250
Computer  449
  Application  450
  Hardware  448
  Software  449
Computer integrated manufacturing  20
Computer numerical control
  (see CNC)
Conductor
  Bad  29
  Good  29
  Semi  29
Control algorithm  308
Control relay  285
Control system  135, 181, 240
  Flowchart  326
Converter, DC  258
Coordinate system  340
Copper plating  115

Cost reduction  4
Couple  81
Current  32
  Characteristics  33
Cutting movement  91
Cyclic program processing  329
Cylindrical roller bearing assembly  199

D-word  345
DC drive
  Brushless  267
  SCR  253
  Servocontrol  249
DC motor  246
  Speed control  251
  Servocontrol  249
Deflection of beam  91
Delivery  4
Design  426
Diagnostics  322
Dial gauges  395
Digital circuit  74
Dimensioning
  Absolute  340
  Incremental  340
Diode  48
  Collector  58
  Emitter  58
  Forward biased  47
  Junction  47
  Light emitting  56
  Opto isolator  57
  P-N junction  47
  Photo  56
  Point contact  46
  Reverse biased  50
  Tunnel  55
  Zener,  54
Diode as rectifier  50
Diode action  48
Direct current (dc)  33
Direct numerical control (DNC)  9, 333
Donors  47
Doping  45
Drive  238–43
  AC, selection  281
  Amplifier  255
  Brushless DC  267
  Optimisation,  276
  Protection  281

## Index

Pulse width modulator  261, 264
SCR-DC  256
Selection  278
Tuning  279
Drive requirements
  Cutting movement  95
  Feed  95
  Feed motor shaft  96
Dwell time  371
Dynamic load  136

Eccentric load  196
Effective switching distance  211
Elastic deformation  228
Elastic limit  88
Electrical cabinet, wiring  287
Electrical panel cooling  289
Electrical standard  289
Electromagnetic force (emf)  38
  Back  39
  Counter  39
  Induced  40
Electromagnetism  246
Electroplating  115
Emitter  58
Encoder
  Incremental rotary  178
  Mounting  207
End milling  93
Ergonomics  438
  Machine tool  439, 440
  Manufacturing system  441

FMC/FMS
  Benefits  22
  Constitutents  21
FMS  22, 27
  Stepwise approach  25–26
  Trends  24
Face milling  93
Farad  37
  Micro  37
  Pico  37
Feed drive  95, 146, 244
Feed motor, AC servo  265
Feedback elements  207
  Encoder  208
  Proximity switch  208
Feedback system
  Position  311

  Velocity  312
Ferrite  109
Finishing allowance  370
Finishing cut depth  366
Fits  123
  Classification  123
  Clearance  123
  Interference  124
  Transition  123
Flame hardening  112
Flexible coupling  167
Flexible manufacturing cell
  (see FMS)
Flexible manufacturing system
  (see FMS)
Force  79, 247
Force of friction  81
Friction  81
  Effect  230
  Torque  84

Gauging  182
Gate
  AND  75
  EXOR  77
  Logic  74
  NAND  76
  NOR  76
  NOT  75
  OR  74
Geometrical accuracy  371, 387–88
Gothic arc  152
Groove width  371
Guideway coatings  140
  Design  138–39
  Selection  147
Guideways  138
  Aerostatic  145
  Antifriction linear motion  141
  Cylindrical  140
  Dovetail  140
  Flat  140
  Friction  138
  Hydrostatic  145
  Vee, 139

Hall effect  273
Heat treatment  109
  Application  109
  Guidelines  107

Principles  109
Henry  39
   Milli  39
Hertz  34
Hydraulics  237
   Troubleshooting  237

Idle passes  360
Inductance  39
Induction hardening  112
Inductor  37
Industrial design  426
Inertia  84
   Cylindrical bodies  98
   Driven shaft  99
   Electrical  38
   Mass along linear axis  98
Infeed angle  368
Input element  283
Insulator  30
Integrated circuit  71
Interfacing  319
Interrupt drive program  329

Just-in-time  4

Keyboard  313
Kinematic accuracy  137
Kinetic friction  81

LM guide  144, 184
   Assembly  185
   Location  185
   Mounting  187
Ladder diagram  326
Laser measuring system  388
Linear scale  179
   Mounting  207
Liquid
   Carburising  113
   Nitriding  113
Load
   Inertia  98
   Torque  97
Locknut  200–201
Lubrication  200, 208, 235–7

Machine accuracy compensation  323
   Backlash  323
   Cutter diameter  324

Leadscrew  323
Pitch error  323
Sag  324
Tool nose  324
Tool offset  324
Tool wear  325
Machine control panel  315
Machine data  322
Machine tool testing
   Axis drive  390
   Functional aspects  379
   Dynamic behaviour  390
   Geometrical accuracy  395
   Idle power  380
   Idle running  380
   Idle torque  380
   Main spindle drive  393
   Metal removal capability  417
   Noise  384
   Positioning accuracy  396
   Process capability  408
   Regidity  424
   Reliability  424
   Repeatability  396
   Safety  427
   Sound power level  385
   Spindle running accuracy  416
   Technical specifications  379
   Thermal effect  385
   Vibration  381
   Working accuracy  406
Machining
   Cycle  364
   Sequence  371
Machining centre  7–10
   Programming  373
   Testing  421
Magnetic effect  32
Material classification  101
   Ferrous  102
   Miscellaneous  102
   Nonferrous  102
Material selection  101
Measuring system  177
   Direct  177
   Indirect  177
   Mounting  178
Mechatronics
   Definition  1
   Factory automation  3

# Index

Key issues 3
Scope 2
Training 3
Metalcutting 91
Miniature circuit breaker 287
Mirror image 361
Modulus of
    Elasticity 88
    Rigidity 88
Moment 80
Motion nomenclature 341
Motor, DC (see DC Motor)

Needle roller and flat cage assembly 191
    Accuracy 191
    Matching 191
Needle roller bearing 199
Newton's law 84
Nickel plating 115
Nitriding 113
Normal switching distance 209
Normalising 111
Numbering system 308
Numerical control (NC) 308
Nut mounting 196

ON/OFF function 285
Ohm 32
Ohm's law 33, 40
Open loop
    Response 276
    System 312
Operator control panel 313

PLC 317
    Programming 325
Parametric programming 363
Part program 338
    Structure 338, 342
Passive components 35
Pearlite 109
Peripheral devices 317
Physiogonomy 435
Plunge cutting cycle 368
Polytetrafluroethylene (PTFE) 140, 147
Potential 31
Potential difference 32
Preparatory function 345
Product
    Design 435

Safety audit 440
Specification 4
Programmable logic controller
    (see PLC)
Proximity switch
    Capacitive 283
    Inductive 283
    Mounting 208

Quality 4
Quick check system 386–87

Radial groove 371
Rake angle 92
Rake surface 92
Reactance
    Capacitive 40
    Inductive 40
Rectifier
    Bridge 52–53
    Centre tap 51–52
    Full wave 51
    Half wave 51
    Output 53
Relay, overload 286
Relief angle 92
Resistance 32, 39
Resistivity 33
Resistor 34
Roller screw 158
    Lead accuracy 162
    Planetary 159
    Recirculating 160
Rolling friction 82
Rolling resistance 83
Root mean square(rms) 34
Roughing cut 368
Runout path 367

SCR 68
    Converter 256
    Firing 69
    Schematic symbol 69
    Structure 69
SCR drive amplifier
    Single phase full wave converter 257
    Three phase full wave converter 257
Self inductance 39
Semiconductor device 46
Semiconductor 30, 42

# Index

Extrinsic 45
Intrinsic 42
N-type 46
P-type 46
Sensor, rotor position 272
Sequence number 344
Servocontrol unit 311
Servodrive 313
   AC 265
   Amplifier 265
   Braking 263
Servomotor 148
   AC 271
   DC 252
   Operating range 245
Silicon controlled rectifier (see SCR)
Sine wave 34
Specific resistance 32
Speed control unit 311
Spindle accuracy
   Axial runout 175
   Radial runout 175
Spindle bearing 169, 214
   Arrangement 175
   Assembly 221
Spindle drive 242
   Digital 270–2
   Optimisation 280
Spindle speed 345
Starting torque
   Process demand 282
   Static friction 282
Statement list 326
Static load 136
Steel 108
Step 5 programming 326
Stock removal cycle 364
Strain 82
   Linear 82
   Shear 82
Strength of material 87
Stress 85
   Comprehensive 85
   Relieving 111
   Shear 85
   Tensile 85
   Yield 85
Structured programming 328
Subroutine 352

Surface defect 228
Surface roughness 127
Surface texture 127
Switching
   Distance 209
   Frequency 214
   Hysteresis 214

Tensile
   Strain 86
   Strength 87
   Stress 85
   Test 87
Thermal
   Deformation 138
   Load 136
   Stability 175
Thread cutting
   Cycle 365
   Depth 366
   End point 368
   Pitch 366
Time controlled program 329
Timing belt 163–7
Tolerance 117
   Geometric 118
   Symbols 120–3
Tool compensation 351
Tool monitoring 17, 182
   Direct 182
   Electric current 18
   Force measurement 18
   Geometric measurement 18
   Indirect 182
   Principles 17
Tool
   Length compensation 351
   number 345
   Radiius compensation 351
Torque transmission element 163
Torsion 90
Transformer 41
Transistor, 58
   Field effect 64
   Junction field effect 64–5
   NPN 58
   Pulse width modulator 258
   Unijunction 63–64
Transistor as amplifier 60
Transistor as switch 63
Transistor configuration, 61

Common base   61
Common collector   62
Common emitter   61–62
Transmission elements   96
Triac   70
Tufftriding   114
Turning centre   10, 17
   CNC cutting condition,   418
   Programming   376
Turning effect   80
Turning moment   90
Twisting moment   90
Tychoway   187

Assembly   188

Unity   436
Usable sensing distance   213

Value portfolio   3
Video display unit   313
Volt   31
Voltage   31
   Induced   250

Watt   33
Wheel friction   82
Word addressed format   344